Structure and Reactivity in Organic Chemistry

This book is dedicated to my parents, Kevin and Cecily.

Soli Deo Gloria

Structure and Reactivity in Organic Chemistry

Mark G. Moloney

Reader, Department of Chemistry,
University of Oxford

and

EP Abraham Fellow and Tutor in Chemistry,
St Peter's College, Oxford

Blackwell
Publishing

Blackwell Publishing editorial offices:
Blackwell Publishing Ltd, 9600 Garsington Road, Oxford OX4 2DQ, UK
Tel: +44 (0)1865 776868
Blackwell Publishing Professional, 2121 State Avenue, Ames, Iowa 50014-8300, USA
Tel: +1 515 292 0140
Blackwell Publishing Asia Pty Ltd, 550 Swanston Street, Carlton, Victoria 3053, Australia
Tel: +61 (0)3 8359 1011

First published 2008 by Blackwell Publishing Ltd

2   2009

ISBN: 978-1-4051-1451-6

Library of Congress Cataloging-in-Publication Data
Moloney, Mark G.
Structure and reactivity in organic chemistry / Mark G. Moloney.
p. cm.
Includes index.
ISBN-13: 978-1-4051-1451-6 (pbk. : acid-free paper)
ISBN-10: 1-4051-1451-7 (pbk. : acid-free paper)   1. Organic compounds – Structure.   2. Organic reaction mechanisms.   3. Reactivity (Chemistry)   I. Title.

QD476.M38 2008
547 – dc22

2007043767

A catalogue record for this title is available from the British Library

Set in 10/12 pt Minion by Aptara Inc., New Delhi, India

For further information on Blackwell Publishing, visit our website:
www.blackwellpublishing.com

Contents

Preface

Structure and Reactivity in Organic Chemistry provides a mechanistic overview of organic chemistry, which is essential for a broader understanding of the subject. The text is suitable for both undergraduate students learning the details of the subject and graduate students who wish to refresh their knowledge or who are tackling organic chemistry having studied a related discipline, such as biochemistry, materials science or medicinal chemistry, during their undergraduate course. It is also designed to be a revision guide for those familiar with organic chemistry, but for whom some of the important fundamentals have become hazy over time! This audience may include graduate students revising for their final doctoral examination or preparing to teach junior students for the first time, or organic chemists working in industry. The book aims to highlight the mechanistic rules which underpin the structure and reactivity of organic molecules and emphasise that there is a high degree of order in chemical processes as a result. Recognition of the value and importance of these rules can be instrumental in making the learning of organic chemistry much more tractable, if not enjoyable.

I acknowledge all my own undergraduate students over the years who have never accepted insufficient answers and as a result enhanced my own understanding. I also acknowledge Dr Josephine Peach whose appetite for the provision of high-quality teaching has been an inspiration.

Mark G. Moloney
Reader in Chemistry, Department of Chemistry, University of Oxford and
E.P. Abraham Fellow and Tutor in Chemistry, St Peter's College, Oxford

Chapter 1

Bonding

Any molecule can exist only if there is a stable bonding arrangement which holds its constituent atoms together. Since bonding fundamentally results from a sharing of electrons between nuclei, it is crucial to consider the electronic nature of a molecule and its constituent atoms in order to understand the reason for both its existence and its structure. Moreover, not only does bonding determine the structure of a molecule, it also determines its reactivity. This chapter outlines an understanding of bonding suitable for an organic chemist. We will see later that the structure and reactivity of organic compounds is determined by the way their bonding and non-bonding electrons are arranged.

1.1 Atomic structure

1.1.1 The chemical bond

We consider that *ionic bonds* form as a result of electrostatic attraction between positively and negatively charged ions (cations and anions respectively). More important for organic chemistry is the *covalent bond*, which results from the sharing of an electron pair between two nuclei, leading to a mutually attractive interaction between the respective nuclei.

1.1.2 The periodic table

The classification of the chemical elements according to their characteristic reactivity reveals certain patterns of similarities, and when they are compiled into the periodic table (the elements of most immediate relevance for introductory organic chemistry are shown in Table 1.1), these trends become readily apparent. Significantly, elements in the same column (or *group*) exhibit similar patterns of reactivity; examples include nitrogen and phosphorus, oxygen, sulfur and selenium, and carbon, silicon, germanium and lead. This is because these elements possess the same numbers of electrons in their outermost or *valence* shell; the number of valence electrons can be easily determined by simply counting along each of the rows until arrival at the element in question. Some important examples of particular relevance to organic chemistry are given in Fig. 1.1.

The most stable (or unreactive) elements are those of group VIIIa, the inert gases (He, Ne, Ar and Kr), possessing *electronic configurations* with eight electrons in their outermost shells. A chemical reaction enables its participating elements to achieve such a stable inert gas configuration; for elements of groups Ia (except hydrogen), IIa and IIIa, this will be of the inert gas of the previous row, and for the elements of groups IVa, Va, VIa and VIIa, this will be of the inert gas configuration of that same row. Because these changes most often lead to the acquisition of eight electrons in the valence shell (an exception is He, for which the electron total is only two), this is usually called the *octet rule*. In order to achieve these inert gas configurations (in this condition, the element is said to have *filled* its valence

Table 1.1 The periodic table containing the elements of most immediate relevance to organic chemistry

Ia	IIa	IIIa	IVa	Va	VIa	VIIa	VIIIa
^{1}H 2.1							^{2}He
^{3}Li 1.0	^{4}Be 1.5	^{5}B 2.0	^{6}C 2.5	^{7}N 3.0	^{8}O 3.5	^{9}F 4.0	^{10}Ne
^{11}Na 0.9	^{12}Mg 1.2	^{13}Al 1.5	^{14}Si 1.8	^{15}P 2.1	^{16}S 2.5	^{17}Cl 3.0	^{18}Ar
^{19}K	^{20}Ca				^{24}Se	^{25}Br 2.8	^{26}Kr
						^{33}I 2.5	

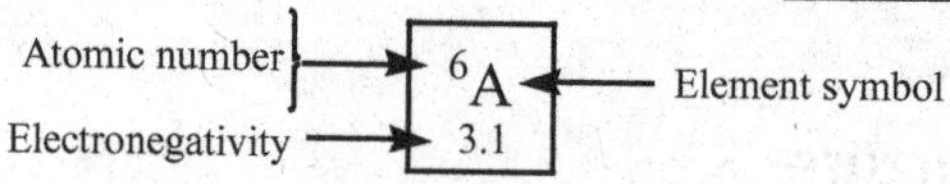

Atomic number ⟶ ^{6}A ⟵ Element symbol
Electronegativity ⟶ 3.1

shell), elements will need to gain or lose electrons, either completely or by sharing them with another nucleus. For example, hydrogen, lithium and beryllium will all achieve the electronic configuration of ^{2}He as a result of a chemical reaction, while carbon, nitrogen, oxygen, fluorine, sodium and magnesium will achieve the inert configuration of ^{10}Ne. In order to achieve these outcomes, hydrogen requires one additional electron, equivalent to a half share in one bond, and lithium needs to lose one electron; both are therefore said to have a *valence* of 1. A carbon atom, on the other hand, needs to gain four electrons and therefore has a valence of 4; this is best achieved by sharing four electrons, since gaining four electrons fully would lead to a tetraanion, which is electrostatically less favourable. Nitrogen, oxygen and fluorine have valencies of 3, 2 and 1, respectively, since acquisition of that number of electrons gives a stable octet; this can be achieved either in ionic bonds or by sharing the required number of electrons in covalent bonds. Exactly how many electrons are needed to

Atom	Shell		
	K	L	M
H	1		
He	2		
B	2	3	
C	2	4	
N	2	5	
O	2	6	
Si	2	8	4
P	2	8	5
S	2	8	6

Figure 1.1 Electronic configurations of some elements.

be gained or lost to arrive at an octet is then readily deduced from the periodic table; the most stable arrangement of electrons of any row is given by the electronic configuration of the inert gas (group 8) in that row. Atoms to the right side of the table can gain the stable octet of the inert gas of that row most easily by gaining one to four electrons; for example, the addition of two electrons to ^8O achieves the same electronic configuration of ^{10}Ne. Atoms to the left side of the table most readily achieve the inert gas configuration of the previous row by losing one to four electrons; for example, the loss of two electrons from ^4Be achieves the stable electronic configuration as that of ^2He. Because organic chemistry is most commonly concerned with compounds between carbon and elements from the first three rows of the periodic table, it is worthwhile to commit rows 1–3 to memory, along with the respective electronic configuration of each of the elements (see Table 1.1).

1.1.3 Valence electrons

We therefore do not need to consider all of the electrons in any atom to be able to understand its chemistry; rather, examination of its *valence electrons*, or those in the highest shell and therefore of highest energy, is generally sufficient. For the first row of the periodic table, this is the K shell; for the second row, the L shell; and for the third row, the M shell (Fig. 1.1). This accounts for the similarity of structure, properties and reactivity of the members of any group in the periodic table, since the lower-energy, closed shells can be ignored.

1.1.4 Lewis structures

We have seen that a fundamental principle central to all chemical understanding is the concept of the stable octet of electrons: an atom is considered to be in its most stable state when it has achieved an octet of electrons in its outer shell, and this can be achieved by gaining, losing or sharing electrons with other atoms. Therefore, the ability to draw correct representations of this situation is crucial to understand the bonding in a molecule. Molecules are most conveniently represented using Lewis structures, as illustrated in Fig. 1.2. Such structures are compiled using the following rules overleaf:

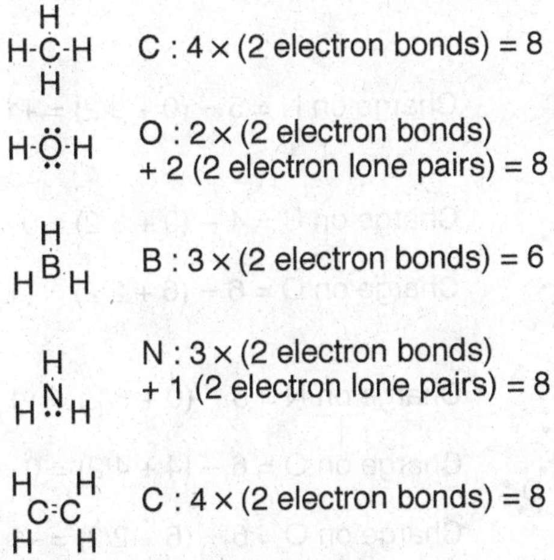

H
H-C-H   C : 4 × (2 electron bonds) = 8
H

H-Ö-H   O : 2 × (2 electron bonds)
          + 2 (2 electron lone pairs) = 8

H
B   B : 3 × (2 electron bonds) = 6
H   H

H
N   N : 3 × (2 electron bonds)
H   H   + 1 (2 electron lone pairs) = 8

H   H
C∷C   C : 4 × (2 electron bonds) = 8
H   H

Figure 1.2 Lewis structures for some simple molecules.

- Each single line (–) representing a covalent bond is a shared pair of electrons and counts two towards the total number of valence electrons for each nucleus.
- Two lines (=) indicate sharing of four electrons between two atoms, called a double bond; and three lines (≡) indicate sharing of six electrons between two atoms, called a triple bond.
- Unshared or lone pairs (represented as '••') count two towards the total number of valence electrons for the nucleus.
- The formal charge on an atom is equal to the number of valence electrons of the atom less the sum of the number of lone-pair electrons and half the number of bonding-pair electrons, as given by Eq. 1.1:

$$
\left\{
\begin{array}{l}
\text{Formal} \\
\text{atomic} \\
\text{charge}
\end{array}
\right\}
=
\left\{
\begin{array}{l}
\text{Number of} \\
\text{valence} \\
\text{electrons} \\
\text{of atom}
\end{array}
\right\}
-
\left(
\left\{
\begin{array}{l}
\text{Number of} \\
\text{lone-pair} \\
\text{electrons}
\end{array}
\right\}
+ 1/2
\left\{
\begin{array}{l}
\text{Number of} \\
\text{bonding-} \\
\text{pair} \\
\text{electrons}
\end{array}
\right\}
\right)
\qquad \text{(Eq. 1.1)}
$$

For the second-row elements C → F, the number of bonds and lone pairs cannot exceed four; that is, the octet rule must not be violated. Although not universally true for higher rows, it is usually also true for the third-row elements Na → Cl. Some charge calculations for the correct Lewis structures for simple molecules and entities commonly encountered in organic chemistry are shown in Fig. 1.3.

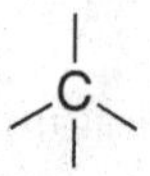

Charge on C = 4 − (0 + 8/2) = 0

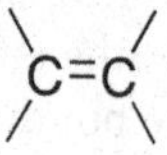

Charge on C = 4 − (0 + 8/2) = 0

Charge on N = 5 − (2 + 6/2) = 0

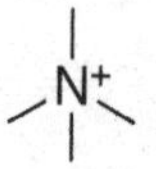

Charge on N = 5 − (0 + 8/2) = +1

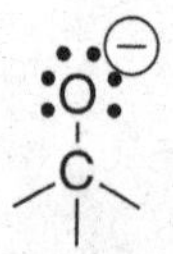

Charge on C = 4 − (0 + 8/2) = 0

Charge on O = 6 − (6 + 2/2) = −1

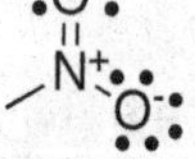

Charge on N = 5 − (0 + 8/2) = +1

Charge on O = 6 − (4 + 4/2) = 0

Charge on O = 6 − (6 + 2/2) = −1

Figure 1.3 Lewis structures for common entities in organic chemistry.

1.1.5 Conventions for drawing structures

Although drawing Lewis structures is the most accurate method for representing the bonding electrons in a molecule, including as it does bonding and non-bonding electrons and electric charge, in practice it can be very cumbersome, especially for organic compounds, and is in fact usually unnecessary, because we need to focus only on those atoms of a molecule which are reacting. In order to simplify drawing structures of organic compounds, an accepted convention is to draw a zigzag line representing the carbon backbone, with each corner denoting a carbon atom, and the required number of hydrogens to make up to a valence of 4 is assumed (Fig. 1.4). This format can be used for linear and cyclic structures, and can also be adapted to include substitution by *heteroatoms* (atoms which are not carbon, and include nitrogen, oxygen, sulfur, phosphorus and the halogens). One exception to this rule is aldehydes, for which the aldehydic hydrogen (RC*HO*) needs to be included.

An important convention concerns charge and lone pairs: anionic species are very important in organic chemistry, and these are usually drawn without lone pairs; it should be

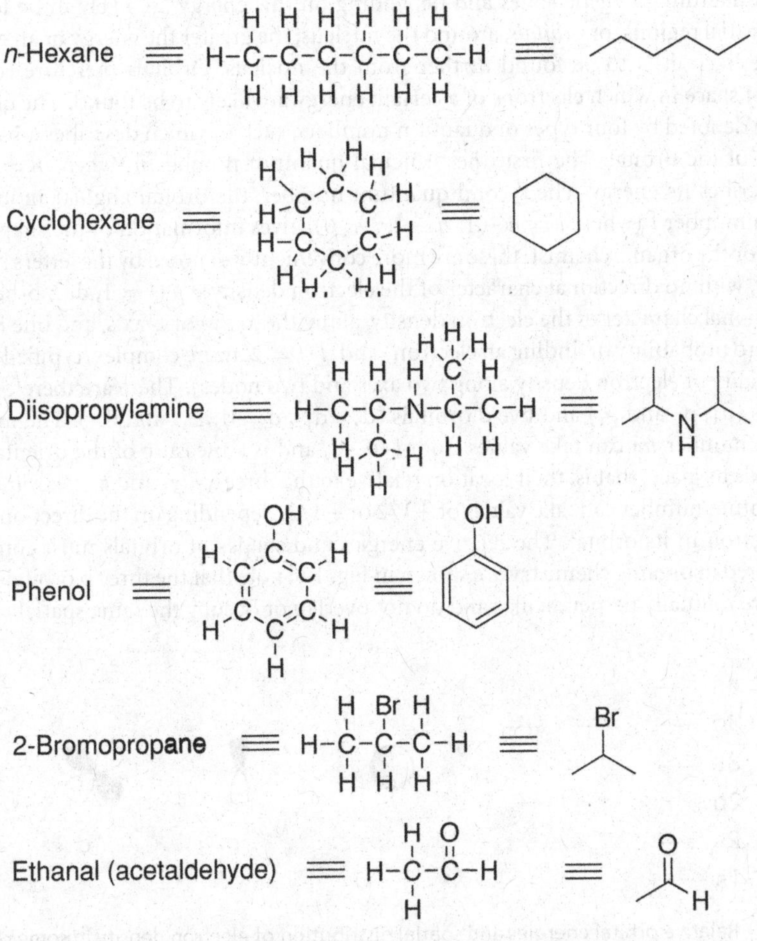

Figure 1.4 Use of zigzag structures for structural representations of organic compounds.

$$\left[R-\ddot{\underset{\cdot\cdot}{O}} : \right]^{\ominus} \equiv RO^{\ominus} \qquad \left[R-\overset{\cdot\cdot}{N}-R \right]^{\ominus} \equiv R-\overset{\ominus}{N}-R$$

Figure 1.5 Representation of anions.

remembered that one unit of negative charge accrues as a result of a nucleus having one extra electron, but that this additional electron will normally comprise part of a lone pair. Thus, a negative charge normally represents a pair of electrons (Fig. 1.5).

1.1.6 Atomic orbital theory

The concepts of the Lewis octet and valence electrons are very useful, but they do not take account of an important additional complication which is that the energies of elections of an atom are not all equal, but differ according to their spatial proximity to the nucleus. Those which are further away from the nucleus are of higher energy. Electrons exhibit discrete quantum energetic states and depending on this energy are likely to be found in certain spatial regions, or *orbitals*, around the nucleus; the greater the energy of an electron, the more likely it is to be found further from the nucleus. Orbitals therefore represent regions of space in which electrons of a certain energy are likely to be found. The quantum states are denoted by four types of quantum numbers, each of which describes a particular property of the orbital. The first, the principal quantum number n (where $n = 1, 2, 3, \dots$), describes its energy. The second quantum number, the orbital angular momentum quantum number l (where $l = n - 1, n - 2, \dots, 0$), gives information of the shape of the orbital. For the organic chemist, these are more conveniently denoted by the letters s ($l = 0$, spherical, with no directional character of the electron density), p ($l = 1$, dumb-bell, with unidirectional character of the electron density, along the x-, y- or z-axis, and one *node*, an area of zero probability of finding an electron) and d ($l = 2$, more complex, typified by high directionality of electron density along two axes and two nodes). There are therefore three p orbitals (p_x, p_y and p_z) and five d orbitals (d_{xy}, d_{yz}, d_{xz}, $d_{x^2-y^2}$ and d_{z^2}). The magnetic quantum number m_l can take values from l to $-l$, and is a measure of the orientation of the orbitals in space, that is, their location relative to the three x-, y- and z-axes. Finally, the spin quantum number can take values of $+1/2$ or $-1/2$, depending on the direction of spin of an electron in its orbital. The relative energies and shapes of orbitals most commonly encountered in organic chemistry are shown in Fig. 1.6; note that the three p orbitals, p_x, p_y and p_z, are mutually perpendicular and do not overlap or occupy the same spatial regions.

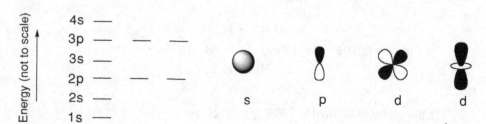

Figure 1.6 Relative orbital energies and spatial distribution of electron density in some common orbitals.

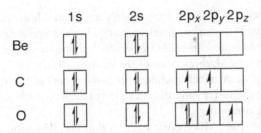

Figure 1.7 Electronic configurations for some elements of the second row.

Using this information, we can construct the electronic configuration and an atomic orbital (AO) diagram for any given atom using three rules:

1. *The Aufbau principle*: Lowest energy orbitals are filled first. Thus, 1s orbitals are filled before 2s before 2p before 3s before 3p, according to the energy profile given in Fig. 1.6.
2. *The Pauli exclusion principle*: A maximum of only two electrons per orbital is allowed and they must have opposite spins.
3. *The Hund's rule*: Empty orbitals of equal energy take one electron each before pairing occurs.

Since elements of the first and second rows are most commonly encountered in organic chemistry, at least at the introductory level, these will be considered in more detail. For example, application of each of these principles to determination of the electron configuration for each of Be, C and O gives the results indicated in Fig. 1.7. Beryllium, with four electrons, will first fill the lowest energy orbital, 1s, with two electrons and then the next highest energy orbital, 2s, with the remaining two; each set of electrons is paired, that is, of opposite spin. Carbon with six electrons will fill the 1s and 2s orbitals with four electrons as for beryllium, and the last two will go into two 2p orbitals, but will be unpaired, as required by the Hund's rule. For oxygen with eight electrons, the 1s and 2s orbitals will be filled with the four electrons as for beryllium, and three electrons will go into each of the 2p orbitals (the Hund's rule); the remaining electron will go into any of those 2p orbitals, but needs to be paired (Pauli exclusion principle). This analysis allows the determination of the electronic configuration of any element to be established, and with this information in hand, it is possible to predict the likely bonding in which it will participate. Thus, for the first-row elements Li and Be, the stable octet is most readily achieved by loss of two electrons to give the $1s^2$ configuration of helium, while for B, C, N, O and F, acquisition of the required number of electrons to achieve the stable octet of neon ($1s^2 2s^2 2p^6$) is most likely. For the second-row elements, Na and Ca will lose one or two electrons respectively to give the neon configuration ($1s^2 2s^2 2p^6$), while Al, Si, P, S and Cl will gain electrons to give the argon configuration ($1s^2 2s^2 2p^6 3s^2 3p^6$). From this analysis, we can predict the number of bonds (valency) any atom is likely to adopt, and for nitrogen and phosphorus, this is three, while for oxygen and sulfur, this is two, and for halogens, one.

1.1.7 Molecular orbital theory

AO theory is very useful for understanding the reactivity of atoms, but does not help us directly with understanding molecules. By simple analogy, we might expect that the bonding in molecules is to be described by molecular orbitals (MOs), but how these might be similar or different from simple AOs is not immediately apparent. However, by making a very simple assumption that MOs can be considered to be a linear combination of atomic orbitals (LCAO), it is possible to extend simple AO theory to describe bonding in molecules; a linear combination is simply the algebraic sum of the electron density of each AO. This simple approach has proven to be extremely useful for the development of a powerful understanding of bonding in organic chemistry and forms the basis of the way we think about the reactions of all types of organic compounds.

Fundamentally, therefore, MO theory assumes that bonding between elements results due to overlap of the valence AOs of those elements, representing a sharing of electron density, and that we can estimate this overlap simply by adding the electron density for each of the constituent AOs. Furthermore, the greater the overlap, the stronger the bond; best overlap is achieved with AOs of similar energy. There is one complication, however: For every AO that interacts to form an MO, the same number of MOs must be generated; thus, the interaction of two AOs leads to the formation of two MOs. This is where the 'linear combination' part is needed: In order to combine AOs, we need both to add *and* to subtract their electron density to generate a so-called *bonding MO* ψ and an *antibonding MO* ψ^*, as shown in Fig. 1.8a; the bonding AOs are each indicated on the right- and left-hand sides of the diagram and the resulting MOs between them. The bonding MO is lower in energy than both the contributing AOs, and the antibonding MO higher in energy. Notice that there is no gain or loss of energy in the system as a whole as required by the conservation of energy. For example, the overlap of two 1s orbitals of two hydrogen atoms creates two new MOs, one of lower energy and therefore bonding in character, and the other of higher energy and therefore antibonding in character (Fig. 1.8b). In a hydrogen molecule, the two bonding electrons taken from each contributing hydrogen atom occupy the bonding orbital as indicated, and there is a net gain in energy relative to the two isolated hydrogen atoms, and so the molecule is a stable entity. In He_2, however, four electrons occupy both the bonding and antibonding orbitals, there is no net gain in energy and no bond is formed, and so He_2 is not a stable entity (Fig. 1.8c). However, there is one problem with this analysis; its application to the bonding in carbon-containing compounds would lead to the erroneous deduction that carbon, on the basis of its electronic configuration ($1s^2 2s^2 2p^2$; see Fig. 1.7), would be divalent, since MOs would result due to overlap of two unfilled 2p orbitals with

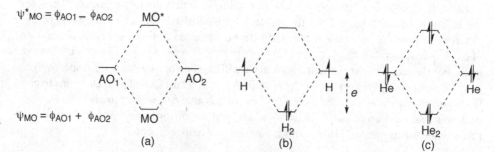

Figure 1.8 (a) AO and MO diagrams for an unspecified interaction, showing the bonding MO (lower) and antibonding MO* (upper), and their derivation from contributing AOs; (b) MO diagram for H_2, showing net stabilisation resulting from the filling only of the lower, bonding MO; and (c) MO diagram for He_2, showing no stabilisation resulting from the filling of both the bonding MO and antibonding MO* orbitals.

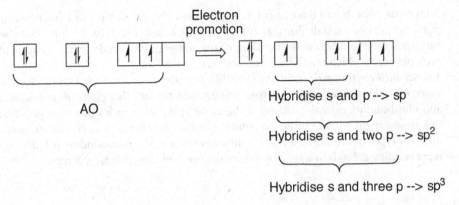

Figure 1.9 Hybridisation of the AOs of carbon to generate three possible hybrid orbitals.

those of another element to give a stable octet of electrons. Of course, we know that carbon is in fact tetravalent, forming, for example, CH_4, and in order to end up with this result, two important modifications to the basic LCAO concept are necessary:

1. *Hybridisation*: It is possible to describe more accurately the bonding in carbon compounds by allowing for the possibility that the AOs of the valence electrons are capable of mixing to form *hybrids*, a process resulting due to *hybridisation*. Three outcomes are common, resulting from mixing an s orbital and one, two or three p orbitals, and these are denoted as sp, sp^2 and sp^3, as illustrated in Fig. 1.9. However, the shapes of s and p orbitals have important consequences for the shapes of their derived hybrids. The sp orbital results from mixing one 2s and one 2p orbital, giving a linear hybrid which looks similar to a p orbital, but differs in the volume of space which it occupies (Fig. 1.10a). The sp^2 orbital results from mixing one 2s and two 2p orbitals, giving a planar trigonal orbital with the three lobes at 120° to each other (Fig. 1.10b). The sp^3 orbital results

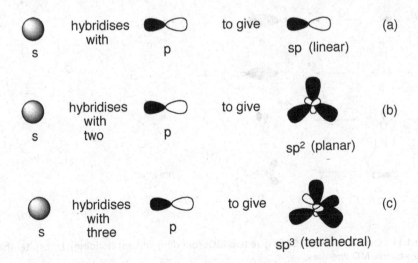

Figure 1.10 Hybridisation of AOs to give hybrid orbitals.

from mixing one 2s and three 2p orbitals, giving an orbital with the four lobes pointing to the vertices of a tetrahedron, at 109°28′ to each other (Fig. 1.10c). Which of these hybrids is used to describe the bonding in any compound depends on the number of multiple bonds which may be present (see Section 1.2.1).

2. *Valence shell electron pair repulsion (VSEPR)*: In order to successfully account for the shape of molecules, we need to invoke VSEPR, which states that electrons in bonding and non-bonding orbitals will seek to be as far apart from each other as is possible; this arrangement will minimise the repulsive interactions between electrons. However, different types of orbitals will repel to different extents, and non-bonding orbitals will repel bonding orbitals to a greater extent than two bonding orbitals will repel.

1.2   Covalent bonding

The overlap of AOs leads to MOs, and reorganisation of the valence electrons of atoms into MOs can lead to a net energy stabilisation. That is, a bonding arrangement provides an energetic advantage for the molecule to exist rather than its constituent atoms. Drawing the appropriate AO overlap for the constituent atoms of a molecule can account for single, double and triple bonds, which are commonly observed in organic compounds; σ- and π-bonds result due to different orientations of overlapping AOs. Thus, the formation of σ-bonds results due to the end-on overlap of s and s, s and p, or p and p orbitals, or of any of the hybridised orbitals sp, sp² or sp³ with s, p or other hybridised orbitals, as indicated in Fig. 1.11. These are denoted as σ (sigma, Greek s) by analogy with the s AO, because when viewed along the interatomic axis, the radial distribution of electron density is spherical,

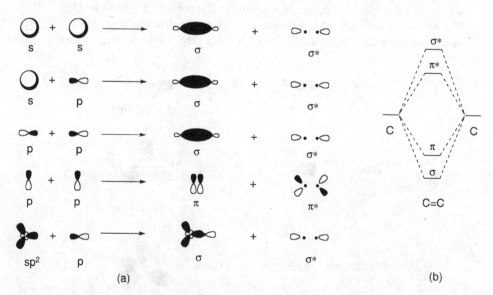

Figure 1.11   Overlap of two AOs to give two MOs (bonding and antibonding): (a) orbital shapes and (b) relative MO energies.

that is, the same as that of an s AO (Fig. 1.11). On the other hand, edge-on overlap of p and d orbitals leads to the formation of π-bonds (pi, Greek p), so named because when viewed along the interatomic axis, the electron density distribution has one node (which parallels that of a p AO). The sum of the AOs leads to a radial distribution of electron density, which is represented in Fig. 1.11. Both σ- and π-orbitals are important because they concentrate electron density between the bonding nuclei; that is, they lead to a net attractive interaction. MO theory of course also requires the existence of antibonding orbitals (σ^* and π^*), and their shape is indicated in Fig. 1.11. Antibonding orbitals are noteworthy because electrons occupying them are likely to be found away from the internuclear bond axis, resulting in a net repulsive interaction between nuclei, that is, an antibonding arrangement. The relative energies of σ-, π-, σ^*- and π^*-orbitals are indicated in Fig. 1.11b.

1.2.1 Bonding in hydrocarbons

For the hydrocarbons methane, ethane, ethene, ethyne and allene, the bonding is illustrated in Fig. 1.12. Thus, for methane, the four C–H bonds are formed by overlap of the C sp^3 and four H 1s orbitals; these σ-bonds point to the corners of a tetrahedron (Fig. 1.12a). For ethane, the situation is nearly identical, except that one of the H 1s orbitals is replaced

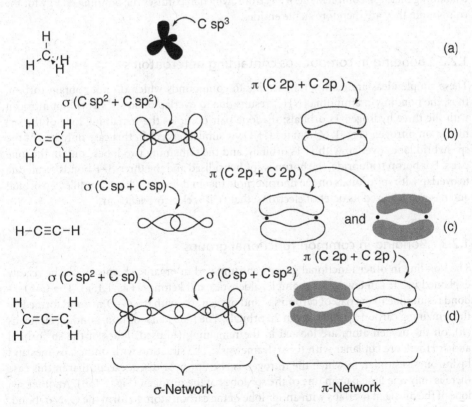

Figure 1.12 σ- and π-bonding networks in methane, ethene, ethyne and allene.

by the C sp^3 of the other carbon atom, leading to a carbon–carbon σ-bond formed by C sp^3–C sp^3 overlap. For ethene, in which there is a carbon–carbon double bond, we need to consider the σ- and π-frameworks separately (Fig. 1.12b). The carbon–carbon σ-bond is formed by C sp^2–C sp^2 overlap, and the four carbon–hydrogen bonds are formed by overlap of the remaining lobes on each of the C sp^2 orbitals with the four H 1s orbitals. The double bond is formed by overlap of the remaining 2p orbitals on each of the carbon atoms, but since the electrons in this bond are not located along the internuclear bond axis, they are less tightly held and therefore the bond is weaker. Of course, along with the bonding orbitals, the corresponding antibonding orbitals are formed, but since the electrons of each of the atoms of ethane can all be accommodated in the bonding orbitals only, the antibonding orbitals remain unfilled, and a strongly bonding arrangement exists. The carbon–carbon σ-bond in ethyne results due to overlap of C sp–C sp and the two double bonds by overlap of orthogonal 2p orbitals on each of the carbons (Fig. 1.12c). The situation in allene ($CH_2{=}C{=}CH_2$), in which there are two cumulative double bonds, is more unusual; for each of the double bonds, carbon–carbon σ-bond formation is via overlap of two sp orbitals, and the π double bonds are formed by mutually orthogonal overlapping sets of the remaining p orbitals, forming two double bonds, each orthogonal to the other (Fig. 1.12d). The orbital energy level diagrams for methane, ethene and ethyne are shown in Figs. 1.13a–1.13c respectively, in which addition of electrons to their respective orbitals results in all bonding orbitals containing their full complement of electrons, which maximises the bonding energy for the compound; they are therefore stable entities.

1.2.2   Bonding in compounds containing heteroatoms

These simple ideas are readily transferable to compounds which do not contain carbon; thus, the bonding in ammonia (NH_3) results due to overlap of an sp^3 hybrid on nitrogen with the three hydrogen 1s orbitals; the lone pair occupies the remaining lobe of the sp^3 hybrid on nitrogen (Fig. 1.14). Water (H_2O) is similar, except in this case only two of the sp^3 hybrid lobes overlap with H 1s orbitals, and the remaining two lobes contain the lone pairs. For boron trifluoride, the boron is sp^2 hybridised and the three B–F bonds result due to overlap with sp^3 orbitals on the fluorine; note that in this case, B has an unfilled 2p orbital and has not achieved its octet of electrons; that is, it is electron deficient.

1.2.3   Bonding in common functional groups

The bonding in other functional groups encountered in organic chemistry can be readily explained in the same way, for example, aldehydes and ketones (Fig. 1.15a). The C–O σ-bond results due to overlap of carbon sp^2 and oxygen sp^2, with the C–O π-bond formed by the remaining carbon 2p and oxygen 2p orbitals. The lone pairs, or non-bonding electrons (n), on the oxygen atom are located in the remaining (unused) lobes of the sp^2 orbital and are therefore coplanar with the σ-framework. The situation with imines, oximes and hydrazones is similar, in which the nitrogen is also sp^2 hybridised, except that in this case there is only one lone pair in one of the sp^2 lobes of the nitrogen (Fig. 1.15b). Another sp^2 lobe of the nitrogen overlaps with an sp^2 lobe of the carbon atom to form the C–N σ-bond, and the third sp^2 lobe overlaps with an H 1s orbital, O sp^3 or N sp^2 to form the C–H, C–O or C–N σ-bond of the imine, oxime or hydrazone respectively. Nitriles are like alkynes; thus,

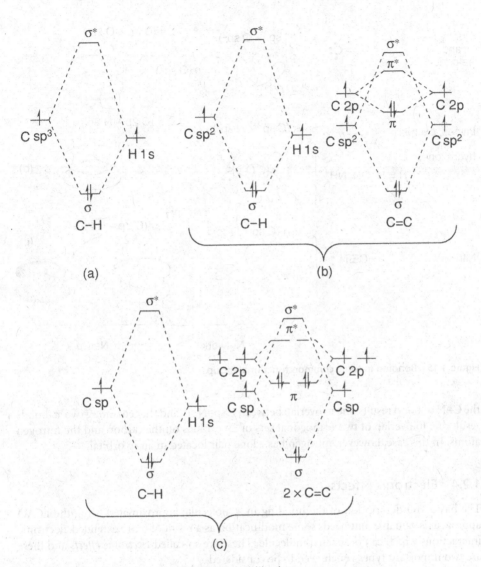

Figure 1.13   MOs in (a) methane, (b) ethene and (c) ethyne.

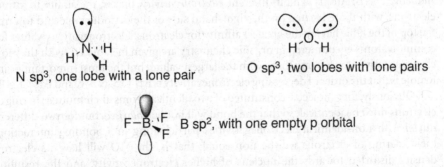

N sp³, one lobe with a lone pair

O sp³, two lobes with lone pairs

F–B   B sp², with one empty p orbital

Figure 1.14   Bonding in BF₃, H₂O and NH₃.

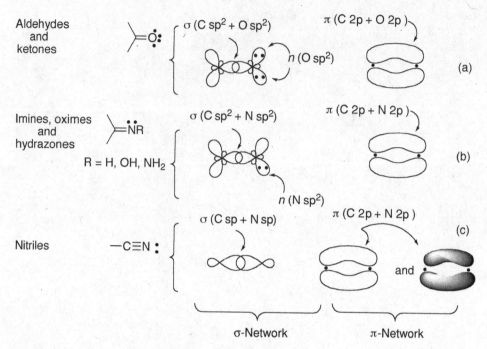

Figure 1.15 Bonding in some common functional groups.

the C–N σ-bond results due to overlap between C sp–N sp, and the remaining two π-bonds result due to overlap of two orthogonal sets of 2p orbitals on the carbon and the nitrogen atoms. In this case, however, nitrogen has a lone pair located in an sp orbital.

1.2.4   Electronic effects

The basic MO description of the bonding in a molecule, approximated using the LCAO approach, is valuable, but needs some modification as a result of more extended electronic interactions which can occur in a molecule. These are so-called *electronic effects* and there are two important types which need to be considered.

1. *Inductive effects*: Atoms exhibit different *electronegativity*, that is, the ability to attract electrons, with those located on the right-hand side of the periodic table and towards the top of the table having the greatest affinity for electrons. Electronegativity values for common atoms encountered in organic chemistry are given in Table 1.1, with the most electronegative atoms being those with the largest values, but the most important point to note is that the order of decreasing electronegativity is F > O > N > C and F > Cl > Br > I. Obviously, for a molecule constituted of two identical atoms, their mutual sharing of electrons must be identical, so that the bond will not be *polarised*, but for two different nuclei with non-identical electronegativities participating in a bonding interaction, their sharing of electrons will be non-equal; that is, the MO will have its electron density distorted towards the nucleus of higher electronegativity, and the resulting bond will be polarised. It is important to note that the difference in electronegativity

$$\begin{array}{cccc} \delta^+ \; \delta^- & \delta^+ \; \delta^- & \delta^+ \; \delta^- & \delta^- \; \delta^+ \\ \text{C–F} & \text{C–O} & \text{C–N} & \text{C–H} \end{array}$$

Figure 1.16 Polarisation in covalent bonds involving carbon.

involving carbon and most other atoms is not large enough to lead to the formation of completely ionic bonds, but the covalent bonding which does occur can experience a non-equal sharing of electrons between the bonding nuclei. We can represent this by considering one end of the bond to be negatively polarised and the other positively polarised. For example, C–F, C–O and C–N bonds are quite strongly polarised, as shown in Fig. 1.16, with the sense given by the electronegativity differences of each of the atoms in the bond. A C–N bond has only slight polarisation as indicated, since the electronegativities of carbon and hydrogen are nearly identical (2.5 and 2.1 respectively). Importantly, this effect can be transmitted along the σ-bonds of a molecule, so that one polarised bond can affect several others immediately adjacent to it. This can withdraw electron density from one part of a molecule to another, for example, in CF_3CO_2H or pentachlorophenol (Fig. 1.17), in which the fluorine or chlorine atoms withdraw electron density from as far as the O–H bond, giving it a partial positive charge. One additional complication is that the electronegativity of an atom can be modified by the type of bonding it participates in; this is particularly important for carbon in organic compounds. For example, the more s character of any hybrid orbital, the shorter and stronger bonds it will form, the larger its bond angles and the more electronegative it will be. Thus, an sp^3 hybrid orbital possesses only about one-fourth s character, but an sp^2 orbital possesses about one-third s, and so is more electronegative and forms stronger bonds. An sp hybrid is about half s in character and is even more electronegative. This is very important for the stabilisation of negative charge in that orbital.

2. *Mesomeric effects*: These effects (also called *resonance effects*) result due to the delocalisation or sharing of electron pairs over more than two atoms, and this leads to enhanced bonding and therefore greater stabilisation. This occurs when more than one Lewis structure can be written to describe the bonding of any one compound, but the true structure is considered to be a hybrid of both of them; more correctly, resonance occurs when orbitals overlap, and this occurs most efficiently when there are alternating single and double bonds, or when a lone pair is adjacent to a double bond (Figs. 1.18a and 1.18b). Resonance structures are linked using a double-headed arrow and enclosed in square brackets. Curved, double-headed arrows indicate the

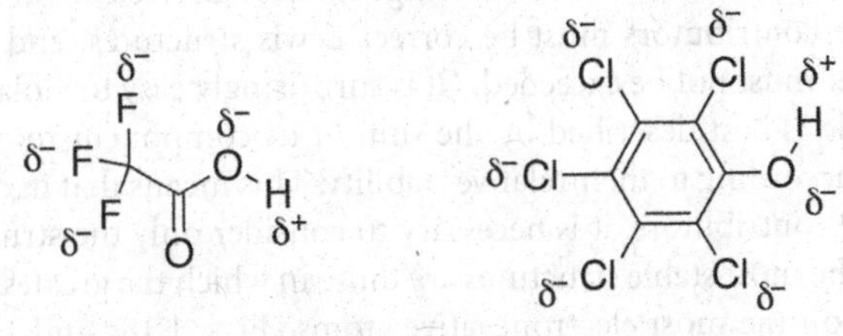

Figure 1.17 Polarisation in a carboxylic acid and an alcohol.

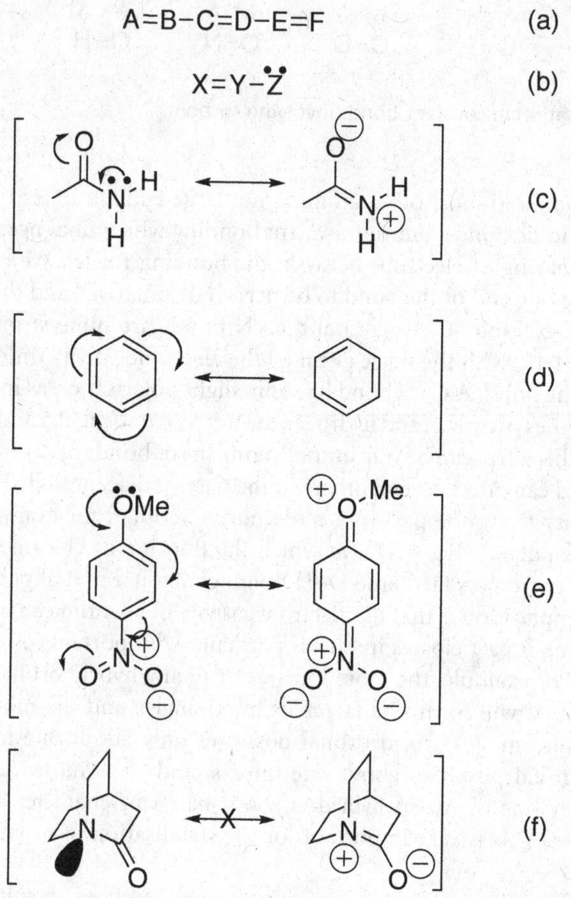

$$A=B-C=D-E=F \qquad \text{(a)}$$

$$X=Y-\overset{\bullet\bullet}{\overset{\bullet}{Z}} \qquad \text{(b)}$$

(c)

(d)

(e)

(f)

Figure 1.18 Examples of resonance in some molecules.

movement of two electrons corresponding to one covalent bond. For example, amides exhibit delocalisation of the electron pair on nitrogen through to oxygen, and there are two possible structures (Fig. 1.18c). Benzene has two so-called Kekulé resonance structures (Fig. 1.18d) and more extended resonance is possible if electron-donating and electron-accepting groups are appropriately placed on the ring (Fig. 1.18e). There are some important rules which must be followed if effective resonance is to occur:

(i) Nuclear positions must remain unchanged, although electrons are allowed to move.
(ii) Resonance contributors must be correct Lewis structures, and in particular the stable octet must not be exceeded. (It is surprisingly easy to violate this criterion.)
(iii) A molecule is best described by the sum of its component resonance structures weighted according to their relative stability. This means that in deciding the most important contributors, it is necessary to consider only the structures of greatest stability; the most stable structures are those in which the greatest electron density is located on the most electronegative atoms (Figs. 1.18c and 1.18e). Equivalent (called 'degenerate') structures contribute equally to the overall structure.

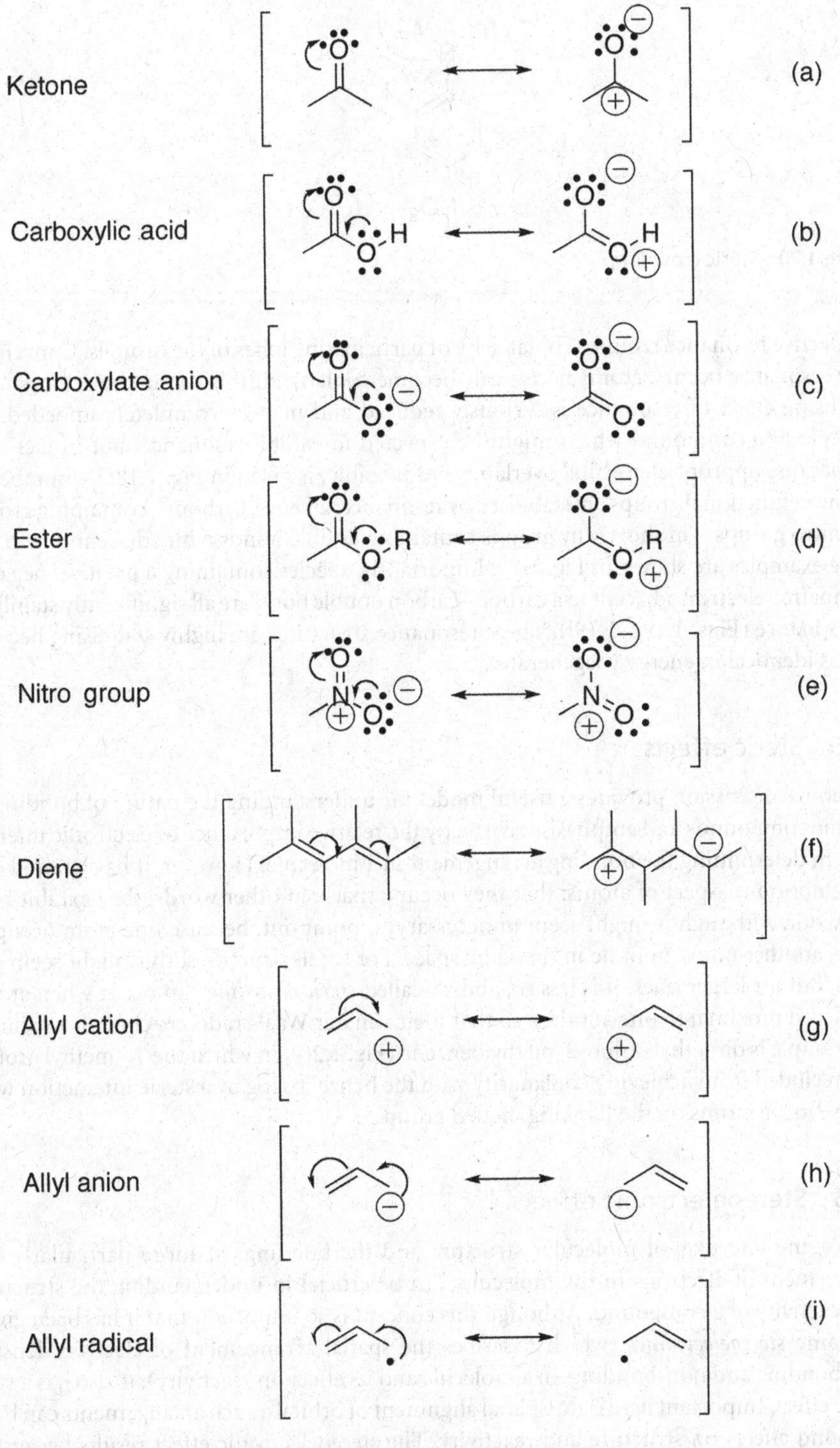

Ketone

Carboxylic acid

Carboxylate anion

Ester

Nitro group

Diene

Allyl cation

Allyl anion

Allyl radical

(a)

(b)

(c)

(d)

(e)

(f)

(g)

(h)

(i)

Figure 1.19 Resonance in common functional groups.

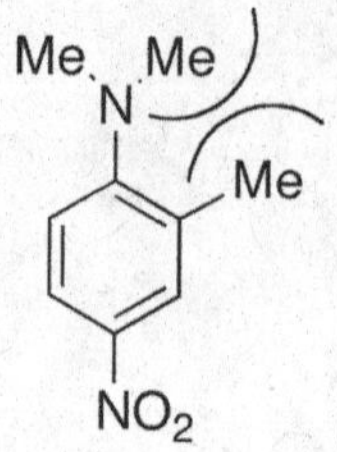

Figure 1.20   Steric crowding.

Effective resonance requires coplanarity of participating lobes of the orbitals. Conversely, once resonance occurs, atoms necessarily become coplanar. If this cannot be achieved, the stabilising effect of resonance is seriously reduced and may be completely impeded. An example of a compound which might be expected to exhibit resonance, but in fact does not because appropriate orbital overlap is not possible, is given in Fig. 1.18f. A number of common functional groups are stabilised by resonance: alkenes, carbonyl-containing groups and nitro groups – in short, any groups containing double bonds with adjacent lone pairs. Some examples are shown in Fig. 1.19. Importantly, species containing a positive, negative or unpaired electron adjacent to a carbon–carbon double bond are all significantly stabilised by resonance (Figs. 1.19g–1.19i). These resonance structures are highly stabilising because each is identical in energy (degenerate).

1.2.5   Steric effects

The above discussion provides a useful model for understanding the nature of bonding in organic compounds and emphasises correctly the relative importance of electronic interactions in determining the bonding arrangement in a molecule. However, it has ignored one very important aspect of atoms: that they occupy space; in other words, they exhibit bulk behaviour. Although it might seem unnecessary to point out, because one atom occupies space, another one cannot be in the same space. For small structures, this might seem obvious, but for larger ones, it is less so, and so-called *steric crowding* can occur when atoms are placed proximal to one another, so that their van der Waals radii are close or touching. An example is dimethylamino-2-methylbenzene (Fig. 1.20), in which the *N*-methyl groups are precluded from achieving coplanarity with the benzene ring by a steric interaction with the hydrogen atoms on the flanking methyl group.

1.2.6   Stereoelectronic effects

In fact, the interplay of molecular structure and the bonding, or more particularly the arrangement of electrons in the molecule, can be crucial in understanding the structure and reactivity of a compound. Although this concept is so important that it has been given the name *stereoelectronics* (which describes the spatial arrangement of electron density, both bonding and non-bonding, in a molecule and its effect on reactivity), it also has a very subtle effect. Important here is the spatial alignment of orbitals; such arrangements can have profound effects on structure and reactivity. The stereoelectronic effect results because a non-bonded electron pair (n) is able to interact with a suitably disposed antibonding orbital,

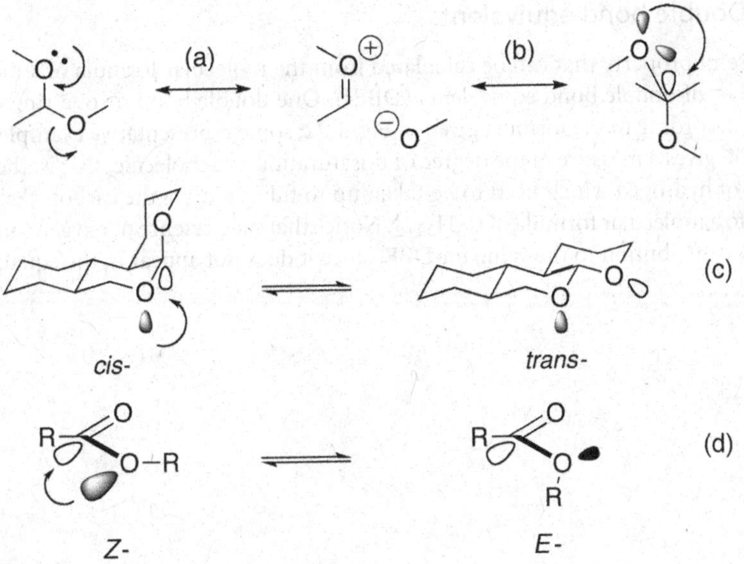

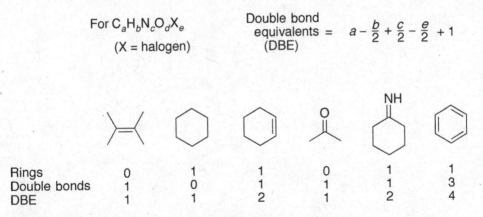

Figure 1.21 Stereoelectronic effects.

leading to what has been called 'bond–no bond' resonance; crucially, this interaction requires the correct spatial arrangement of orbitals. For example, an acetal can be considered to be stabilised as shown in Figs. 1.21a and 1.21b; the first of these indicates the 'classical' valence representation, and the second provides the more modern orbital explanation. In the latter case, coplanarity of the (filled) non-bonded orbital and the (unfilled) σ^*-orbital permits the transfer of electron density and the formation of a partial double bond; no other orbitals are so suitably aligned. An example of the importance of this phenomenon is that the *cis*-acetal (Fig. 1.21c) is more stable than the *trans*-acetal by about 7 kJ mol^{-1}. In a similar way, the *Z*-arrangement of an ester is preferred over the *E*-arrangement (Fig. 1.21d) due to suitable *n*–σ interaction, which is possible only in the *Z*-ester.

For $C_aH_bN_cO_dX_e$
(X = halogen)

Double bond
equivalents = $a - \dfrac{b}{2} + \dfrac{c}{2} - \dfrac{e}{2} + 1$
(DBE)

Rings	0	1	1	0	1	1
Double bonds	1	0	1	1	1	3
DBE	1	1	2	1	2	4

Figure 1.22 Calculations of double bond equivalents.

1.2.7   Double bond equivalents

A very useful property that can be calculated from the molecular formula of a molecule is the number of double bond equivalents (DBEs). One double bond or one ring counts as one DBE according to the formula given in Fig. 1.22; some representative examples are also given. DBE gives a measure of the degree of unsaturation in a molecule, that is, the number of moles of hydrogen which need to be taken up to fully saturate the carbon skeleton and return it to a molecular formula of C_nH_{2n+2}. Notice that the presence of oxygen (and sulfur) makes no contribution to the value for DBE, since it does not appear in the equation.

Chapter 2
Structure

The bonding which unites the constituent atoms of a molecule can be conveniently understood using the molecular orbital (MO) approach, which in turn is most simply considered to be a linear combination of atomic orbitals (AOs), as described in Chapter 1. Crucially, in determining the most appropriate types of atomic orbitals which contribute to an MO, we need to recall that hybridisation of AOs on first- and second-row elements is important, and the derived sp hybrids are linear, sp^2 hybrids are trigonal and sp^3 hybrids are tetrahedral; thus, these AOs induce well-defined structural constraints on bonds in which they participate and therefore on the overall shape of the molecule. The bonding of a molecule therefore directly determines its structure, and since the study of this aspect of chemistry involves a description of the shape of the molecule in three dimensions, it is called *stereochemistry*.

It is possible for compounds with the same molecular formula to have different structures, resulting from the different ordering of bonds. These are called *isomers* or more particularly, *constitutional isomers*. For example, for the molecular formula C_4H_{10}, there are two possible structures, namely, *n*-butane and *t*-butane, and for C_6H_{12}, there are the positional isomers 1-hexene, 2-hexene and 3-hexene, along with cyclic structures, such as cyclohexane and methylcyclopentane amongst others (Fig. 2.1). An additional level of complexity arises because tetrahedral carbons with four different substituents (a, b, c and d) can exist in two different spatial arrangements as shown in Fig. 2.2. These structures, although possessing a similar ordering of bonds, are not identical in three dimensions and are mirror images (provided a $\neq$ b $\neq$ c $\neq$ d) that are not superimposable. In this arrangement, the carbons are called *stereogenic*, and the molecule as a whole is *chiral*; that is, it is not superimposable on its mirror image, being related in the same way as left and right hands are related. The handedness of a stereogenic carbon atom is transferred to a molecule of which it is a part. This, and the fact that linear, trigonal and tetrahedral carbons are common bonding motifs, leads to some important complications when we consider the structure of organic compounds, and a satisfactory description of their structure relies on two additional key concepts *configuration* and *conformation*, and these are discussed in detail in the remainder of this chapter.

2.1 Configuration

The existence of tetravalent carbons as key linear, trigonal and tetrahedral bonding partners often leads to their derived organic compounds having well-defined three-dimensional structures and shapes. The three-dimensional structure of organic compounds is called the *absolute configuration*. Although this is most reliably determined by X-ray crystallographic analysis, it is often not possible, since the compound may not be crystalline or may not form suitable crystals. If crystallographic analysis is not possible, then indirect methods must be used, although these will not be discussed here. Significantly, configuration cannot be

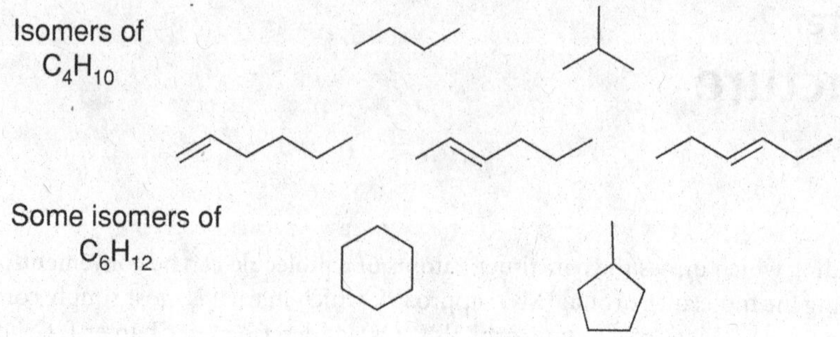

Isomers of
C_4H_{10}

Some isomers of
C_6H_{12}

Figure 2.1 Isomeric hydrocarbons.

altered without breaking and making bonds; that is, a chemical reaction is required; there is therefore a significant barrier to be overcome. Compounds which differ in their absolute configuration are called *stereoisomers*. This stereoisomerism can be of two types, *geometrical isomerism* and *optical isomerism*.

For the description of configuration, the Cahn–Ingold–Prelog (C–I–P) system was devised; for a given carbon atom, its substituents are assigned priorities based on their atomic number. (Isotopes of the same atom are ordered on the basis of increasing atomic mass; thus, $T(^3H) > D(^2H) > H$.) Atoms of higher atomic number are of higher priority. On this basis, a hierarchy of functional groups in descending order of precedence can be compiled: $-SMe > -SH > -OC(O)CH_3 > -OCH_2CH_3 > -OCH_3 > -OH > -NO_2 > -NMe_2 > -NH_2 > -CO_2H > -C(O)CH_3 > -CH(O) > -CH_2OH > -CN > -CH_2NH_2 > -Ph > -CMe_2CH_2CH_3 > -C(Me)=CH_2 > -CHMeCH_2CH_3 > -CH_2CH_2CH_3 > -CH_2CH_3 > -CH_3$. This hierarchy can be used to define the stereochemistry of a stereoisomer, as described in the following sections.

2.1.1 Geometrical isomerism

The restricted rotation around a carbon–carbon double bond leads to the existence of E/Z (also called *trans/cis*) isomers. The assignment of stereochemistry is determined by ordering the substituents of the double bond according to their atomic number, and the two substituents of highest priority on each of the carbons decide the configuration; if they lie on the same side of the double bond, they are *cis* or (Z) (for *zusammen*), and if on opposite sides, *trans* or (E) (for *entgegen*) (Fig. 2.3). This phenomenon also applies for C=N systems, although in this case the isomers are called *syn* (when the highest priority substituents on carbon and the nitrogen are located on the same side) and *anti* (when on the opposite sides). In this case, the lone pair is always of lowest priority.

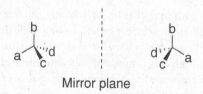

Mirror plane

Figure 2.2 Chiral tetrahedra have non-superimposable mirror images.

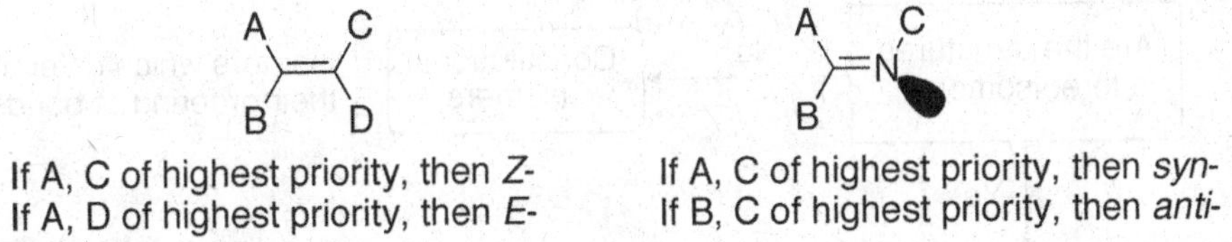

If A, C of highest priority, then *Z*-
If A, D of highest priority, then *E*-

If A, C of highest priority, then *syn*-
If B, C of highest priority, then *anti*-

Figure 2.3 Assignment of geometric isomers.

2.1.2 Optical isomerism

The different absolute configurations of stereogenic carbons can lead to the existence of different stereostructures, called enantiomers and diastereomers. Any molecule which cannot be superimposed on its mirror image is an *enantiomer* (Fig. 2.2). Stereoisomers which are not enantiomers are called *diastereomers*. Compounds which are not chiral are called *achiral*. Although straightforward to state, the full implications of these definitions are not immediately apparent and require some careful thought: Notice that both of the terms – enantiomers and diastereomers – are defined in terms of what it is they are not and that both terms describe a relationship between two structures, not one in isolation. Depending on what it is being compared with, any one structure could be both enantiomeric and diastereomeric. In order to reliably assign stereoisomers as enantiomers or diastereomers, there are several key questions to be answered:

1. 'Are the structures stereoisomers?' That is, do they differ only in the arrangement of atoms in space? If not, they are constitutional isomers, and the following analysis is not applicable.
2. 'Are the structures mirror images?' The ability to answer this question depends on the ability to accurately draw mirror images; some people find this easy and others difficult, so some practice may be necessary. One helpful technique is to use a small mirror to see the mirror image and to transcribe this (accurately!) on to a page.
3. The third is 'Are the structures non-superimposable?' This is probably the most difficult question of all to answer, but is crucial to the overall analysis.

A flow chart which illustrates this procedure is given in Fig. 2.4, and application of this sequence permits discrimination among constitutional isomers, diastereomers and enantiomers; a fourth category which can also arise, *meso* isomers, will be considered in Section 2.2.6. The following examples will serve to clarify the assignment of stereoisomers as enantiomers or diastereomers using this sequence (Fig. 2.5). Thus, the hydrocarbons of Fig. 2.5a are constitutional isomers, whose bond connectivity is fundamentally different; therefore they cannot be stereoisomers. Although the derivatives shown in Fig. 2.5b have tetrahedral carbon atoms, they are not stereoisomers, since their mirror images are identical (because two substituents are identical), whilst those shown in Fig. 2.5c are chiral, since each carbon has four different substituents and the two structures are related as non-superimposable mirror images; these structures are called enantiomers. The structures shown in Fig. 2.5d are also enantiomers, although in this case each molecule has two stereogenic carbon atoms. However, in Fig. 2.5e, the structures are diastereomers, since they are stereoisomers that are not enantiomers. Having established the existence of enantiomers, it is possible to assign the configuration using the C–I–P rules. By placing the atom of lowest atomic number furthest away from the viewer, the carbon is assigned priority (*R*) if the

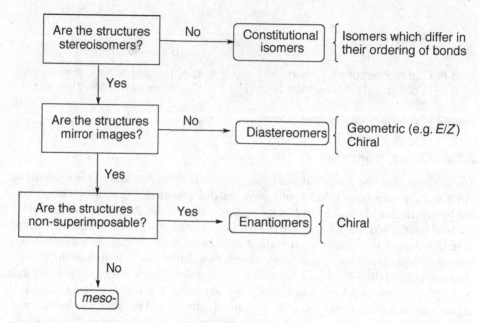

Figure 2.4   Route map for the assignment of isomers.

ordering (atomic number decreasing highest to lowest) of the remaining three is clockwise (*R = rectus = right*) or (*S*) if their ordering is anticlockwise (*S = sinister = left*). This is illustrated in Fig. 2.6 along with some examples. If assignment is not possible by considering the first four substituents, then prioritisation based on atomic number is applied successively at further points away from the stereogenic centre until a decision can be made; if this

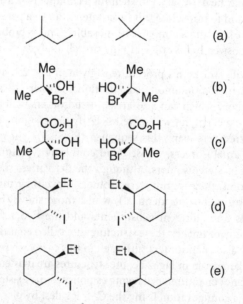

Figure 2.5   Examples of isomers.

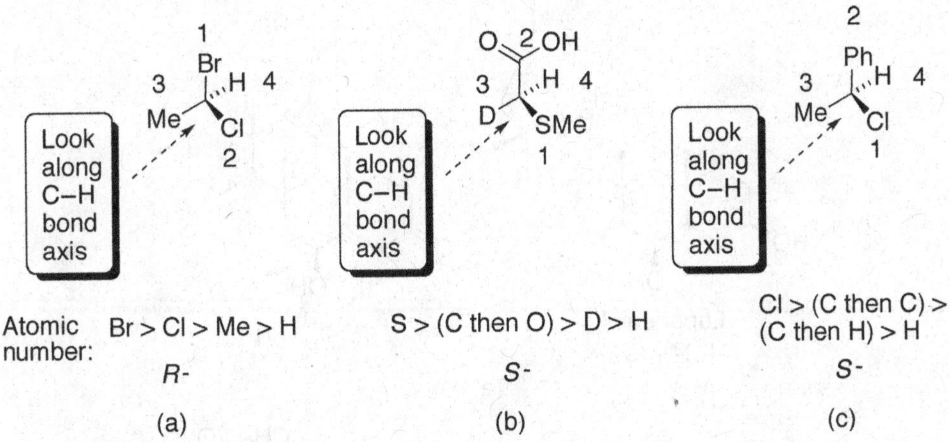

Figure 2.6 Examples of the assignment of C–I–P rules.

is not possible after the third iteration, then special rules apply, which will not be discussed here. In order to determine the absolute configuration of a molecule, this analysis needs to be applied to each one of its stereogenic centres.

Experimentally, optical isomerism is manifested as *optical activity*: This is the ability of a compound to rotate the plane of plane-polarised light, expressed as the *specific rotation* $[\alpha]_D$. Rotation to the right is expressed as a positive value $(+)$ and rotation to the left as negative $(-)$; numerically, the value of $[\alpha]_D$ can vary from 0 to several thousands. This technique can be used to determine the *optical purity* of a compound; this in turn can be used to estimate its *enantiomeric excess* (excess of one enantiomer over the other). All optical isomers are chiral, but all compounds with stereogenic centres do not necessarily lead to an observable optical rotation, since this value depends on the structure of the compound, its concentration and the nature of the solvent. The $[\alpha]_D$ value cannot be easily predicted from the structure of the compound, nor from its C–I–P descriptor, and can be determined only empirically. This point is illustrated in Fig. 2.7a; lupenediol and its relative, which differ only in having alcohol and alkene functionality replaced by ketone and alcohol functions, have specific rotations that differ in both magnitude and sign. Similarly, in Fig. 2.7b, an apparently small change in structure of (R,R)-1-methyl-1,2-cyclohexanedicarboxylic acid (diacid → anhydride) gives a very significant change in magnitude and sign of the specific rotation. (R)-3-Methylheptanedioic acid, on the other hand, virtually exhibits no change among the free acid, dimethyl ester and dianilide (Fig. 2.7c).

2.1.3 Representations of stereoisomers

That organic compounds exist as three-dimensional structures causes us an immediate problem: How do we represent these structures on a two-dimensional page? There have been a number of solutions to this problem, and each has its merits, depending on the complexity of the structure in question. The simplest approach for a single tetrahedral carbon is to use wedge–dash diagrams of the type shown in Figs. 2.2 and 2.6. For molecules with more than one stereogenic centre, though, this representation becomes cumbersome, and different conventions can be used (Fig. 2.8). Fischer projections are particularly useful in this case, but some important rules for their use apply. Firstly, horizontal bonds are assumed

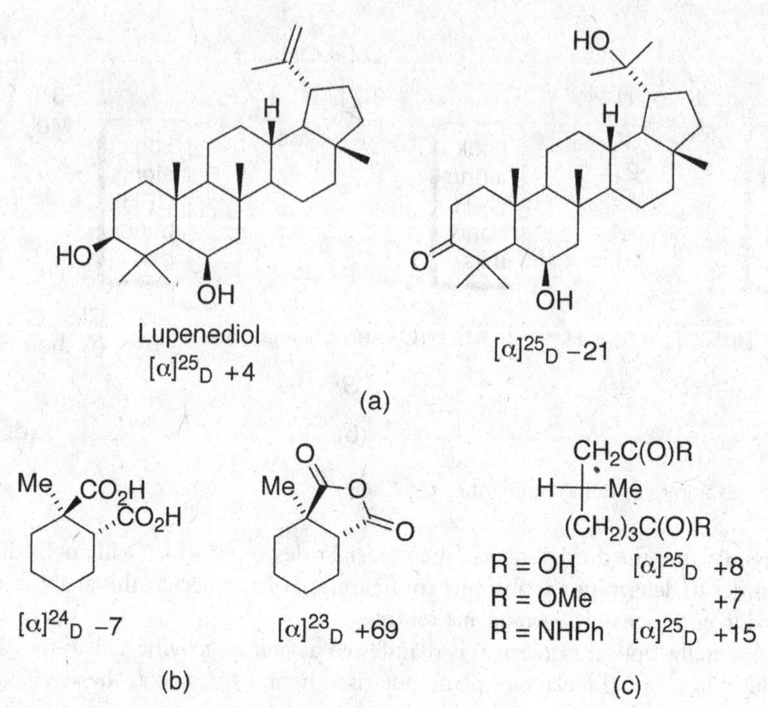

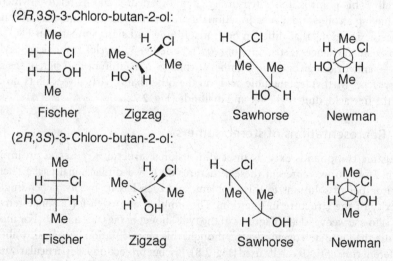

Figure 2.7   Examples of specific rotation.

to come out of the page, and secondly, the structures may be translated or rotated in the plane of the page, but must not be taken out of that plane. Sawhorse projections are useful, too, as they give a clear indication of relative stereochemistry as well as a sense of perspective. Newman projections are useful to visualise the structure of a compound looking along a C–C bond axis.

(2*R*,3*S*)-3-Chloro-butan-2-ol:

Fischer   Zigzag   Sawhorse   Newman

(2*R*,3*S*)-3-Chloro-butan-2-ol:

Fischer   Zigzag   Sawhorse   Newman

Figure 2.8   Representations of stereoisomers (Fischer, zigzag, sawhorse and Newman projections).

2.1.4 Molecules with one stereogenic centre

For molecules with a single stereogenic centre (i.e. a tetrahedral atom bearing four different substituents), there is the possibility of only one sort of stereoisomerism, namely, that of enantiomerism, since the mirror images of the molecule are non-superimposable, as discussed in Section 2.1.2 and shown in Fig. 2.2. Compounds which consist of only one type of enantiomer are said to be *enantiomerically* (and therefore *optically*) *pure*, but those consisting of an equal mixture (50:50) of each type of enantiomer are *racemic mixtures* (*racemates*); for the latter, the $[\alpha]_D$ will be zero, since there are an equal number of left- and right-hand rotating stereocentres. The process of converting a pure enantiomer to a racemate is called *racemisation*. The separation of a racemate into its enantiomerically pure components is called *resolution*.

Enantiomers possess identical chemical and physical properties except in a chiral environment. Thus, they have identical melting points, boiling points, solubilities and nuclear magnetic resonance (NMR) spectra, but different $[\alpha]_D$ values and different chemical behaviour with other chiral compounds.

2.1.5 Molecules with more than one stereogenic centre

For molecules with more than one stereogenic centre, enantiomerism as well as diastereomerism becomes possible. The definitions of these terms as given above do not change, but their implications in this case become more subtle and sophisticated. Thus, enantiomers arise if there are non-superimposable mirror images. All other types of stereoisomers are diastereomers. This is best illustrated using monomethyl tartarate as an example (Fig. 2.9). Structures A and B and C and D are mirror images and non-superimposable, so are enantiomers. However, A relates to both C and D as a stereoisomer which is not an enantiomer, that is, a diastereomer. Similarly for B → C and D, C → A and B, and D → A and B. This emphasises the point made earlier that a compound may be both an enantiomer and a diastereomer, depending on what it is being compared with. A useful rule of thumb is that for a compound with *n* stereocentres, there will be a maximum of 2^n stereoisomers; in this case, two stereogenic centres generate four stereoisomers. Compounds A and B, in which the two identical substituents are on opposite sides of the main carbon backbone, are

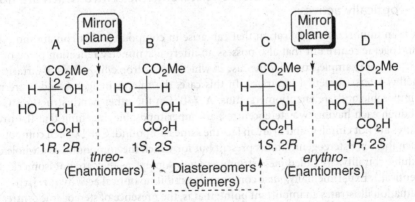

Figure 2.9 Stereoisomers with two stereogenic centres.

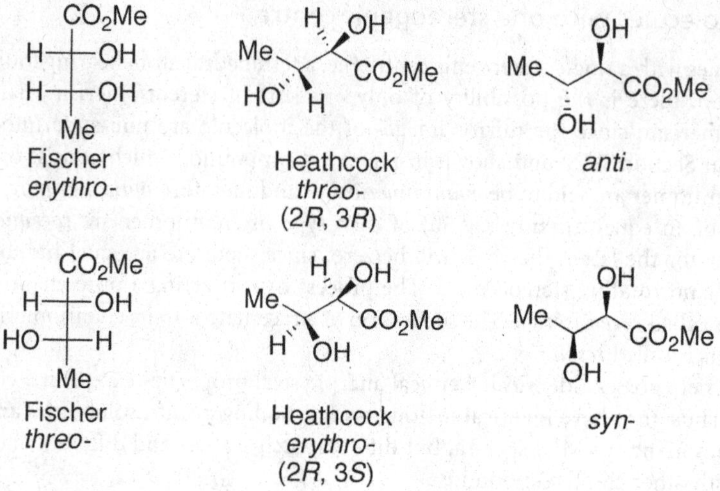

Figure 2.10   Representations of stereoisomers.

called *threo* isomers, and C and D, in which the identical substituents are on the same sides of the main carbon backbone, are called *erythro* isomers. Diastereoisomers which differ in configuration at just one carbon are called *epimers*. Unlike enantiomers, the chemical and physical properties of diastereomers are different. *Relative configuration* describes the relative disposition of all of the substituents but does not differentiate between the two enantiomeric forms. However, a cautionary note is worth making at this stage: Traditionally, the Fischer projection was used to assign *erythro* and *threo* to acyclic structures. More recently, the same descriptors were applied to the more widely used zigzag line representations, and so there is now a situation in which these two terms can mean exactly the opposite, depending on the convention which has been followed (and this is not always made clear!). The only reliable descriptor in use now is *syn/anti*, as shown in Fig. 2.10. In this scheme, the relationship of substituents is described relative to the main carbon backbone: If on the same side of the zigzag projection, they are called *syn*, and if on different sides, *anti*.

2.1.6   Molecules with more than one stereogenic centre which are not optically active

There is an important complication that can arise in compounds that contain more than one stereogenic centre but that also possess an internal symmetry reflection plane or axis (Fig. 2.11). For example, consider the case in which both carboxylic groups of tartaric acid are methyl esters (dimethyl tartrate). In this case, the mirror images A and B are non-superimposable and are enantiomers; thus, A $\neq$ B. On the other hand, structures C and D, although each having two stereocentres, are superimposable, meaning that the overall structures are not chiral; C and D are in fact the same compound: C = D. Such compounds, in which there are stereogenic atoms present but for which the compound as a whole does not exhibit chirality, are called *meso* structures (*not* isomers, as they are not isomeric, they are identical!). Experimentally, *meso* compounds exhibit no optical activity (i.e. $[\alpha]_D = 0$). This situation illustrates an important point; that is, the presence of stereogenic centres in a molecule does not automatically confer chirality on that molecule; for chirality to exist, the

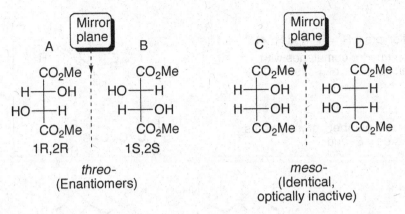

Figure 2.11 Stereoisomerism in dimethyl tartarates.

key requirement is that the molecule *as a whole* must be non-superimposable on its mirror image.

2.1.7 Optically active molecules without stereogenic centres: molecular asymmetry

Not all compounds which exhibit bulk chirality have stereogenic centres. As we have seen, the key criterion for the presence of optical activity is the fact that mirror images of a molecule should be non-superimposable. Some examples are considered in more detail in the following sections.

2.1.7.1 Molecules with structural orthogonality

(a) *Allenes*: Compounds in which there are two cumulated double bonds (Fig. 2.12a) have an unusual structure, in which the substituents at each end of each double bond are orthogonal, and provided that $R^1 \neq R^2$ and $R^3 \neq R^4$, it is not superimposable on its mirror image; that is, the compound exists as enantiomers. A similar situation also exists for even-numbered cumulated double bonds ('*cumulenes*'), but for odd-numbered cumulenes, only geometrical isomerism is possible, with *E*- and *Z*-possibilities.

(b) *Spiranes*: A similar situation also exists with spirane systems, in which one sp^3-hybridised centre is part of two rings (Fig. 2.12b); these rings are necessarily orthogonal and provided that the rings are appropriately substituted will have non-superimposable mirror images; that is, they will exist as enantiomers.

(c) *Hindered biphenyls*: Two aromatic rings linked as indicated in Fig. 2.12c and substituted at the *ortho* positions suffer from restricted rotation. Provided that the rotation is sufficiently slow at room temperature, the mirror images are non-superimposable, and the compound exists as enantiomers.

2.1.7.2 Molecules with structural helicity

(a) *Helicenes*: Helical structures exhibit chirality, depending on whether it is a left-handed or right-handed twist. The polyaromatic structure in Fig. 2.13a is an example, and of course so is DNA and RNA.

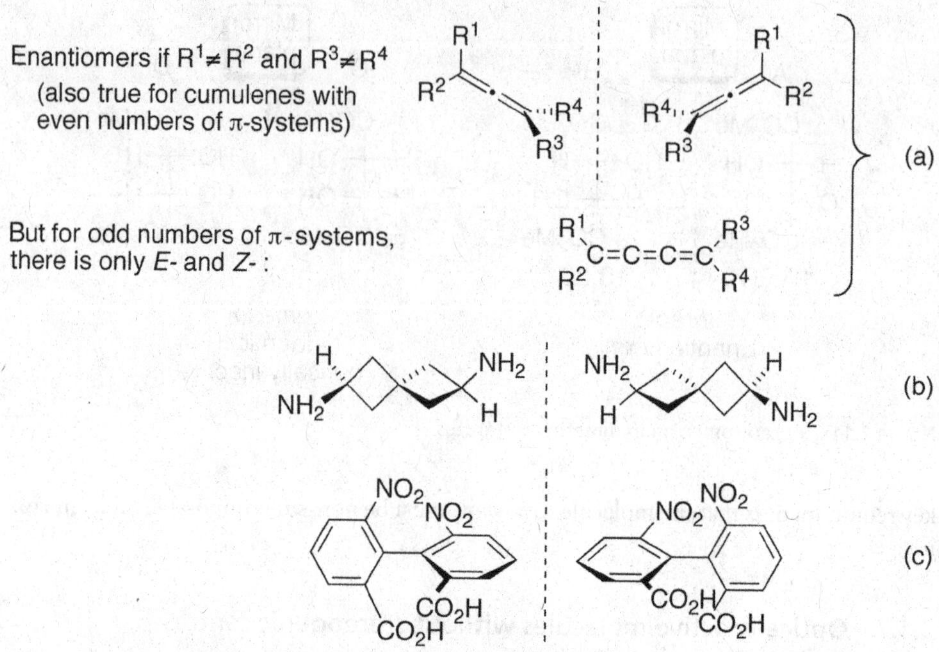

Enantiomers if $R^1 \neq R^2$ and $R^3 \neq R^4$
(also true for cumulenes with
even numbers of π-systems)

But for odd numbers of π-systems,
there is only *E*- and *Z*- :

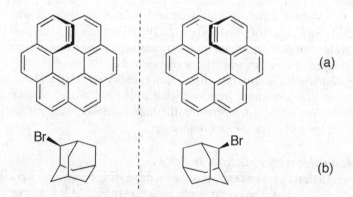

Figure 2.12   Chiral compounds without stereogenic centres.

(b) *Substituted adamantanes*: The rigid structure of adamantane leads to the possibility of non-superimposable mirror images, again provided that there is appropriate substitution on the ring structure (Fig. 2.13b).

2.1.8   Asymmetric heteroatoms

It is not only tetrahedral carbon atoms that can be stereogenic, but other heteroatoms in appropriate molecular arrangements may also exhibit chirality.

Figure 2.13   Chiral compounds without stereogenic centres.

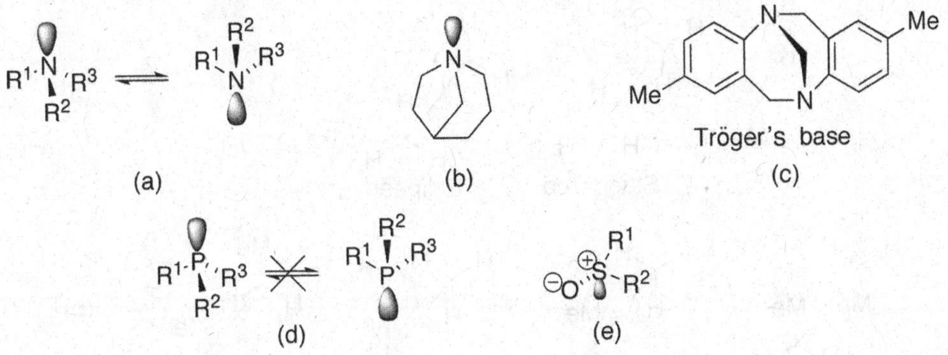

Figure 2.14 Stereogenic heteroatoms.

2.1.8.1 Nitrogen

Most amines are unresolvable at room temperature due to rapid Walden inversion, which converts one isomer into another (Fig. 2.14a); however, if the substituents are constrained, such as in the bicyclic structure indicated, then enantiomers are, in principle, resolvable (Fig. 2.14b). Tröger's base is an example of such a system (Fig. 2.14c).

2.1.8.2 Phosphorus and sulfur

Chiral phosphines (R_3P) are resolvable, as the inversion occurs only slowly even at 200°C (Fig. 2.14d). Sulfoxides are also configurationally stable at room temperature (Fig. 2.14e).

2.2 Conformation

So far, we have discussed the importance of absolute configuration in the assignment of structure to a molecule. Molecules, however, are not static entities, and possess translational, rotational and vibrational energy, as well as the capacity for rotation around the internuclear bond axis of single bonds. Different molecular structures which are accessible by rotation around single bonds are called *conformers* or *rotamers*, and phenomena resulting from this rotation are called *conformational effects*. Generally speaking, the activation energy for rotation around bonds in simple alkanes is very small ($12\,\text{kJ}\,\text{mol}^{-1}$) and can be easily surmounted at ambient temperature. Moreover, because the energy involved in interconverting different conformers is so small (unlike the interconversion of configurational isomers), involving as it does only bond rotation, it is very difficult, if not impossible, to separate conformers.

2.2.1 Representation of conformers

Newman and sawhorse projections are particularly useful for drawing the conformers of any molecular structure (see Figs. 2.15 and 2.16).

2.2.2 Open-chain compounds

For ethane, there are two limiting conformational structures, in which all the hydrogen substituents are *eclipsed*, representing an energy maximum, and *staggered*, representing an

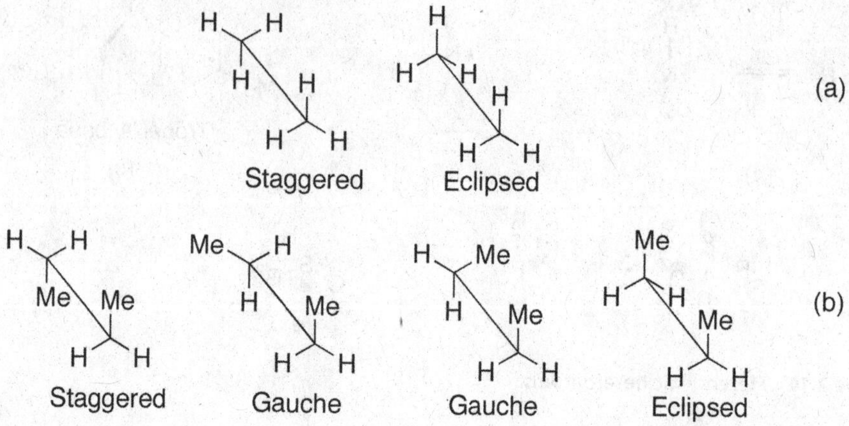

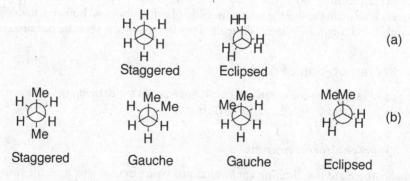

Figure 2.15 Sawhorse representation of conformation in (a) ethane and (b) butane.

energy minimum, although joining these two extremes gives an infinity of possible structures which are accessed by incremental changes in the *torsional angle* describing the intersection of the two planes through the H–C–C–H bond network (Figs. 2.15a and 2.16a). For butane, these staggered and eclipsed conformers also exist, but it is possible to buttress either a hydrogen and a hydrogen or a carbon and a hydrogen together (Figs. 2.15b and 2.16b). This leads to the existence of several intermediate stages, called *gauche*. More generally, substituents which are located in the same hemispheric segment are called *syn*, and those located in opposite halves are *anti*. If they are located in the same plane, they are called *coplanar*, and it is possible to use these terms in combination, thus *syn-periplanar* and *anti-periplanar* relationships (Fig. 2.17).

However, it is important to note that apparently conformationally mobile systems are not always as freely mobile as might first appear. A good example relates to the conformational mobility in alkenes. In the monosubstituted alkene shown in Fig. 2.18a, there is a modest energetic preference for the conformer in which the 1,3-related hydrogen substituents are coplanar; this avoids the steric hindrance which accrues in the alternative conformer in which the 1,3-hydrogen and methyl substituents are coplanar. For a disubstituted alkene

Figure 2.16 Newman projections of conformation in (a) ethane and (b) butane.

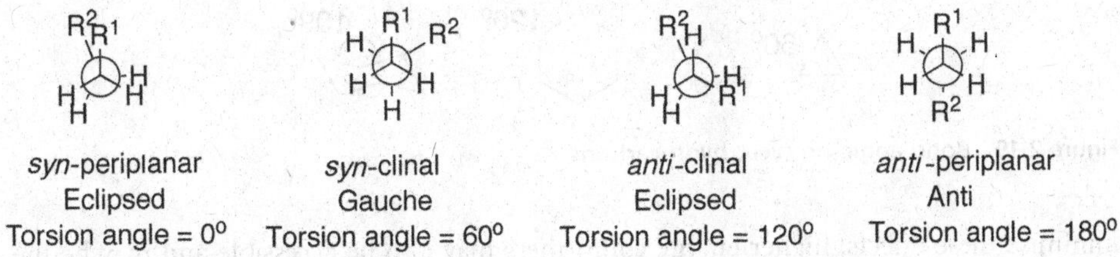

syn-periplanar	syn-clinal	anti-clinal	anti-periplanar
Eclipsed	Gauche	Eclipsed	Anti
Torsion angle = 0°	Torsion angle = 60°	Torsion angle = 120°	Torsion angle = 180°

Figure 2.17 Conformational relationships.

(Fig. 2.18b), the situation becomes more strongly biased and coplanar 1,3-dimethyl groups are strongly destabilised on the basis of steric effects and so the preferred conformer is the alternative in which the 1,3-related hydrogen and methyl groups are coplanar. Obviously, this type of conformational preference becomes even more important with more bulky groups instead of methyls. This phenomenon has been realised to be of considerable importance, since it influences the reactivity of such systems on steric grounds, and is called $A^{1,3}$ strain. A similar preference is observed in more highly functionalised systems, and has been called $A^{1,2}$ strain, which leads to the preference for the conformer indicated (Fig. 2.18).

2.2.3　Ring compounds

In the case of cyclic compounds, there can be a severe restriction of available conformers due to their ring structures. The conformational behaviour of rings is dependent on their size; in some cases the conformational behaviour is very well defined and is controlled by the interplay of several factors. However, conformations are not fixed, with several structures existing together, and the position of the equilibrium can be altered by other external factors.

2.2.3.1　Factors controlling ring conformations

The conformation of molecules can be determined by a combination of the effects of angle strain, torsional strain, non-bonding interactions, hydrogen bonding and dipole effects, and stereoelectronic effects. However, although a molecule will adopt a conformation so as to

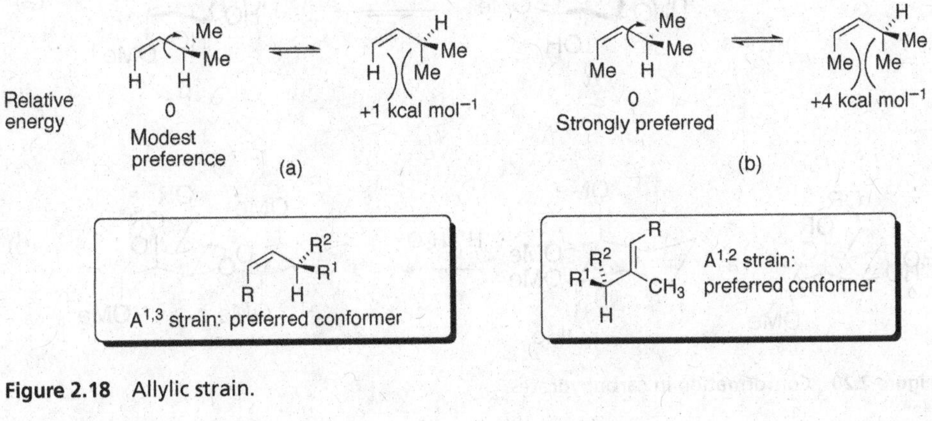

Relative energy	0	+1 kcal mol⁻¹	0	+4 kcal mol⁻¹
	Modest preference		Strongly preferred	
	(a)		(b)	

$A^{1,3}$ strain: preferred conformer

$A^{1,2}$ strain: preferred conformer

Figure 2.18 Allylic strain.

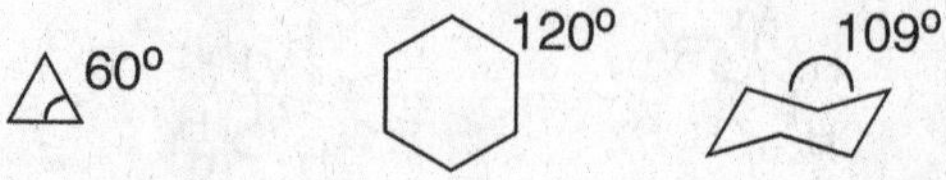

Figure 2.19 Bond angles in cyclic hydrocarbons.

minimise these effects, higher energy conformers may also be accessible and may be the more reactive entity. In detail, there are:

1. *Angle strain*: Deviation of bond angles from the ideal values of sp^3 (109°28′)- and sp^2 (120°)-hybridised carbons leads to angle strain. In cyclopropane, for example, bond angles of only 60° lead to significant angle strain, and a similar situation would exist for a cyclohexane if the ring were planar with angles of 120°, but puckering into a chair conformation allows the more normal sp^3 angle to be achieved (Fig. 2.19).

2. *Torsional strain*: Rotation around single bonds can give rise to conformers in which eclipsing interactions occur; this can give rise to *torsional* strain. A molecule will exist in a rotamer such as to minimise torsional strain as far as possible.

3. *Non-bonded interactions*: When ring substituents become sufficiently close in space that their separation approaches the sum of their van der Waals radii, steric interactions occur. Obviously, two atoms cannot occupy the same volume of space, and such interactions can be minimised by conformational modification.

4. *Hydrogen bonding and dipole effects*: H-bonding effects, linking electropositive hydrogen with electron-donor atoms such as O, N and S, and other dipolar effects can be important.

5. *Stereoelectronic effects*: The interplay of lone pairs of electrons with polar σ-bonds can lead to important conformational effects. For example, in glucose, an equilibrium between the α- and β-forms exists, in which the preferred conformer is the β (axial)-isomer (Fig. 2.20a). This conformer is favoured because of stabilising orbital overlap between one of the oxygen lone pairs and the σ* (C—O) orbital. This effect can operate in more elaborate systems, and as a result, polycyclic spiroacetal systems, for example, exhibit very well defined conformational preferences (Fig. 2.20b).

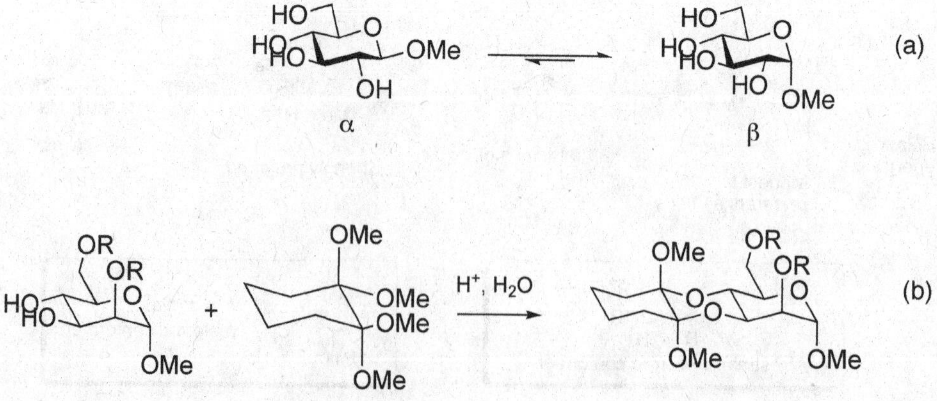

Figure 2.20 Conformation in carbohydrates.

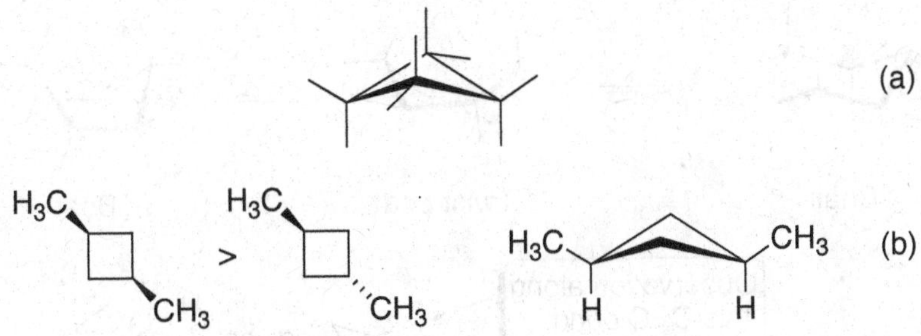

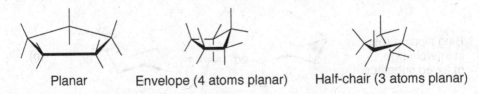

(a)

(b)

Figure 2.21 Conformation in cyclobutanes.

2.2.3.2 Conformation and ring size

In cyclic systems, the effects described in Section 2.2.3.1 can play particularly important roles:

1. *Cyclopropane*: Because there are only three atoms in this ring, it is necessarily planar. The resulting angle and torsional strains make the ring highly susceptible to ring-opening reactions.
2. *Cyclobutane*: The four-membered ring is in fact not planar but puckered; that is, one carbon is out of the plane of the other three. This arrangement helps to avoid eclipsing interactions between substituents on adjacent ring carbons, which would be inevitable in a fully planar structure (Fig. 2.21a). *syn*-1,3-Disubstituted cyclobutanes are more stable than their *anti*-isomers, since the substituents can adopt a pseudo-diequatorial arrangement, which minimises steric interactions across the ring (Fig. 2.21b).
3. *Cyclopentane*: If this ring was planar, although angle strain would be very small (due to an angle of $108°$ compared to the ideal of $109°28'$), torsional strain would be substantial as a result of eclipsing interactions. In order to minimise these interactions, the ring puckers to give an envelope or half-chair form (Fig. 2.22). However, the energy differences between possible conformers are small, and *pseudo-rotation*, in which each carbon is alternatively out of the plane of the ring, is especially facile and leads to an averaging of stereochemical environments at each position around the ring. Thus, cyclopentanes do not usually have conformationally well-defined structures.
4. *Cyclohexane*: In cyclohexane, there are several forms with minimum angle strain, namely, *chair* and *boat*, which interconvert via a twist-boat form, but which possess different torsional strains. The lowest energy conformer is the chair form, and in this all ring carbons benefit from a staggered arrangement of substituents, unlike the alternative boat conformer (Fig. 2.23). In the chair conformer, there are two distinct

Planar   Envelope (4 atoms planar)   Half-chair (3 atoms planar)

Figure 2.22 Conformation in cyclopentanes.

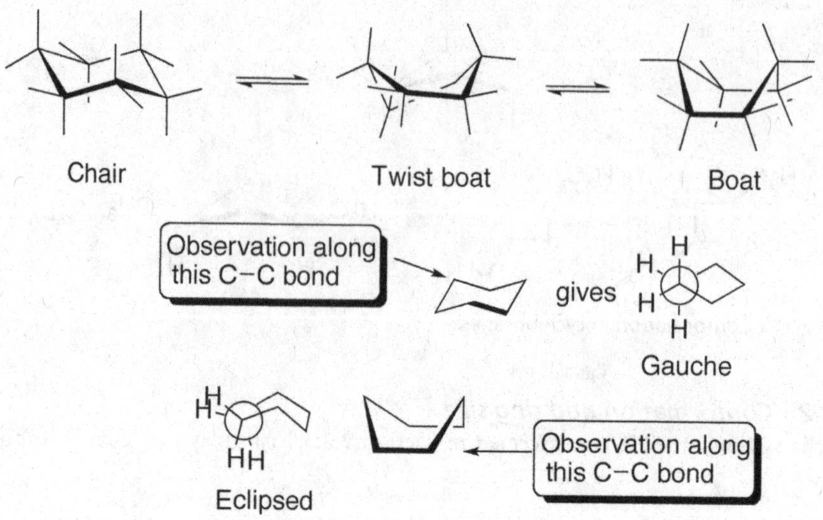

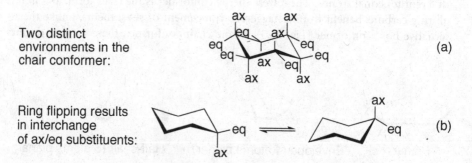

Figure 2.23   Conformation in cyclohexanes.

orientations of the ring substituents – *axial* (approximately perpendicular) and *equatorial* (approximately coplanar) – relative to the (distorted) plane of the ring carbons (Fig. 2.24a). If the ring undergoes a complete flip, an alternative chair form can be accessed in which all axial and equatorial substituents of the original form are interchanged (Fig. 2.24b). Diaxial interactions between the 1- and 3-positions are particularly destabilising (Fig. 2.25a), and for this reason, bulky substituents prefer to be equatorial if this can be achieved by ring flipping (Fig. 2.25b). The *t*-butyl group is very effective at achieving this conformational restriction by residing in an equatorial position, and is often called a conformational lock because it preferentially remains in this position. However, other less stable conformers are accessible by a ring flip, and these include half-chair, twist-boat and boat conformers; the latter is the most unstable, since the ring carbons suffer from eclipsing interactions as well as 1,4-bowsprit interactions (Fig. 2.25c). In the case of two bulky groups, a twist-boat conformation may be adopted and preferred, allowing both substituents to adopt a pseudo-equatorial position, for example, *trans*-1,3-di-*t*-butylcyclohexane (Fig. 2.26). However, these preferences can be modified

Two distinct
environments in the
chair conformer: (a)

Ring flipping results
in interchange
of ax/eq substituents: (b)

Figure 2.24   Conformational relationships in cyclohexanes.

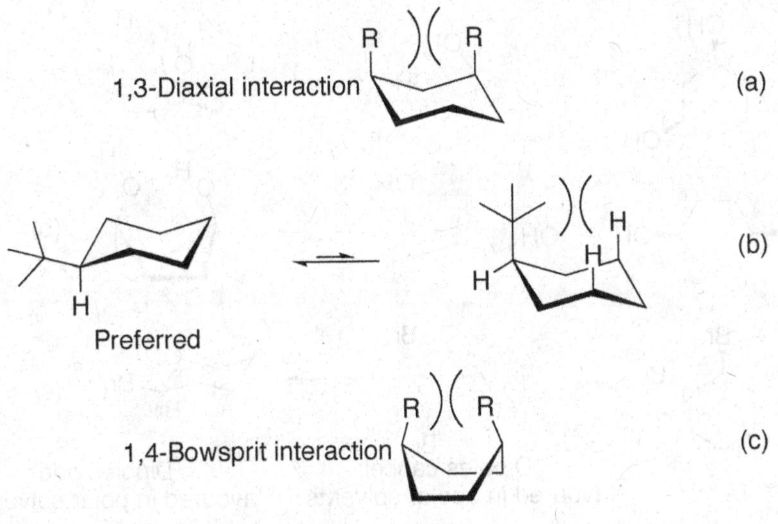

1,3-Diaxial interaction (a)

Preferred (b)

1,4-Bowsprit interaction (c)

Figure 2.25 Steric interactions in cyclohexanes.

by other effects; for example, hydrogen bonding can lead to a preferred diaxial chair (Fig. 2.27a) or boat (Fig. 2.27b) conformer which would normally be unexpected, or in the case of polar bonds as in 1,2-dibromocyclohexane, the conformational preference can depend on the polarity of the solvent (Fig. 2.27c).

5. *Cyclohexene*: The presence of a carbon–carbon double bond in a ring imposes a severe steric constraint on the ring, in which four atoms in the ring are coplanar and the remaining two lie above and below the ring plane (Fig. 2.28a).
6. *Cycloheptane*: The larger ring is more flexible than cyclohexane, and consequently the conformational behaviour is less well defined (Fig. 2.28b).
7. *Cyclooctane*: This ring is very flexible, giving rise to many more energetically accessible conformations, although one or two of these are notable because of their high level of symmetry and minimisation of steric interactions (Fig. 2.28c).

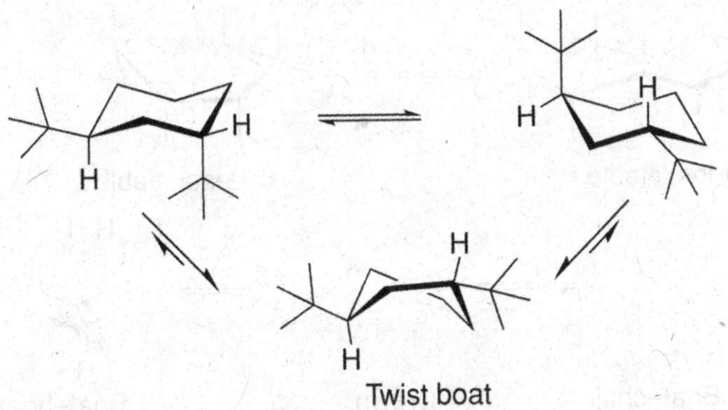

Twist boat

Figure 2.26 Conformation in 1,3-disubstituted cyclohexanes.

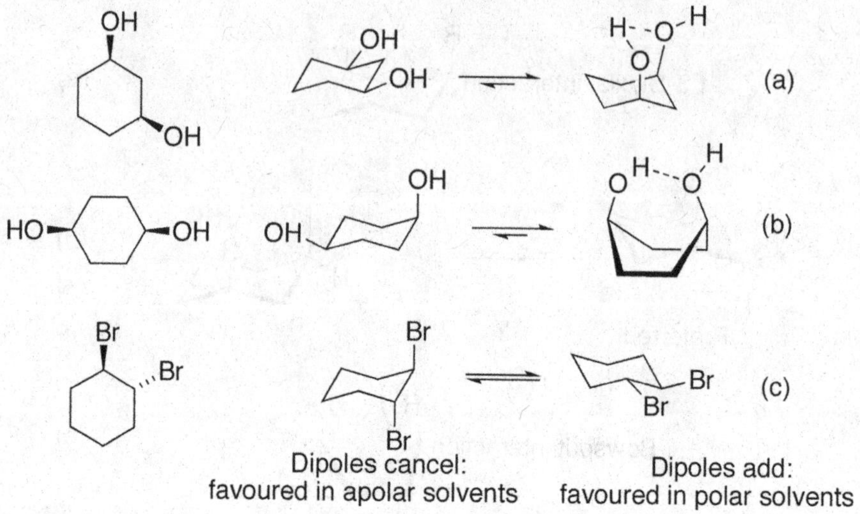

Figure 2.27   Conformation in 1,2-, 1,3- and 1,4-disubstituted cyclohexanes.

8. *Fused-ring systems*: When two rings are fused together, the number of available bond rotations is often substantially reduced and conformational behaviour can be very limited. *trans*-Decalins (Fig. 2.29a), in which there are two six-membered rings joined with a common bond in which the two axial substituents are *trans*-related, are conformationally rigid, and there is no ring flipping and therefore no interconversion of axial and equatorial substituents. By contrast, *cis*-decalins are conformationally flexible, and ring flipping provides access to the two possible structures shown in Fig. 2.29b.

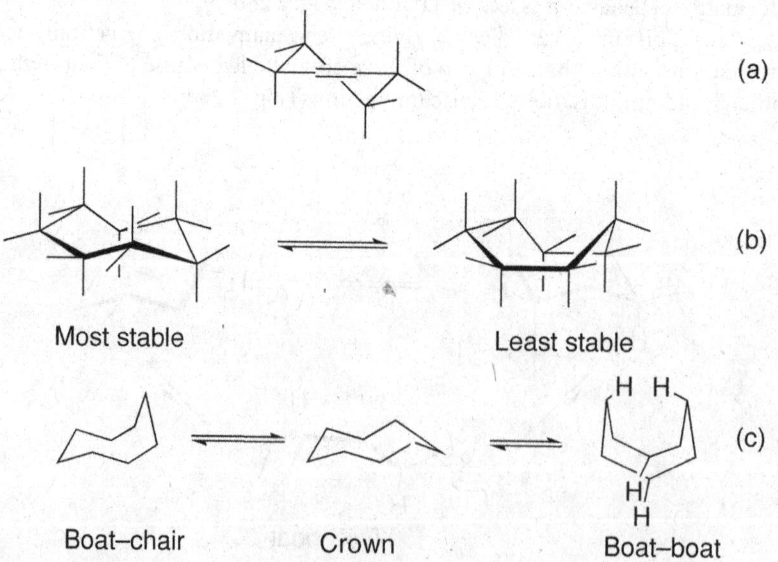

Figure 2.28   Conformation in cyclic hydrocarbons.

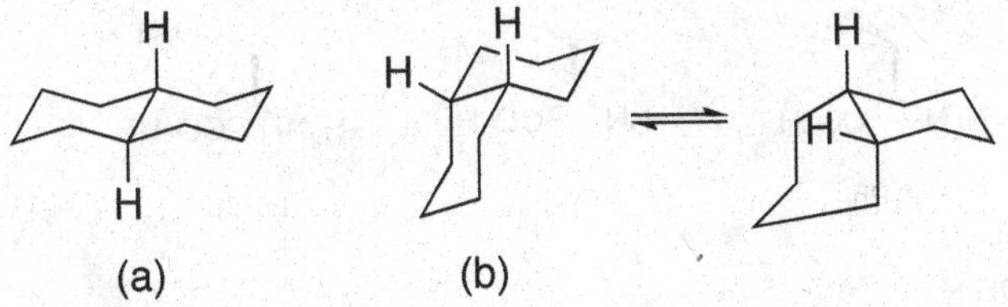

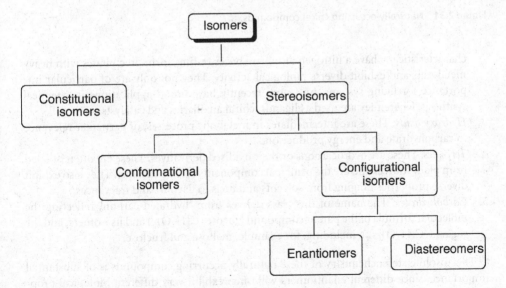

Figure 2.29 Conformation in decalins.

2.3 Summary of stereochemical relationships

A number of relationships have been described in this chapter, and these are summarised in Fig. 2.30.

2.4 Naturally occurring chiral compounds

A wide diversity of chiral chemical compounds occur in nature and in fact are required for the successful operation of many metabolic processes in plants and animals. Many of these compounds are isolable and comprise the so-called chiral pool. These can be subdivided into five classes of compounds (Fig. 2.31):

1. *α-Amino acids*: These are the constituent parts of peptides and proteins, of which there are 24 examples. However, other non-proteinogenic examples are known, and some of these have very important biological activity.
2. *Alkaloids*: These are isolated from a diversity of plant and animal sources, and are so called because they produce an alkaline solution when dissolved in water. They

Figure 2.30 A summary of isomeric relationships.

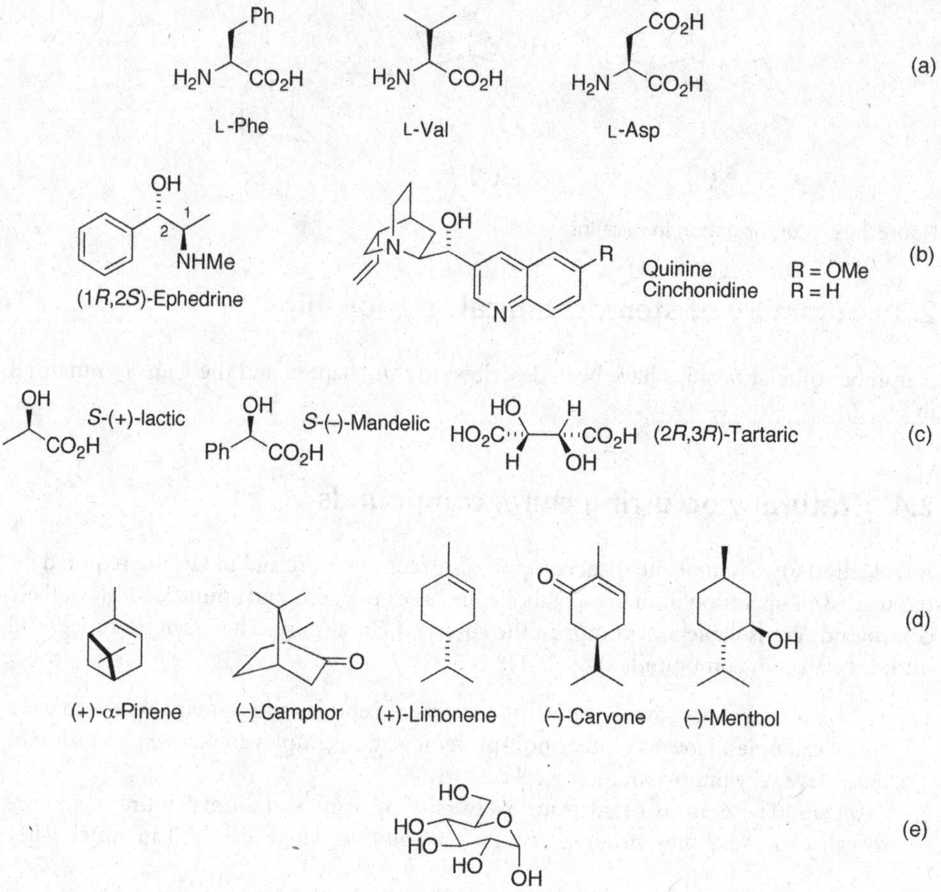

Figure 2.31　Naturally occurring chiral compounds.

characteristically have a nitrogen atom, are bitter tasting, form precipitates with heavy metal salts and exhibit diverse biological activity. They not only are of particular importance for living systems, but more recently have found applications in chemical synthesis, where they are used as ligands, chiral auxiliaries and catalysts.

3. *Hydroxy acids*: These are intermediates in metabolic processes, of particular relevance to carbohydrate and energy production.

4. *Terpenes*: These are hydrocarbons or their oxidised derivatives. These are often isolated from plant sources and are the principal components of 'essential oils'. They also exhibit diverse properties ranging from solvents and fuels to flavours and fragrances.

5. *Carbohydrates*: The name of this class derives from 'hydrated carbon', reflecting the molecular formula of the parent compound glucose ($C(H_2O)_6$) and its isomers, and the sugars ($(C(H_2O)_6)_n$), including, for example, maltose and fructose.

The absolute stereochemistry of these naturally occurring compounds is of substantial importance, since different enantiomers will often exhibit very different biological properties. For example, depending on the absolute stereochemistry of limonene, it exhibits

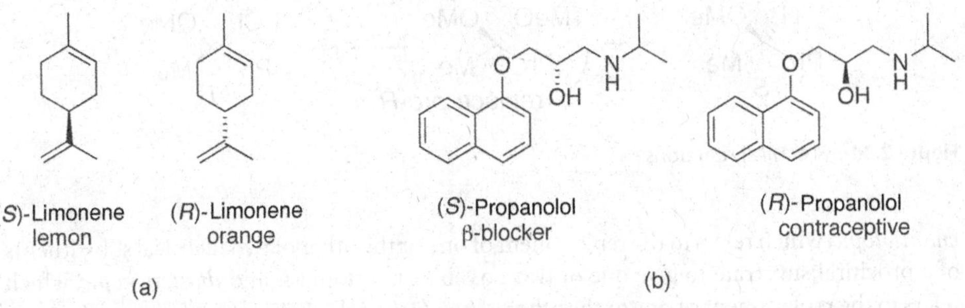

Figure 2.32 Different bioactivity of enantiomers.

(S)-Limonene   (R)-Limonene   (S)-Propanolol   (R)-Propanolol
lemon   orange   β-blocker   contraceptive

(a)   (b)

completely different flavours (Fig. 2.32a). Similarly, enantiomers of pharmaceutically active compounds prepared for use as therapeutic drugs behave differently, and it is crucial that only one enantiomer is prepared and used for the appropriate purpose; for example, propanolol exhibits different therapeutic actions depending on its absolute stereochemistry (Fig. 2.32b). For this reason, the selective synthesis of enantiopure compounds is a very important activity in modern organic chemistry.

2.5 Asymmetric synthesis

Stereoselective synthesis is the synthesis of chiral compounds enriched in one diastereomer. *Asymmetric* or *enantioselective synthesis* is the synthesis of chiral compounds enriched in one enantiomer. It can be defined as the conversion of an achiral unit of a substrate molecule into a chiral unit in such a way that the possible stereoisomeric products are formed in unequal amounts. Such stereocontrol can be achieved using either chiral starting materials or chiral reagents (or both). However, these terms are used very loosely and are often interchanged. In fact, there are two types of stereoselectivities, *diastereoselectivity* and *enantioselectivity*; these relate to the control of absolute stereochemistry and relative stereochemistry, respectively, in a reaction (Fig. 2.33).

2.5.1 Enantioselective synthesis

An important concept is that of *prochirality*. This refers to the fact that two protons or other identical groups located on the same carbon might lead to different enantiomers or diastereomers if one or the other substituent is chemically altered. Related to this are the terms

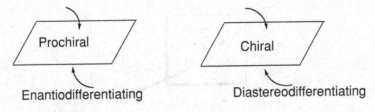

Figure 2.33 Stereoselective reactions.

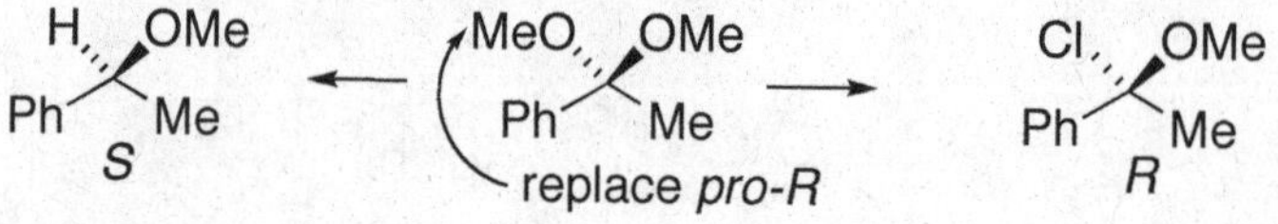

H̥   OMe   MeO̧   OMe   Cl̥   OMe
Ph   Me   ◄──   (Ph   Me) ──►   Ph   Me
   S   replace *pro-R*   *R*

Figure 2.34 Prochiral reactions.

enantiotopic, which refers to the replacement of one or the other of two identical substituents of a prochiral substrate to give one of two possible enantiomers, and *diastereotopic*, which refers to the replacement of one or the other of two identical substituents of an enantiomeric substrate to give one of two possible diastereoisomers. A carbon atom which can be converted to a chiral carbon atom in one chemical step is *prochiral*; this allows the possibility for the synthesis of a single enantiomer. Such a carbon can be:

1. *Tetrahedral*: In this case, the carbon bears one pair of identical substituents; replacement of one of these substituents will give a chiral carbon. Each of the substituents are labelled *pro-R* or *pro-S*, and this is assigned by assuming that the atom to be replaced has the higher priority of the two identical atoms. Note that this does not necessarily correspond to the configuration assignments on the products. For example, in the case of the dimethyl acetal derived from acetophenone (Fig. 2.34), the *pro-R* methoxy substituent is the one behind the plane of the page and if this is replaced by a hydrogen, the *S*-configured product is obtained. If the same *pro-R* methoxy substituent is replaced by a chlorine, the *R*-enantiomer is obtained.

2. *Planar*: An unsymmetrically substituted planar functional group can be converted to a chiral product depending on the face on which it is attacked (Fig. 2.35). In this case, the prochirality is described as *Re* or *Si*: Substituents of the planar functional group are assigned priority using the normal rules, and the classification as *Re* or *Si* depends on whether the substituents appear clockwise (*Re*) or anticlockwise (*Si*), respectively, in decreasing order of priority. Note that just like *pro-R* and *pro-S*, the *Re* or *Si* descriptor has no correlation with the absolute configuration of the product formed from substitution of the prochiral atom; thus, for example, addition of hydride to the *pro-S* face of methyl ethyl ketone leads to the *R*-alcohol, but addition of thiolate gives the *S*-product (Fig. 2.36).

In general, such enantioselectivity is possible only if a chiral reagent, chiral derivatising agent or a chiral environment is applied during the course of the reaction. This area of organic chemistry is of tremendous importance, particularly from a commercial perspective, and significant developments in this area have been achieved over the last 20 years. Although a

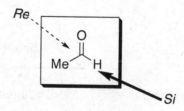

Figure 2.35 Prochiral addition to a carbonyl group.

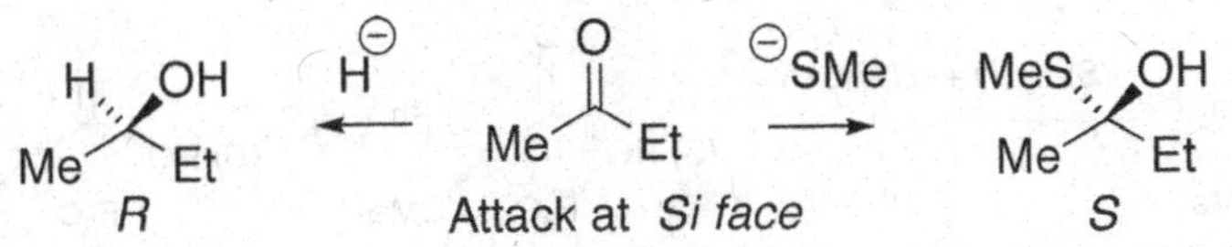

Figure 2.36 Prochiral addition to a carbonyl group.

number of important aspects of this will be discussed in passing, a detailed discussion of this area falls outside the scope of this book.

2.5.2 Diastereoselective synthesis

Practically much simpler than enantioselective synthesis is diastereoselective synthesis. In this, a chiral starting material is converted to a chiral product, but in such a way that one diastereomer is formed in favour over another. Diastereocontrol relies on pre-existing chirality in a molecule to direct the stereochemical outcome of a reaction such that only one diastereomer is favoured; in the most ideal case, it would be the *only* one that is formed.

2.5.3 Methods for the determination of enantiomeric purity

Crucial to the consideration of any discussion of enantioselectivity is a measure of enantiomeric purity. This is best reported as the *enantiomeric excess* (e.e.) and is defined as:

$$\text{Enantiomeric excess} = (\text{Proportion of major enantiomer})$$
$$- (\text{Proportion of minor enantiomer})$$

The measurement of e.e. is not usually straightforward, and this difficulty arises because, as discussed in Section 2.1, enantiomers are identical in physical and chemical properties, except in a chiral environment or in a reaction with another chiral reagent. In the latter situation, diastereomeric complexes or products are formed, which are readily distinguishable, in particular because of their physical properties. On this basis, the enantiomeric excess can be measured using a variety of methods:

1. *Polarimetry*: This crucially depends on the measurement of the optical rotation of a compound (reported as $[\alpha]^{20}_D$). The limitations to the technique, particularly due to a lack of sensitivity, which leads to the requirement for large sample sizes, mean that e.e. determination by polarimetry is falling into disuse.
2. *Chiral chromatography*: Passage of the compound over a column of material (*stationary phase*) which has had its surface modified with an enantiopure agent leads to the transient formation of diastereomeric adducts. Selective desorption of the less stable one permits the faster elution of one enantiomer over the other, and these can be collected separately. Although in principle a very attractive process, practical difficulties, including small sample sizes and small differences in elution times, place considerable limitations on this process.
3. *Diastereomer formation*: Conversion of the analyte to a diastereomeric mixture, using an enantiopure reagent, allows direct compositional analysis of this mixture with a variety of physical methods. A number of reagents have been developed for this purpose and include Mosher's acid chloride (Fig. 2.37a). The reaction of Mosher's acid chloride with

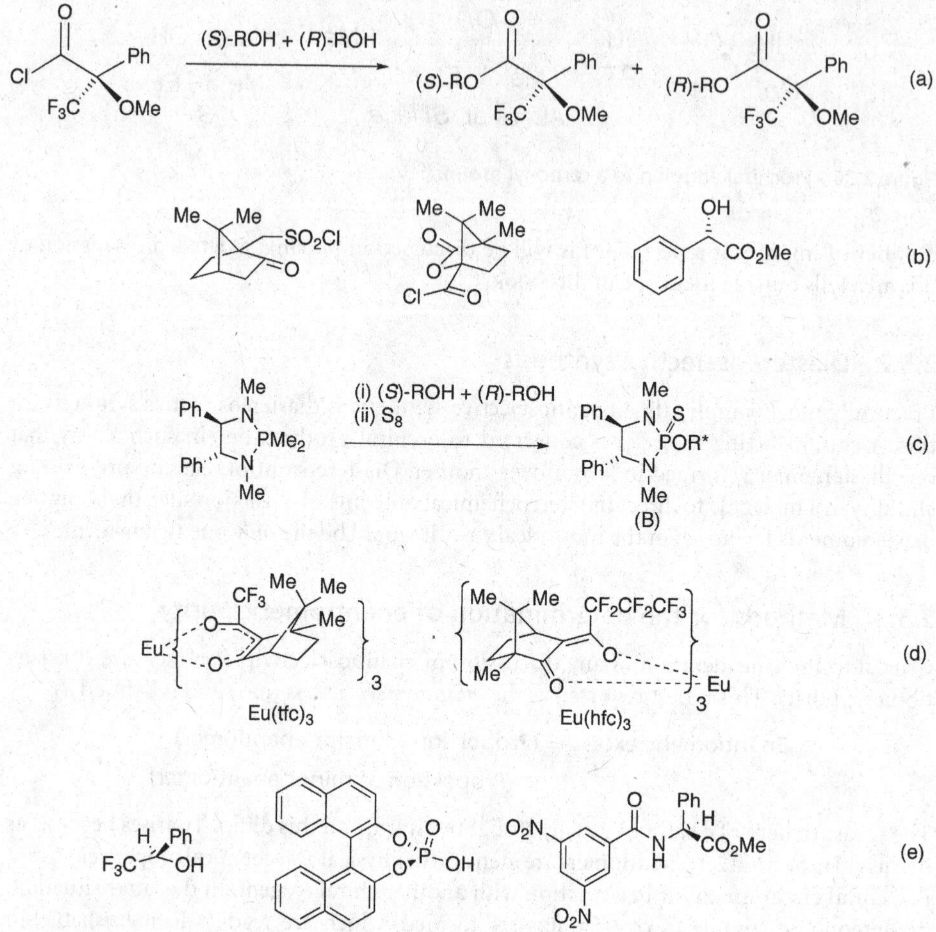

Figure 2.37 Reagents for the determination of e.e. values.

a non-enantiopure chiral alcohol gives a mixture of two diastereomeric products, which can be readily distinguished by chromatography, or most conveniently, by ^{1}H, ^{13}C or ^{19}F NMR resonances of the OMe and the CF$_3$ groups (Fig. 2.37a). Other reactive chiral reagents that can be used in a similar way include chiral acid chlorides (Fig. 2.37b) amongst many others. Organophosphorus derivatives can also be used, for example, to assess the enantiopurity by converting a chiral alcohol to a diastereoemeric mixture of thiophosphoramidates and measuring the diastereomeric ratio using ^{31}P NMR analysis (Fig. 2.37c).

4. *Chiral shift reagents*: Rather than forming a diastereomeric compound, it is also possible to form a transient adduct using a chiral shift reagent. In this case, chelation of a highly oxophilic transition metal cation carrying a chiral ligand (Fig. 2.35c) to a suitable analyte leads to the in situ formation of diastereomeric complexes; these can be readily distinguished by NMR analysis and the diastereomeric ratio calculated by peak integration.

5. *Chiral solvents*: Similarly to chiral shift reagents, it is possible to use chiral solvents to form transient diastereomeric adducts with the analyte, and so distinguish them by NMR analysis (Fig. 2.35d). Although attractive, this approach suffers from the complication of identifying compounds which are synthetically accessible, are good solvents and have sufficient appropriate functionality to facilitate the formation of the required diastereomeric adducts.

Chapter 3
Reactivity

The consideration of structure and bonding described in the earlier chapters allows the development of a rationalisation for understanding the reactivity of compounds. Although the determination of the structure of organic compounds is by no means always trivial, probably the most important aspect of organic chemistry is the study of the way in which compounds react with each other. We will now move on to consider a more dynamic aspect of the chemistry of organic compounds: Why and how they react.

It is well known that some reactions, such as the reaction of hydroxide with chloromethane (3.1), are spontaneous and occur immediately upon mixing of reagents. Some occur spontaneously only if suitable initiation is provided (such as a spark, the input of heat or addition of a chemical catalyst), the combustion of methane and the reduction of ethene with hydrogen gas in the presence of a Pd or Pt catalyst being examples, as shown in Eqs. 3.2 and 3.3. On the other hand, some reactions never occur under any circumstance, for example, the reaction of hydrogen and carbon monoxide to give carbon and water (Eq. 3.4).

$$HO^- + CH_3Cl \rightarrow CH_3OH + Cl^- \qquad \text{(Eq. 3.1)}$$

$$CH_4 + 2O_2 \rightarrow CO_2 + 2H_2O \qquad \text{(Eq. 3.2)}$$

$$CH_2{=}CH_2 + H_2 \rightarrow CH_3CH_3 \qquad \text{(Eq. 3.3)}$$

$$H_2 + CO \rightarrow C(graphite) + H_2O \qquad \text{(Eq. 3.4)}$$

A key goal is therefore the ability to explain and predict when a chemical reaction will proceed spontaneously, and ideally, this would make use of known or measurable fundamental physical properties of the compounds involved. In fact, it is possible to do this, and there are two key parameters to be considered: The first is the extent to which a reaction proceeds ('how far'), and this is under the control of the so-called *thermodynamic* parameters of the reaction. The second is how fast it proceeds, that is, the *kinetic* dimension of the reaction. In fact, thermodynamic and kinetic parameters are the key defining characteristics of any reaction, and although they both derive from a common source, namely, the changes in energy in the course of a reaction, they must always be considered separately.

3.1 Thermodynamics

3.1.1 Gibbs free energy

In order to understand reactions, it has been necessary to identify the 'driving force' or 'affinity' of a chemical reaction which converts starting material A into product B (Eq. 3.5) to allow us to decide if the reaction is possible and to determine the relative amounts of A and B when a reaction is finished.

$$A \rightarrow B \qquad \text{(Eq. 3.5)}$$

Fortunately, a study of thermodynamics, that is, energy and the way it changes as a result of a reaction, allows this to be done quite simply. This can be done by defining the Gibbs free energy (G), which is characteristic of any given molecule and which is itself a function of *enthalpy* and *entropy*. Enthalpy is a measure of the heat content of a compound and will be discussed further in Section 3.1.2, while entropy is a measure of the disorder of a compound and will be discussed further in Section 3.1.3. The change in Gibbs free energy (ΔG) during the course of a reaction can be defined as the difference in free energy of the products and reactants, which is also related to the change in enthalpy (ΔH), the change in entropy (ΔS) and the temperature of reaction (T, in Kelvin), as shown in Eq. 3.6.

$$\Delta G = G_{products} - G_{reactants} = \Delta H - T\Delta S \qquad \text{(Eq. 3.6)}$$

We define a reaction to be *spontaneous* if and only if $\Delta G < 0$; that is, the change in free energy is negative, and the larger the value of $|\Delta G|$, the more favourable the reaction. If $\Delta G < 0$, then energy is released during the reaction of starting materials to give the products and the process is called *exergonic*, and if $\Delta G > 0$, then energy is required for the reaction to proceed and the process is called *endergonic*. From the equation, it can be seen that for a large decrease in ΔG (i.e. $\Delta G < 0$), we need a large decrease in enthalpy and a large increase in entropy. Thus, appropriate free-energy change in determining how far a reaction proceeds is paramount. Under standard conditions ($P = 1$ bar, 1 M solutions, 293 K), we define G as $G°$, in which case the parameter is called the *standard molar Gibbs free energy*, although it should be noted that it is rare for reactions to be run at these conditions in practice.

Paradoxically, although the change in the Gibbs free energy (ΔG) is the defining parameter for the progress of a reaction, it is very difficult to measure, and it therefore becomes necessary to look in more detail at its component parts, that is, enthalpy and entropy.

3.1.2　Enthalpy

Enthalpy is a measure of the energy content of a compound, made up from the energy of its chemical bonds, inter- and intramolecular forces, and vibrational and rotational energy. It is therefore a measure of the stability of a compound, and the change in enthalpy of any reaction can be readily determined by calorimetry, that is, by measuring the heat released or absorbed during the course of a reaction. If $\Delta H > 0$, energy is absorbed and the reaction is said to be *endothermic*, and if $\Delta H < 0$, energy is released and the reaction is *exothermic*.

By measuring the enthalpy changes in a series of reactions at standard temperature and pressure, it has been possible to determine standard average enthalpies of formation ($\Delta H°$) for various types of bonds, and values for those commonly encountered in organic chemistry are given in Tables 3.1 and 3.2. Examination of these tables shows the following notable aspects: All types of single bonds of heteroatoms to themselves (e.g. N–N, O–O, F–F, Cl–Cl, Br–Br, I–I and Si–Si) are amongst the weakest of all bonds, but C–H and C–C bonds are amongst the strongest, with the strongest of all being a Si–F bond. A strong bond is considered to be $\Delta H° > 250$ kJ mol^{-1} and a weak bond $\Delta H° < 250$ kJ mol^{-1}. The order of bond strength for halogens bonded to carbon is C–F > C–Cl > C–Br > C–I, with C–I bonds being particularly weak. Amongst multiple bonds, notice that the $\Delta H°$ for O=O is more than double that of O–O, and for N≡N it is more than triple and for N=N more than double that of N–N; multiply-bonded heteroatoms are therefore particularly stable. By contrast, C≡C is not quite triple and C=C not quite double that of C–C, making carbon–carbon

Table 3.1 Standard average enthalpies of formation ($\Delta H°$, kJ mol^{-1}) for various types of single bonds

	C	N	O	F	Cl	Br	I	Si
H	420	391	462	571	432	370	298	302
C	340	290	353	441	332	281	239	290
N		160	181	273	201			
O			138	210	210	223	239	433
F				159	252	252	281	592
Cl					242	223	210	403
Br						192	181	290
I							149	210
Si								189

multiple bonds weaker than their single-bond counterpart (we saw this earlier with π-bonds being weaker than σ-bonds); that is, they are the more reactive component of C–C multiple bonds. For a C=O, the double-bond component is worth $722 - 353 = 369$ kJ mol^{-1}, only just more than the C–O single-bond energy, and for C–N bonds, the triple-bond part is worth 272 and the double-bond part 332, indicating again that higher multiple bonds for carbon–heteroatom bonds are similar in energy to their single-bond counterparts.

It is important to note that the values given in Tables 3.1 and 3.2 are average values; in fact, bond enthalpies are not independent of other bonding arrangements in any given compound, and they do therefore vary between compounds. This is illustrated by the data for different types of carbonyl groups (Fig. 3.1).

Changes in enthalpy in any reaction can be estimated from standard enthalpies of formation ($\Delta H°$) by simple algebra. For example, in order to assess the likelihood of the reaction of hydrogen bromide with ethene, we can calculate the difference between the energy consumed from breaking bonds in the starting material and the energy released by forming bonds in the product (of course, we can afford to ignore those bonds which are unchanged during the course of the reaction) to be –49 kJ mol^{-1}, as shown in Fig 3.2a, indicating an energetically favourable reaction proceeding from reactants to products. A similar analysis for the reaction of hydrogen cyanide with benzaldehyde gives a difference of –113 kJ mol^{-1} (Fig. 3.2b), indicating a favourable reaction. For the transesterification reaction of ethyl acetate with butanol, there is no thermodynamic advantage over the formation of products or reversion to reactants ($\Delta H = 0$; Fig. 3.2c), and in the case where $\Delta H < 0$ (Fig. 3.2d), the process is thermodynamically unfavourable.

Table 3.2 Standard average enthalpies of formation ($\Delta H°$, kJ mol^{-1}) for single and multiple bonds

Elements	Single	Double	Triple
O–O	139	403	—
N–N	160	420	949
C–C	340	622	815
C–O	353	722	—
C–N	290	622	894

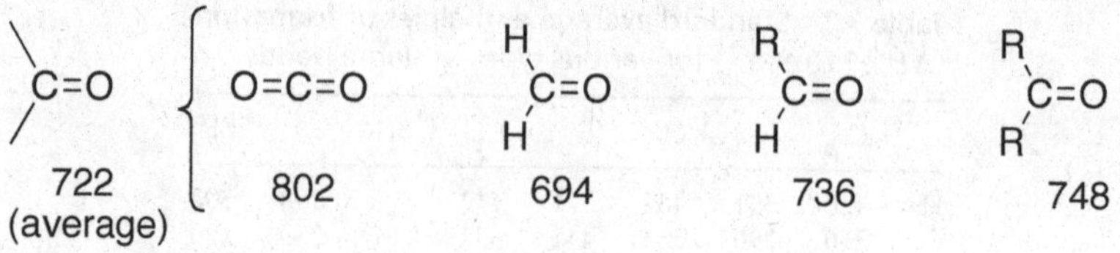

$$\begin{array}{ccccc}
\overset{\diagdown}{\underset{\diagup}{C}}{=}O & O{=}C{=}O & \overset{H}{\underset{H}{\diagdown}}C{=}O & \overset{R}{\underset{H}{\diagdown}}C{=}O & \overset{R}{\underset{R}{\diagdown}}C{=}O \\
722 & 802 & 694 & 736 & 748 \\
\text{(average)} & & & &
\end{array}$$

Figure 3.1 Bond enthalpies in kJ mol^{-1} for different types of carbonyl compounds compared to the calculated average value.

3.1.3 Entropy

The second term on which the Gibbs free energy depends is entropy (S), which is a measure of the disorder in a system. Entropy increases with increasing randomness or disorder. The entropy content of a molecule depends on its molecular motion and therefore will not surprisingly diminish with temperature (T) and S tends to 0 at 0 K. Also, transition to more condensed phases leads to a significant reduction in entropy ($S_{gas} > S_{liquid} > S_{solid}$), but it increases with increasing complexity of a molecule, since more intramolecular motions become available. Open-chain molecules have more entropy than cyclic ones, because the former has more available conformations.

The change in entropy (ΔS) during the course of a reaction is favourable ($\Delta S > 0$) if the number of molecules increases and is unfavourable ($\Delta S < 0$) if the number of molecules decreases. Note that the signs of ΔH and ΔS are opposite, that is, $\Delta H < 0$ and $\Delta S > 0$, for a favourable reaction. Unless values for ΔH and ΔS are known for a given reaction, ΔG cannot be easily determined. Unfortunately, entropy is difficult to measure, but for many commonly encountered organic reactions, entropy changes are small and can be ignored, particularly if there are no changes in the number of reactant and product molecules in the stoichiometric equation; thus, we can assume that $\Delta S \approx 0$. This allows for a very useful approximation that the change in Gibbs free energy is equivalent to the change in enthalpy (Eq. 3.7), and since ΔH can be easily estimated from bond enthalpy data (see Tables 3.1 and 3.2), as described in Section 3.1.2, it is also possible to estimate ΔG.

$$\Delta G \approx \Delta H \tag{Eq. 3.7}$$

However, it is unwise to automatically assume that entropy effects are always small; in fact, they can be significant. One example is the interconversion of cyclopentadiene to dicyclopentadiene (Fig. 3.3). For this reaction, thermodynamic data have been determined as follows: $\Delta H° = -72$ kJ mol^{-1} and $\Delta S° = -1.5 \times 10^{-1}$ kJ mol^{-1}. Using this information, $\Delta G°$ values at 298 and 488 K can be calculated:

At 298 K $\Delta G° = -72 - 298(-0.15) = -27$ kJ mol^{-1} $K = 50\,000$

At 488 K $\Delta G° = -72 - 488(-0.15) = 0$ kJ mol^{-1} $K = 0$

Thus, at room temperature, cyclopentadiene exists as a dimer but can be cracked to the monomer by slow distillation at 488 K (the boiling point of dimer).

Any reaction that leads to a reduction in the available degrees of freedom also leads to a decrease in entropy; a good example is of cyclisation reactions, for example, the intramolecular cyclisation of an ω-bromoamine, as shown in Fig. 3.4.

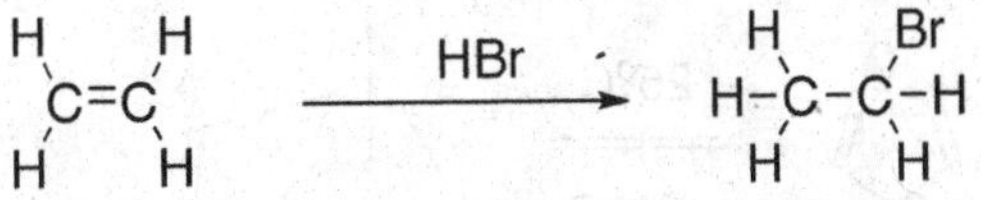

H, H ... H, Br
C=C → HBr → H-C-C-H   (a)
H, H ... H, H

π-bond (C–C) 282   C H   420
H–Br   370   C–Br   281
              ___   ___
              652   701

ΔH = Enthalpy change for (bonds broken – bonds formed)
   = 652 – 701 = –49 kJ mol⁻¹

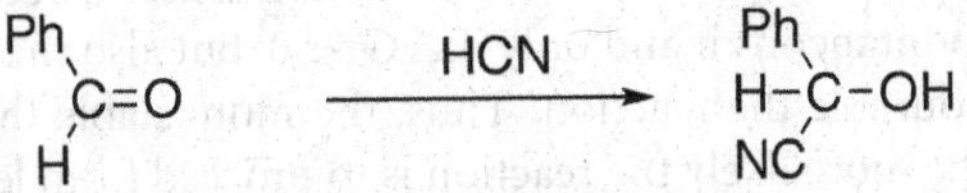

Ph,   Ph,
C=O → HCN → H-C-OH   (b)
H   NC

π-bond (C–O) 269   C–C   340
H–CN   420   C–H   462
              ___   ___
              689   802

ΔH = Enthalpy change for (bonds broken – bonds formed)
   = 689 – 802 = –113 kJ mol⁻¹

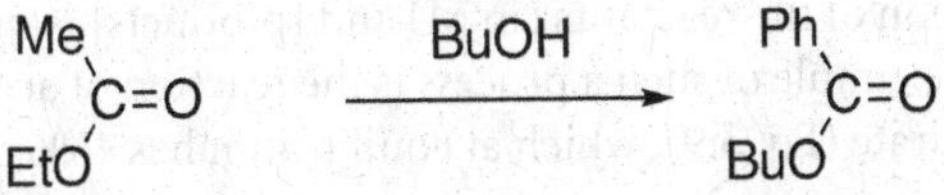

Me,   Ph,
C=O → BuOH → C=O   (c)
EtO   BuO

C–O   353   C–O   353
              ___   ___
              353   353

ΔH = Enthalpy change for (bonds broken – bonds formed)
   = 353 – 353 = 0 kJ mol⁻¹

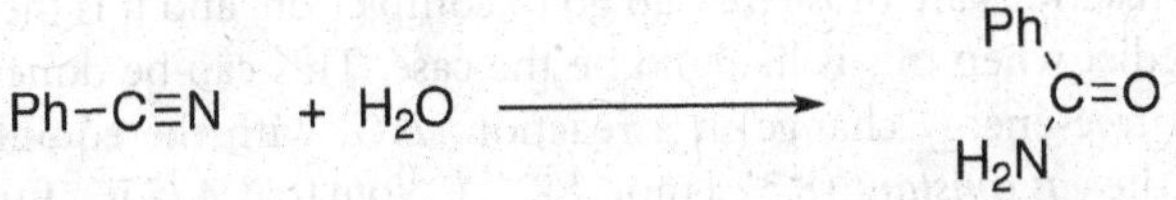

                                    Ph,
Ph–C≡N + H₂O →   C=O   (d)
                                    H₂N

2 × π-bond (C–N)   604   π-bond (C–O) 369
2 × O–H   924   σ-bond (C–O) 353
                   ____   2 × H–N   782
                   1528   ____
                                         1504

ΔH = Enthalpy change for (bonds broken – bonds formed)
   = 1528 – 1504 = +24 kJ mol⁻¹

Figure 3.2 Estimated enthalpy changes in some reactions.

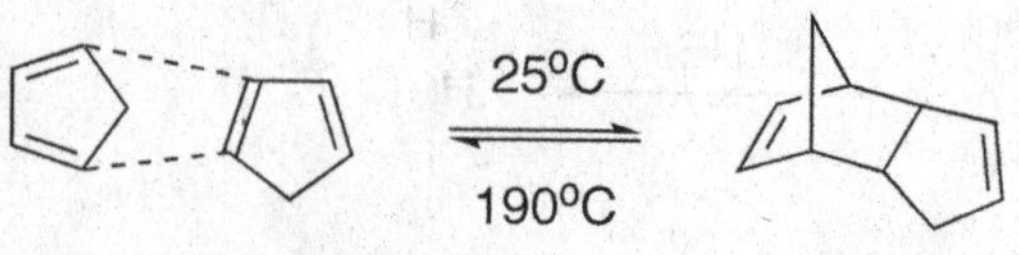

Figure 3.3 A reaction leading to a loss of entropy.

3.1.4 Chemical equilibrium

We have seen that $\Delta G = \Delta H - T\Delta S$ and that we can afford to ignore the $T\Delta S$ term in a majority of cases, since the entropy change in reactions is small, provided that we have the same number of reactant and product molecules. This equation is useful because it tells us not only that the reaction is spontaneous if and only if $\Delta G < 0$, but also that the larger the value of $|\Delta G|$, the more favourable the reaction. Thus, the more stable the product, and the less stable the reactant, the more likely the reaction is to proceed from left to right. The corollary to this is that depending on the value of ΔG, not all reactants in reactions are completely consumed to give the corresponding products. In fact many reactions stop while there are still some reactants present. At this point, the forward and reverse reactions proceed at the same rate, and there is an *equilibrium*, which can be defined by the *equilibrium constant K*, as given in Eq. 3.8.

$$K = [products]/[reactants]$$
(Eq. 3.8)

where [reactants] is the concentration of the reactants (in M) and [products] is the concentration of the products (in M). An example of such a process is the reaction of acetaldehyde with water to give acetaldehyde hydrate (Eq. 3.9), which at equilibrium has 46% of reactant and 54% of product present.

$$CH_3CHO + H_2O \text{ (excess)} \rightleftharpoons CH_3CH(OH)_2$$
46%   54%
(Eq. 3.9)

Processes like these tend to be troublesome for the synthetic organic chemist, since there is the problem of the separation of reactants and products, which is generally both difficult and expensive. The most useful reactions are those that do go to completion, and it is therefore important to be able to predict when this is likely to be the case. This can be done using Eq. 3.10, which relates the free-energy change in a reaction, ΔG, with the equilibrium constant K and where R is the *gas constant* ($8.32 \text{ J mol}^{-1} \text{K}^{-1}$). Note that ΔG is a function of both K and T.

$$\Delta G = \Delta H - T\Delta S = -RT \ln K$$
(Eq. 3.10)

The magnitude of the values for ΔG and K is illustrated by considering a generalised reaction in which A and B are at equilibrium and where A is less stable than B (Table 3.3).

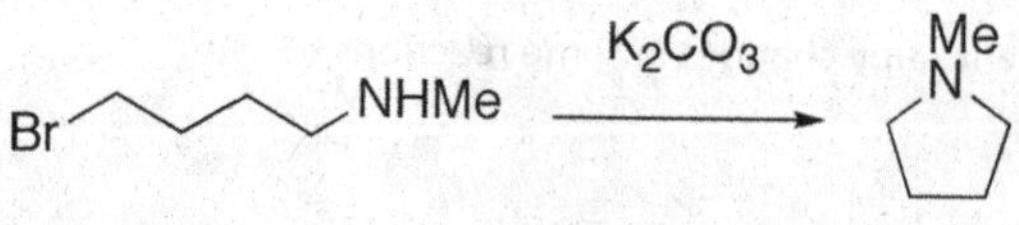

Figure 3.4 A reaction leading to a loss of entropy.

Table 3.3 ΔG and K for the reaction A $\rightleftharpoons$ B at 298 K

$-\Delta G^\circ$ (kJ mol^{-1})	K	Percentage of A	Percentage of B
0	1	50	50
1.00	1.5	40	60
2.1	2.33	30	70
3.42	4	20	80
5.44	9	10	90
11.35	99	1	99
17.0	999	0.1	99.9
22.8	9999	0.01	99.99

Note: A is less stable than B.

When there are equal amounts of A and B present, we have $[B]/[A] = 1$ and so $\Delta G = 0$. For increasing proportions of B, the value of $[B]/[A]$ steadily increases so that at $[B]/[A] = 9$, the value of ΔG° is -5.44. An 11-fold increase in $[B]/[A]$ to a value of 99 only doubles the free-energy difference (-11.35 kJ mol^{-1}), and a further 10-fold increase to $[B]/[A] = 999$ only produces a 1.5-fold increase in free energy ($\Delta G^\circ = -17.0$ kJ mol^{-1}) as does another 10-fold increase ($[B]/[A] = 9999$, $\Delta G^\circ = -22.8$ kJ mol^{-1}). Thus, small changes in ΔG° lead to large changes in the position of equilibrium, and that these changes are indeed small is best appreciated in the context of a typical C—C bond energy, of about 340 kJ mol^{-1}.

The position of equilibrium can be altered by application of the *Le Chatelier principle*: A system at equilibrium adjusts so as to oppose any applied change. This could be done by increasing the concentration of reactants, decreasing that of products (e.g. by removing them) or changing the temperature of the reaction. For example, for an endothermic reaction, increasing the temperature leads to products being favoured over reactants.

Equilibrium processes do not only involve chemical reactions; an equilibrium involving a conformational interchange by ring flipping in substituted cyclohexanes is extremely important, and it is strongly dependent on the steric bulk of the substituent (Fig. 3.5 and Table 3.4). As the bulk of ring substituent increases, the proportion of equatorial conformer increases in the series H < Me < Et < i-Pr < t-Bu. For the t-Bu-substituted system, there is effectively no axial conformer present and the ring is unable to flip between chair conformers and is therefore 'locked'. The same can be seen in the halogen series for increasing atomic radius, in which the proportion of equatorial conformer follows the sequence F < Cl < Br ~ I. The axial isomer is disfavoured due to the presence of highly destabilising 1,3-diaxial interactions.

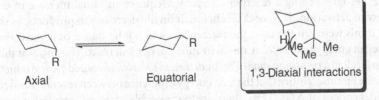

Axial   Equatorial   1,3-Diaxial interactions

Figure 3.5 Energy change in a conformational equilibrium.

Table 3.4 Equilibrium constants and axial:equatorial ratio for monosubstituted cyclohexanes

R	K_{eq}	ax:eq
H	1	1:1
Me	18	5:95
Et	23	4:96
i-Pr	38	3:97
t-Bu	4000	0.02:99.98
F	1.5	40:60
Cl	2.4	29:71
Br	2.2	31:69
I	2.2	31:69
OH	5.4	16:84

3.2 Kinetics

3.2.1 Rates of reaction

Calculation of the change in Gibbs free energy (ΔG) between the reactants and the products of a reaction enables us to determine if a reaction is *possible*; however, this calculation does not consider all the changes in free energy during the course of the reaction and therefore whether the reaction actually *occurs* at an observable rate. Reaction occurs only if a suitable low-energy pathway links reactants and products. Thus, a negative ΔG is a necessary but not a sufficient reason for a reaction to take place; more careful consideration of intermediate energy changes in the course of a reaction is required to determine the likely reaction rate. The pathway taken by reacting molecules and leading to the products is called the *reaction mechanism*. For a generalised reaction given in Eq. 3.11, the rate of reaction is equal to the rate of formation of a product (e.g. $d[X]/dt$) or the rate of disappearance of a reactant (e.g. $-d[A]/dt$). More generally, it is a function of the concentration of reactants, given by Eq. 3.12. The proportionality constant k is the *rate constant*, and the values m and n must be determined by experiment; they are most certainly not determined by the stoichiometry of reaction between A and B. A very fast reaction would have rate constant k (s^{-1}) of $>10^4$, a fast one of between 1 and 10^4, a slow one of between 10^{-4} and 1 and a very slow one $< 10^{-4}$.

$$A + B \rightarrow X + Y \qquad \text{(Eq. 3.11)}$$

$$\text{Rate of reaction} = k[A]^m[B]^n \qquad \text{(Eq. 3.12)}$$

In order to understand how the rate of reaction is determined by free energy G, it is necessary to consider changes in G as a function of the extent of reaction (the *reaction coordinate*). Progress along a reaction coordinate requires an initial increase in energy G (the energy of activation, E_{act}, or $\Delta G^{\ddagger}$) before it finally decreases to products, as shown in Fig. 3.6a. If this were not the case, the reactants would not be stable; only those molecules with sufficient energy to overcome this activation barrier will react. The entity at the energy maximum is called a *transition state* (TS), or *activated complex* (denoted $\ddagger$), and is metastable, with no finite lifetime. Notice that the free-energy difference between reactants and products $\Delta G°$ is independent of $\Delta G^{\ddagger}$. Thus, there are four possible cases: an exergonic reaction (i.e. spontaneous) which has a low value of $\Delta G^{\ddagger}$ (a fast reaction) or a high value of $\Delta G^{\ddagger}$ (a slow reaction), or an endergonic reaction which has a low value of $\Delta G^{\ddagger}$ (a fast reaction) or a

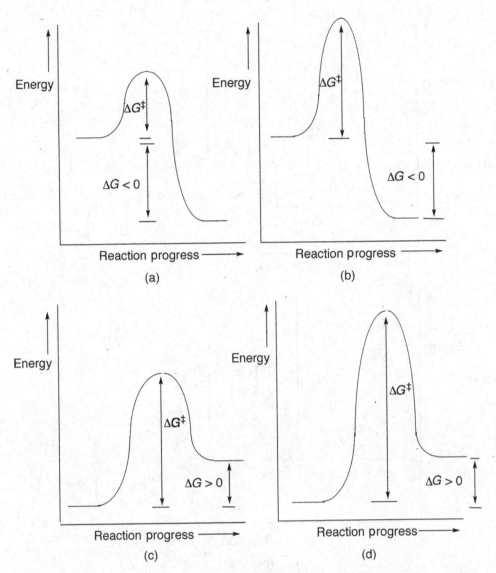

Figure 3.6 Reaction coordinate diagrams illustrating four possible cases with differing values for ΔG and $\Delta G^{\ddagger}$.

high value of $\Delta G^{\ddagger}$ (a slow reaction); all these cases are illustrated in Figs. 3.6a–3.6d. Each of these possibilities is an example of a single-step process. It is important to remember what information each energy difference provides: As $\Delta G^{\ddagger}$ increases, the rate of reaction decreases and can in fact go to zero, but when $\Delta G°$ becomes more negative, the reaction becomes more irreversible.

Such single-step processes more commonly occur in sequence as do multi-step processes. In this case, in addition to the existence of transition states, at a relative energy minimum there are *intermediates*. Again, depending on the relative magnitude of $\Delta G^{\ddagger}$ and $\Delta G°$, several possibilities can occur and two of these are illustrated in Fig. 3.7. An initial fast equilibrium

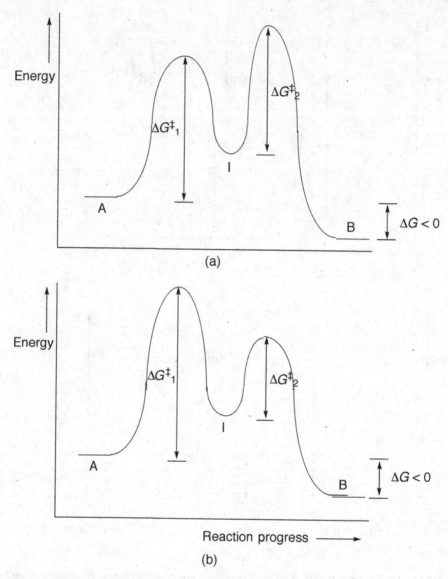

Figure 3.7 Reaction coordinate diagrams for reactions involving successive steps.

(i.e. small $\Delta G_1^{\ddagger}$ and small ΔG) from starting material A leads to intermediate I in a relative energy minimum, which is followed by a much slower (i.e. large $\Delta G_2^{\ddagger}$) conversion to product B (Fig. 3.7a); since the second step is the slowest (it is the step with the largest $\Delta G^{\ddagger}$), it is therefore the *rate-determining step*. In an alternative sequence, a slow conversion of starting material A (i.e. large $\Delta G_1^{\ddagger}$) proceeds to product B in a fast second step (i.e. small $\Delta G_2^{\ddagger}$); in this process, the rate of the overall reaction is governed by the slowest step, that is, the one with the highest $\Delta G^{\ddagger}$, in this case the first one. These types of multi-step sequences are very common.

However, the rate constant k can vary significantly with temperature T, since an increase in temperature can increase the proportion of molecules with the required $\Delta G^{\ddagger}$. This increases the number of molecules with enough energy to surmount the activation barrier and proceeds to product. The dependence of k on T is given by the Arrhenius equation (3.13), which has been determined empirically.

$$k = A \exp(-E_{act}/RT) \qquad \text{(Eq. 3.13)}$$

where A is the Arrhenius constant and R is the universal gas constant. From this equation, it can be calculated that T has a marked effect on reaction rate, and in fact reaction rate approximately doubles for a $10°$ rise in temperature. It is for this reason that reactions are often heated to the boiling point of the solvent in which they are conducted.

There are some important fast reactions. Most notable amongst these are proton transfers between oxygen and nitrogen, which often occur as fast as possible and have no activation energy barrier. This means that rapid reversible proton transfer to oxygen or nitrogen before or after the rate-determining step is common in acid- or base-catalysed reactions. Conversely, proton transfers to and from carbon are often slow with significant energy barriers. Data which illustrate the size of the energy involved, for examples, *o*-chlorophenol and 2-nitropropane, with almost identical acidity, are given in Fig. 3.8.

It is perhaps worth emphasising that when we talk about the stability of a compound, it is important to emphasise whether we are referring to kinetic stability (unreactive due to a slow rate of reaction) or thermodynamic stability (unreactive due to an unfavourable energy difference between it and its products).

3.2.2 Reactions with competing steps

We have seen earlier that reactions can proceed via a sequence of steps, one after the other (Section 3.2.1). However, it is possible for reactions to work in opposition to each other, and the overall outcome is based on the competition between the two. Depending on whether each process is controlled by kinetic or thermodynamic parameters, a different outcome is possible. If we consider two reactions, starting from a common reactant A, and leading to either products B or C, there are several possible reaction pathways of interest (Fig. 3.9). If both $\Delta G_C^{\ddagger} > \Delta G_B^{\ddagger}$ and $\Delta G_C > \Delta G_B$, then product B will be formed faster, although it will be less stable. If the reaction is allowed to proceed for an extended time, or at elevated temperature, product B, formed initially more rapidly, will be able to revert to starting material A and then product C will be formed as the more stable product, but only slowly. The extended reaction time required to achieve this result is called *equilibration*, since the system is brought to an equilibrium. Provided that the energy differences are large enough, product C may be formed exclusively over the alternative B. The faster formation of product

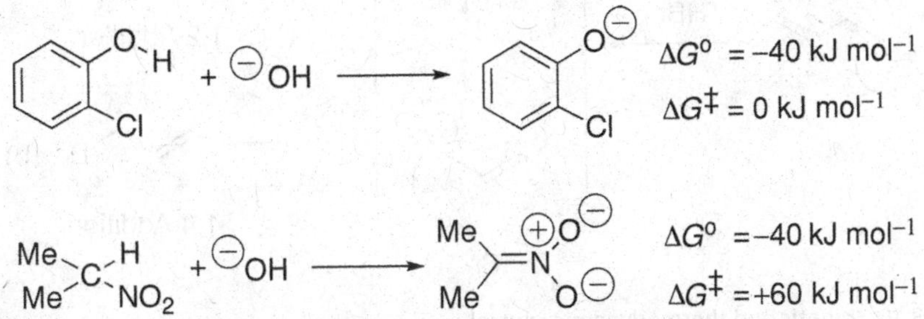

Figure 3.8 Free energy and activation parameters for proton-transfer reactions.

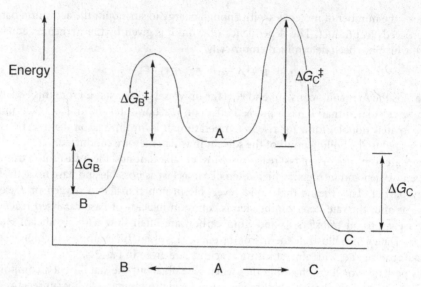

Figure 3.9 Reactions with competing steps.

B under less forcing conditions is called *kinetic control* and that of product C under extended reaction time or higher temperature is called *thermodynamic control*. An example of kinetic and thermodynamic control is given in Fig. 3.10. In the reaction of HBr with butadiene, the starting material proceeds first to a common carbocation intermediate, and at this point divergence to two possible products is possible. At ambient conditions, the 1,2-addition product is favoured (Fig. 3.10a), but at elevated temperature, the 1,4-addition product is formed (Fig. 3.10b).

3.2.3 Overcoming activation energy barriers

We have seen that reactions frequently have energy activation barriers which must be surmounted if a productive outcome is to be achieved. This energy can be introduced into

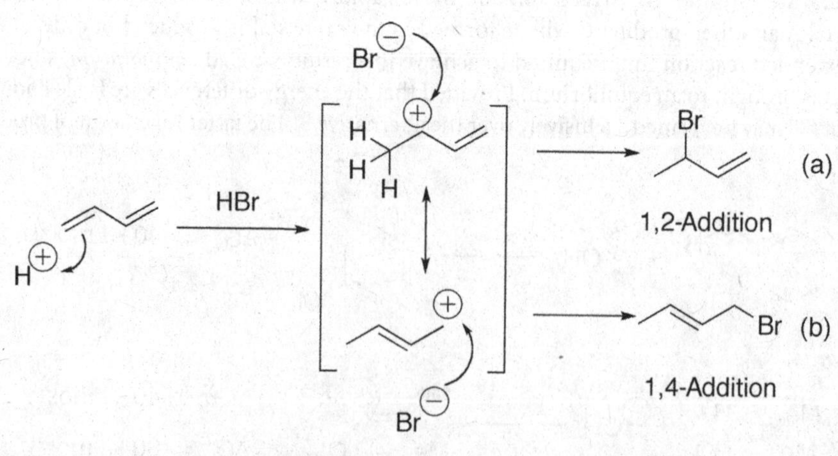

Figure 3.10 Kinetic and thermodynamic control.

Table 3.5 Some commonly used solvents for reactions at elevated temperature

Solvent	Structure	Abbreviation	b.p. ($^\circ$C)	Dielectric constant ε	tan δ
Hexane	$CH_3(CH_2)_4CH_3$	—	69	1.9	0.02
1,4-Dioxane		—	101	2.2	—
Carbon tetrachloride	CCl_4	—	76	2.2	
Benzene	C_6H_6	—	80	2.3	—
Toluene	C_6H_5Me	—	111	2.4	0.04
Diethyl ether	$CH_3CH_2OCH_2CH_3$	Et_2O	35	4.3	—
Chloroform	$CHCl_3$	—	61	4.8	0.091
Ethyl acetate	$CH_3CH_2OC(O)CH_3$	EtOAc	78	6.0	0.059
Acetic acid	$CH_3C(O)OH$	AcOH	118	6.2	0.174
Tetrahydrofuran		THF	66	7.4	0.047
Dichloromethane	CH_2Cl_2	DCM	40	9.1	0.042
Pyridine		py	115	12	—
Acetone	$CH_3C(O)CH_3$	Me_2CO	56	20.7	0.054
Ethanol	CH_3CH_2OH	EtOH	78	24.3	0.94
2,2,2-Trifluoroethanol	CF_3CH_2OH	TFE	78	26	—
Hexamethylphosphoric triamide	$[(CH_3)_2N]_3P{=}O$	HMPA	235	30	—
N-Methylpyrrolidine		NMP	77	32.2	0.275
Methanol	CH_3OH	MeOH	68	32.6	0.66
Nitromethane	CH_3NO_2	$MeNO_2$	103	36.0	0.064
N,N-dimethylformamide	$HC(O)NMe_2$	DMF	153	37.7	0.161
Acetonitrile	CH_3CN	MeCN	81	37.5	0.062
Dimethylsulfoxide	$CH_3S(O)CH_3$	DMSO	189	45.0	0.825
1,2-Dichlorobenzene		—	180	10.4	0.280
Water	H_2O	—	100	80.4	0.123

the system using heat or light, but normally thermal processes relying on convectional and conductional transfer of heat, using solvents at reflux temperature, are preferred because of the convenience of such a procedure in the laboratory. For an effective thermal process, a solvent is required which dissolves the reacting species, but which also boils at a temperature appropriate to the energy input required, which is inert under the reaction conditions and which can be removed once the reaction is complete. Some commonly used solvents for reactions at reflux, and their boiling points, are shown in Table 3.5. Of interest has been a recent realisation that mechanical forces can act selectively to activate covalent bond breakage, and it appears that this is of considerable importance in polymer degradation, but it is not of general utility in chemical processes.

An alternative approach for the introduction of energy is by the use of microwave irradiation. Although the precise mode of action of microwave irradiation is not clear, it has been found that such heating efficiently leads to significant reaction rate enhancements, so much so that reactions not amenable to conventional heating can become possible; this has obvious and important implications in the development of novel reaction processes. Such irradiation relies on the fact that polar reagents, solvents (and, indeed, products), interact with the electrical and magnetic oscillations of the microwave irradiation to induce molecular rotation, which is manifested as heat; the efficiency with which this is done is indicated by the dielectric loss factor, tan δ, and values for solvents commonly used in microwave-mediated reactions are shown in Table 3.5. It can be seen that more polar solvents are usually more effective as microwave absorbers, but this is not always essential for reaction, since the reacting species will also frequently act in this regard, even in a non-polar solvent. The use of microwave heating combined with careful temperature, power and time control permits reactions to be conducted rapidly and reproducibly; additionally, if reactions are conducted in sealed containers, there is the added benefit of pressurisation during the course of the reaction. Of interest is the application of water as a solvent for reactions of organic substrates; although normally not suitable in organic chemistry because of of its poor solvating power for non-polar substrates, at elevated pressure water can be an effective solvent for many organics and has been found to be very suitable for the conduct of many reactions, especially also given that it may act as a weak acid or base, it is inexpensive and, unlike organic solvents, it does not have a direct undesirable environmental impact.

3.3   Reaction mechanism

3.3.1   What is reactivity?

Up until this stage, we have been principally concerned with the structure of organic compounds and reaction intermediates, but have not considered in detail how reactants are converted to products in the course of a reaction. In order to do this, we need to consider the sequence and timing of bond-making and bond-breaking processes. Since the bonding in a molecule is determined by the way its electrons are shared between its constituent atoms, describing changes in electron-sharing arrangements during the course of a reaction enables us to understand in detail the processes which occur in a chemical transformation. In fact, there are an infinity of possible intermediate structures on the path from starting materials to products, but it is nonetheless possible to provide a broad picture of the way in which bonding electrons are redistributed; such a description is called the *reaction mechanism*. Implicit in any mechanism is that it describes the changes in bonding arrangements which lead from reactant to product *along the lowest energy pathway*; any other pathway will be more energetically costly and will therefore not be preferred. Furthermore, this pathway also corresponds to an orderly movement or flow of electrons and to the least movement of atoms required to reorganise reactants into products (*the principle of least motion*). The most satisfactory explanation for the processes which occur during a reaction assumes that organic compounds and reagents have polar characters, and the interaction – attraction or repulsion – of partial or complete charge is a key factor which initiates and controls the bonding reorganisation during a reaction. This approach is applicable to a wide variety of processes, and this chapter will concentrate on such cases in detail.

3.3.2   Lewis acids and bases: 'philicity'

Reactivity is understood and rationalised using a reaction mechanism. Our understanding of the reorganisation of bonding and non-bonding electrons in the course of a reaction depends on the fact that some molecules are depleted of electrons and others can be

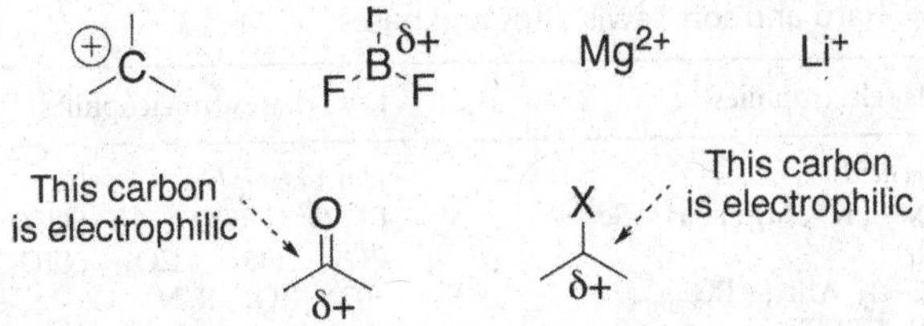

Figure 3.11 Some examples of Lewis acids.

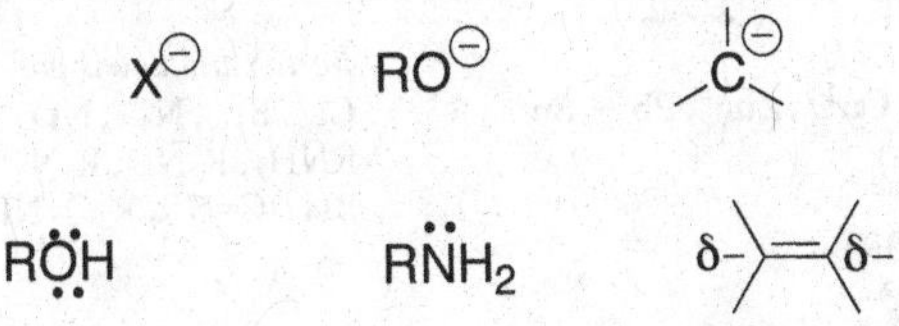

Figure 3.12 Some examples of Lewis bases.

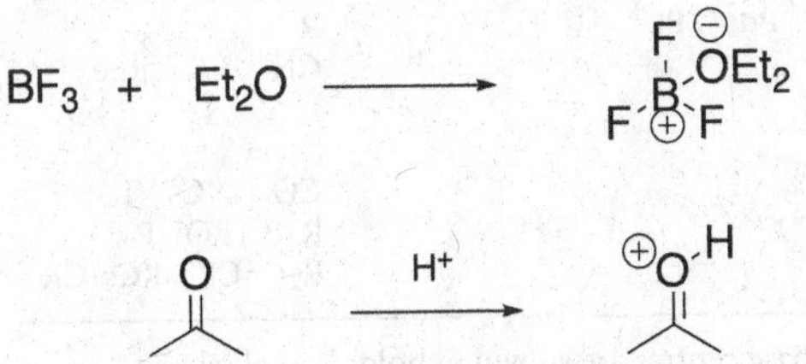

Figure 3.13 Some examples of Lewis acid–Lewis base adducts.

considered to have a surplus of electrons. This means that some molecules can accept electrons and others can give them away in order to achieve the most stable arrangement, that is, a stable octet of electrons. Lewis acids are electron-pair acceptors, and Lewis bases are electron-pair donors. In organic chemistry, we often call Lewis acids *electrophiles* ('electron loving'), which are species that are relatively electron deficient, are fully or partially positively charged and seek areas of electron density. Lewis bases are called *nucleophiles* ('nucleus loving'), which are species that are relatively electron rich, are fully or partially negatively charged and seek positive charge. This concept, which might be called 'philicity', describes the ability of any given chemical entity to accept or release electrons. Common examples of electrophiles which possess a full positive charge include protons, carbocations and metal cations, and those with a partial positive charge include carbonyl groups and carbon–halogen bonds (Fig. 3.11). Examples of nucleophiles which possess a full negative charge include halide anions, hydroxide, alkoxides and carbanions (i.e. organometallic derivatives), and those with a partial negative charge include amines, alcohols and alkenes (Fig. 3.12). Reaction of Lewis acids and bases can form stable adducts, and some examples are given in Fig. 3.13.

3.3.3 Polarisability effects: Hard–Soft Acid–Base Theory

The simple rationalisation of reactivity in organic chemistry using the concepts of electrophiles and nucleophiles is very useful, but does not explain the fact that not all

Table 3.6   Hard and soft Lewis acids and bases

Lewis acids (electrophiles)	Lewis bases (nucleophiles)
Hard Lewis acids	*Hard Lewis bases*
H^+, Li^+, Na^+, K^+, Mg^+, La^{3+}, Sc^{3+}	F^-, O^{2-}, $RCO_2{}^-$, $CF_3SO_2O^-$
BX_3, $B(OR)_3$	$PO_4{}^{3-}$, $SO_4{}^{2-}$, $CO_3{}^{2-}$, $ClO_4{}^-$, $NO_3{}^-$
AlH_3, $Al(OR)_3$, AlR_3, AlX_3	HO^-, RO^-, R_2N^-
R_3Si^+, R_3Sn^+, R_3Ti^+	H_2O, R_2O
$SiCl_4$, $TiCl_4$, $Ti(OR)_4$	NH_3, H_2NNH_2, RNH_2
$RC(O)^+$	
$RMgX$	
Borderline Lewis acids	*Borderline Lewis bases*
Fe^{2+}, Co^{2+}, Ni^{2+}, Cu^{2+}, Zn^{2+}, Pb^{2+}, Sn^{2+}, R^+	Cl^-, Br^-, $N_3{}^-$, $NO_2{}^-$, $SO_3{}^{2-}$, $AlH_4{}^-$
RZn^+, XZn^+	RNH_2, R_2NH, R_3N
ScX_3, BR_3	$R(H)C{=}NR$, $R_2C{=}NR$
$SiCl_4$, SnX_4, $Sn(OR)_4$	
$R(H)C{=}O$, $R_2C{=}O$	
$R(H)C{=}NR$, $R_2C{=}NR$	
Soft Lewis acids	*Soft Lewis bases*
RHg^+, XHg^+, Hg^{2+}	H^-, I^-, SCN^-, $BH_4{}^-$
Cu^{2+}, Ag^+, Au^+, Pd^{2+}, Pt^{2+}	R^-
Cu^+, I^+, Br^+,	CN^-, RS^-, RSe^-
I_2, Br_2, ICN	R_2S
H_3B, $Sn(SR)_4$	
RO^+, RS^+, RSe^+	CO, RNC
Metal atoms (M^0)	R_3P, $(RO)_3P$
R_2C (carbenes)	$R_2C{=}CR_2$, $RC{\equiv}CR$

Note: The acid/base centres are shown in bold; R = alkyl.

electrophiles react equally well with all nucleophiles. This additional level of complexity was rationalised by the hard–soft acid–base (HSAB) theory, introduced by Pearson in 1963. It is most simply espoused as 'soft-likes-soft, hard-likes-hard' and the relevant classification of hardness/softness is given in Table 3.6; this classification can be applied to both inorganic and organic entities. Overall, it is found that hard electrophiles react faster with, and form stronger bonds to, hard nucleophiles and that soft electrophiles react faster with, and form stronger bonds to, soft nucleophiles. Thus, a matched pair participate in a reaction which is both thermodynamically and kinetically favoured.

Hard electrophiles are usually positively charged and have a high-energy lowest unoccupied molecular orbital (LUMO), while hard nucleophiles are usually negatively charged and have a low-energy highest occupied molecular orbital (HOMO). Soft electrophiles are not necessarily positively charged and have a low-energy LUMO, while soft nucleophiles are not necessarily negatively charged and have a high-energy HOMO. In a hard–hard interaction, the reacting orbitals are far apart in energy, the reaction is favoured by Coulombic attraction and the resulting bond is more ionic in character, while in a soft–soft interaction, the reacting orbitals are closer in energy and the bond is more covalent.

The hardness of carbocations depends on their hybridisation; sp^3 carbocations are softer than sp^2 which are in turn softer than sp, consistent with their potential for polarisation, and for carbanions, the softest are sp hybridised followed by sp^2 and then sp^3, consistent with their polarisability.

3.3.4 Curly ('curved') arrows

In order to describe reaction mechanisms, we need to be able to keep track of the changes which occur to the bonding and non-bonding electrons during the course of a reaction; this requires the use of a standard formalism. This is called the *curved-arrow* or *curly-arrow* convention and is very valuable and superficially trivial, but the implication of many of its conventions are not often fully understood and the power of the formalism is therefore lost. Curved arrows are used to denote polarisation and polarisability phenomena, and bond-making and bond-breaking processes.

3.3.4.1 General conventions

So far, we have used a number of conventions, and it is useful to compile the more important into a short glossary. The following is a useful guide to general conventions used in descriptions of structure and reactivity:

1. *Bonds*: A single line signifies that *two* electrons are present between the atoms so joined, a double line *four* electrons and a triple line *six* electrons.
2. *Symbols between structures*: A → B means that a reactant (A) is converted to a product (B). A ⇌ B means that an equilibrium exists between (A) and (B). A ↔ B is used only between valence-bond (canonical) structures that are contributors to a hybrid.
3. *The inductive (I) effect*: For a molecule X–Y, the representation X → Y shows that Y has a $^-$I effect or X has a $^+$I effect, or both.
4. *Dipole arrows*: It should thus be shown ⊢→.
5. *Partial charge*: It is represented by placing the symbol δ^+ or δ^- adjacent to the relevant atom.
6. *Stereochemistry*: A line of ordinary thickness denotes either 'no stereochemical implication' or a bond in the plane of the paper. Lines of extra thickness or a wedge denote a bond above the plane of the paper. Dotted or broken line denotes a bond below the plane of the paper.

3.3.4.2 Curved-arrow conventions

The following rules must be consistently applied to correctly use curved-arrow notation:

1. Arrows show movement of electrons, *never* of nuclei or groups of nuclei.
2. Double-headed arrows (⌒➤) indicate the direction of movement of two electrons. A single-headed arrow or 'fish hook' (⌒➤) indicates the direction of movement of one electron.
3. A set of arrows must always lead from one correct Lewis structure to another correct Lewis structure, including (a) the relationship between one valence-bond structure (canonical form) to another, for example

or (b) the conversion of reactants to intermediates (or transition states) and of these in turn to products, for example

or (c) the conversion of reactants directly into products, for example

4. Arrows must be chemically correct; that is, electrons must flow from a nucleophilic (Lewis basic) to an electrophilic (Lewis acidic) centre. The diagram H⊕ ⇁ is a very common mistake and confuses the notion that the arrow is used to indicate electron movement but not atomic movement; a proton cannot donate electrons because it has none.

5. Special care must be taken in representing molecular rearrangements (e.g. of carbonium ions) to clearly show that a group or atom migrates *with* its pair of bonding electrons.

The application of curly arrow formalism is best appreciated with some examples. Consider the deprotonation of an acid by an amine (Fig. 3.14a). Removal of a proton from an alcohol requires the reverse movement of electrons: Sodium hydride, a source of H⁻, will remove the acidic proton according to the movement of electrons indicated (Fig. 3.14b). Similarly, the reaction of a carbanion with an acid gives the corresponding alkane; electrons move in this case from the (basic) carbanion to the acid (Fig. 3.14c). More complex examples include the reaction of the nucleophile PhS⁻ with an alkyl halide (an electrophile), leading to the product of substitution (Fig. 3.14d); in this case, simultaneous formation of one bond and collapse of another are required in order to prevent the carbon centre from acquiring ten electrons. Simultaneous bond formation and cleavage is also required in the addition of cyanide (a nucleophile) to a carbonyl group (Fig. 3.14e). Finally, in the reaction of an acid HX with an alkene, the first step involves interaction such that the electrophile H⁺ receives two electrons from the (nucleophilic) alkene to generate a carbocation. In a second step, the (electrophilic) carbocation intermediate is intercepted by the remaining nucleophile, X⁻, to give the product (Fig. 3.14f). Notice that the curved arrow describes the movement of the electrons and at no point indicates the movement of atoms in course of any of the reactions. Note that it is, however, convenient and quite in order to use a single valence-bond structure (canonical form) to describe a reaction of a particular resonance-stabilised molecule or species. The only reason that one particular valence-bond structure is used is that the description of a particular reaction requires the use of fewer arrows, and hence the outcome is more quickly realised. As an example, consider the bromination of a ketone catalysed by alkali (Fig. 3.15). The initial deprotonation leads to an enolate anion, which can be represented by either of the two indicated canonical structures. (The second of these is more important, however, since the negative charge is located on the oxygen atom). Either valence-bond structure of the enolate anion may then be used to attack bromine, and both

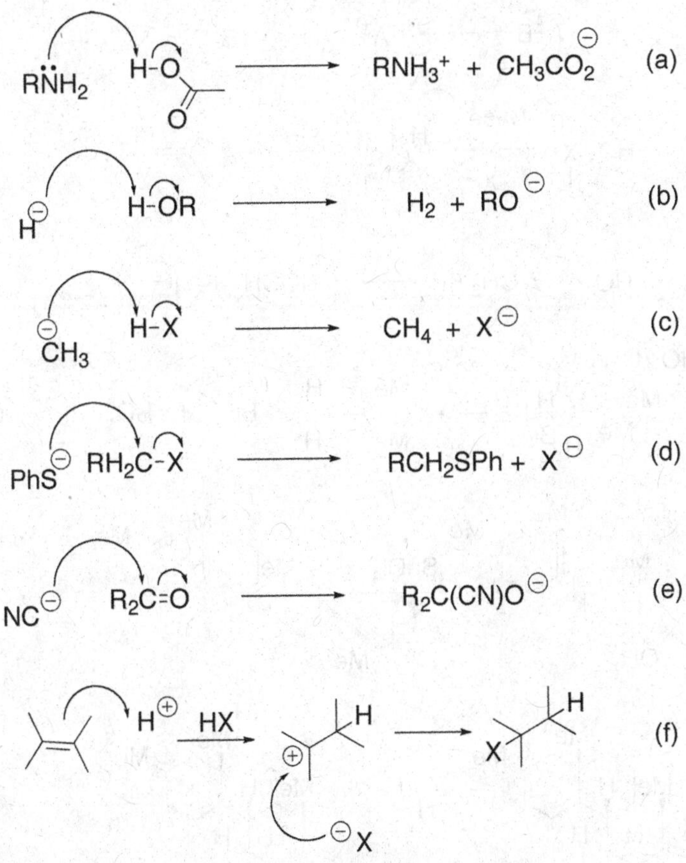

Figure 3.14 The use of curly (curved) arrows.

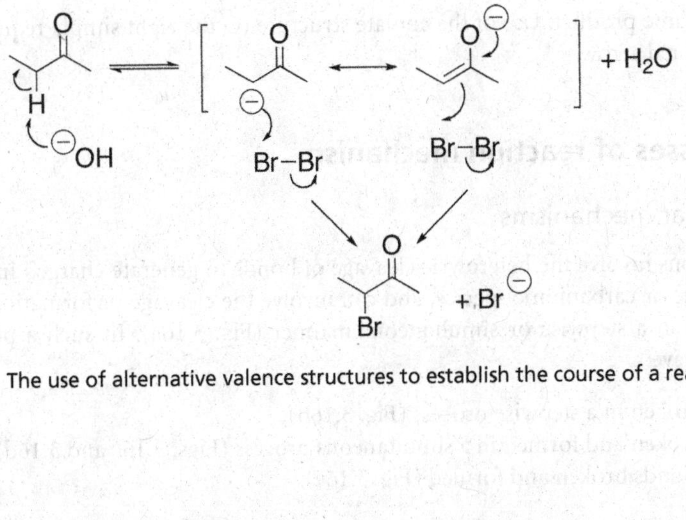

Figure 3.15 The use of alternative valence structures to establish the course of a reaction.

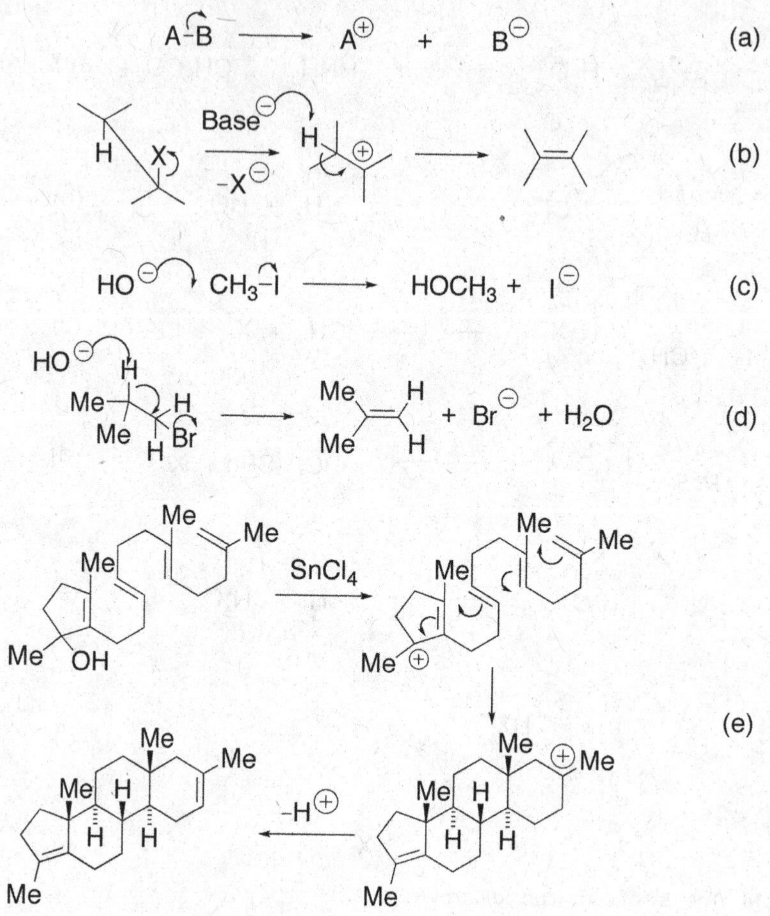

Figure 3.16 Examples of polar reactions.

lead to the same product. Use of the enolate structure on the right simply requires the use of an additional arrow.

3.4 Classes of reaction mechanism

3.4.1 Polar mechanisms

Polar reactions involve the heterolytic cleavage of bonds to generate charged intermediate carbocationic or carbanionic species, and can involve the cleavage or formation of one or more bonds in a stepwise or simultaneous manner (Fig. 3.16a). In such a process, it is possible to have:

1. bonds broken in a stepwise process (Fig. 3.16b);
2. bonds broken and formed in a simultaneous process (Figs. 3.16c and 3.16d);
3. several bonds broken and formed (Fig. 3.16e).

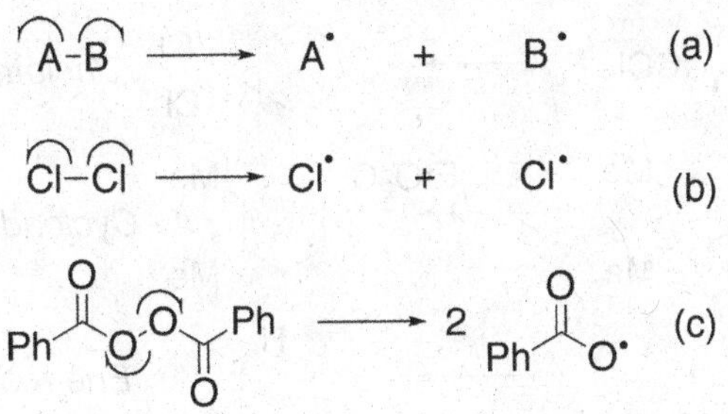

Figure 3.17 Formation of radicals.

3.4.2 Radical mechanisms

Radical intermediates are generated in reactions that involve homolytic cleavage of bonds (Fig. 3.17a); 'fish-hook' arrows indicate the movement of one electron. Two commonly encountered examples are the homolysis (or homolytic cleavage) of molecular chlorine, to generate two chlorine radicals (Fig. 3.17b), and of benzoyl peroxide (Fig. 3.17c). Once formed, radicals usually react by abstraction reactions to generate another radical and a neutral molecule (Fig. 3.18a), or by addition, to another multiple bond, to generate another radical (Fig. 3.18b).

3.4.3 Pericyclic mechanisms

Reactions that proceed by a redistribution of bonding electrons but without the formation of charged intermediates – sometimes erroneously referred to as 'no mechanism' reactions – are *pericyclic*. They are characterised by a mechanism in which there is concerted (or simultaneous) bond breakage and formation, and this is facilitated by a cyclic movement of electrons. Such processes should be compared with the stepwise examples which were considered in Section 3.4.1 and 3.4.2, and in which intermediates were formed. There are several different types, the most important of which are cycloadditions, ene reactions and chelotropic reactions (Fig. 3.19).

3.4.4 Ligand coupling reaction mechanisms

Certain transition and main group metals and metalloids are capable of coupling organic residues which they are carrying as ligands; the reactions are driven by the simultaneous

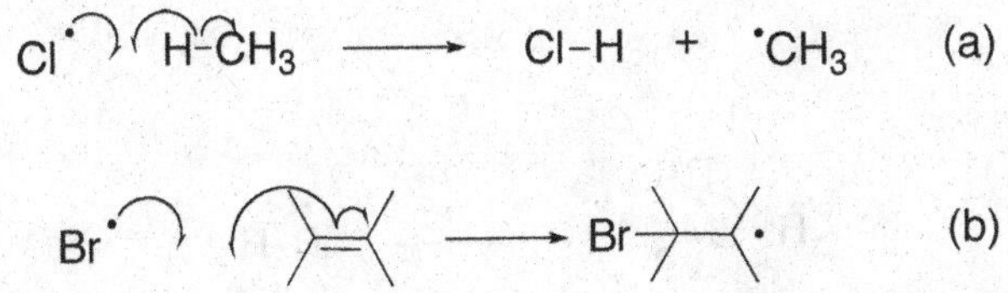

Figure 3.18 Examples of radical reactions.

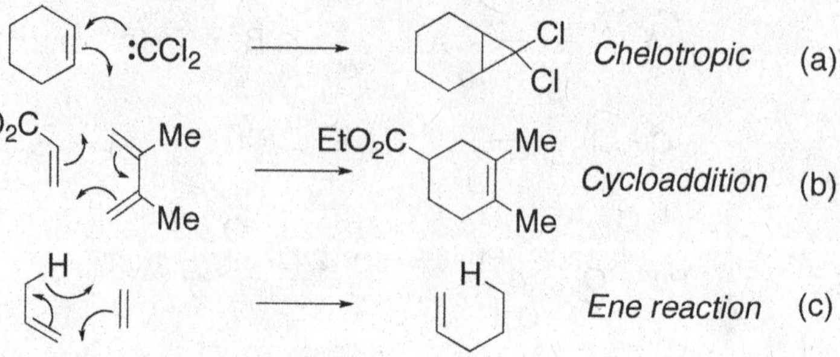

Figure 3.19   Examples of pericyclic reactions.

reduction of the metal and formation of a carbon–carbon bond (Fig. 3.20). The reaction is very general and is successful for R and R' = alkyl, vinyl, alkenyl and aryl, and suitable metals and metalloids include Pd, Cu, Ni, Pb, Bi and I. For transitions metals, the reoxidation of the metal to the higher oxidation state is facile, and this allows the reactions to be run in a catalytic mode.

3.5   Principle of microscopic reversibility

The mechanism of a reaction in the reverse direction must retrace each step of the reaction in the forward direction in microscopic detail (i.e. they must be identical), and this requirement arises as a result of the need for the conservation of energy; this is the *principle of microscopic reversibility*. Remembering that a reaction will proceed in the forward direction via the lowest energy pathway, it is entirely logical that the lowest energy pathway in the reverse direction will be the exact reverse of the forward pathway. However, although the pathway must be identical in both directions, both are not necessarily likely to be of equal rate, since each may not have identical activation energies.

For example, in the addition of methanol to isobutylene under acidic catalysis (Fig. 3.21), the reaction proceeds by (slow and rate-determining) initial protonation of the double bond to give a tertiary carbocation, which is then intercepted by methanol to give an oxonium species, which then loses a proton (fast) to give the ether product (Fig. 3.21a). The reverse reaction, acid-catalysed elimination of methanol from *t*-butyl methyl ether, according to the principle of microscopic reversibility, must proceed by the exact reverse mechanism; that is, initial protonation of the ether oxygen is followed by departure of the good leaving group, methanol, and then loss of a proton to give isobutylene (Fig. 3.21b).

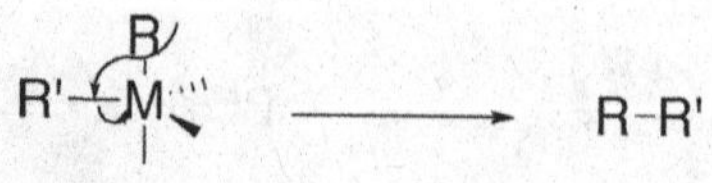

Figure 3.20   Ligand coupling reaction.

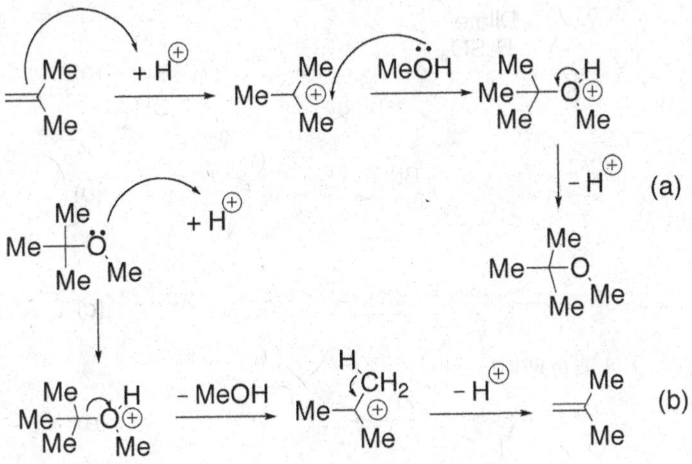

Figure 3.21 Application of the principle of microscopic reversibility.

3.6 Selectivity of reactions

Thus far, we have ignored the complication that comes if there are multiple possible sites for reaction or in which there is more than one possible outcome of the reaction. If a reaction can have more than one possible outcome, but only one is observed in practice because not all reactions are equally favoured, the reaction is said to be *selective*. Such an outcome will arise because the fastest reaction will proceed through a transition state of lowest energy or because the thermodynamically most favoured reaction will lead to the most stable product. For example, if more than one functional group might have reacted, but only one did so because it is the more intrinsically reactive, the reaction is said to be *chemoselective*. Examples of this include the reduction of ketones in favour of esters (Fig. 3.22a), the hydrolysis of

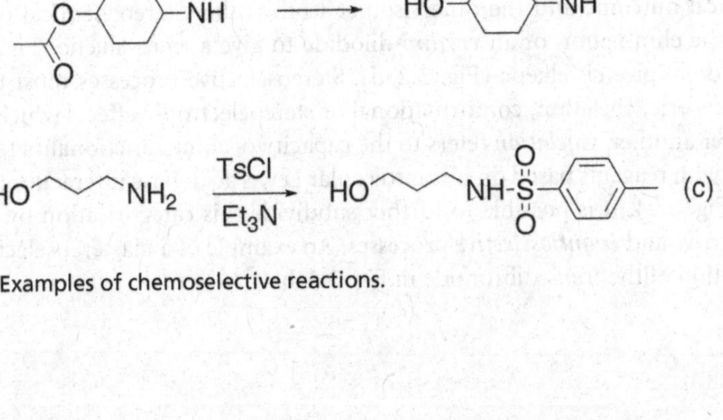

Figure 3.22 Examples of chemoselective reactions.

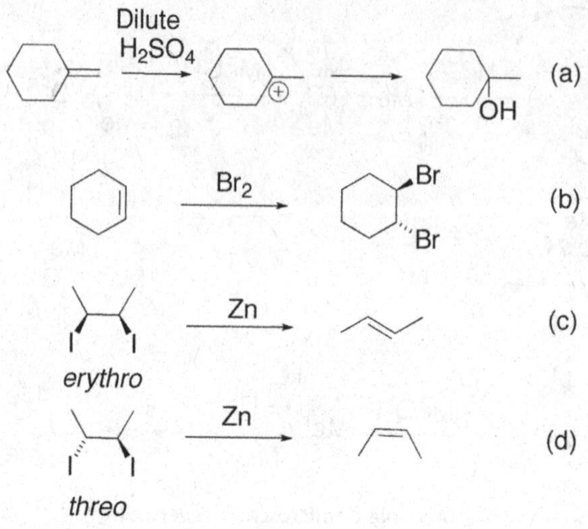

Figure 3.23 Examples of stereoselective reactions.

esters in favour of amides (Fig. 3.22b) and the sulfonylation of amines in favour of alcohols (Fig. 3.22c). Chemoselectivity can also be achieved by careful choice of reagents; for example, $ZnBH_4$ selectively reduces aldehydes over ketones, ketones over enones and aliphatic esters over aromatic esters, although lithium aluminium hydride does not show any selectivity for these groups. However, sometimes complete chemoselection cannot be achieved because the difference in reactivity is not large enough, and in this case *protection* of the relevant functional groups is used. For example, by converting a group into another of different reactivity, it is possible to mask or block undesired reactivity; this strategy works well only if such a conversion is easily reversible, enabling the original functional group to be regenerated under mild and convenient conditions.

    If for a given functional group, reaction can occur at more than one site but only one outcome is obtained, it is said to be *regioselective*. An example is the Markovnikov hydration of alkenes, which proceeds through the most stable carbocationic intermediate (Fig. 3.23a), although hydroboration proceeds with the opposite regioselectivity. And if, of several possible stereochemical outcomes, only one is observed, the reaction is said to be *stereoselective*; an example is the bromination of alkenes, which leads selectively to the *trans*-dibromide rather than the *cis*-alternative (Fig. 3.23b). However, sometimes this selectivity is not complete, and both products are obtained, but in a ratio in which one predominates. A *stereospecific* reaction is one in which one isomer of starting material is converted cleanly to one stereochemical outcome and the other isomer to the other stereochemical outcome. An example is the elimination of an *erythro*-diiodide to give a *trans*-alkene (Fig. 3.23c) or a *threo*-diiodide to give *cis*-alkene (Fig. 3.23d). Stereoselective processes most usually arise as a result of steric, chelation, conformational or stereoelectronic effects which favour one outcome over another. *Chelation* refers to the capacity of some functionality to form transient bonds with reagents based on intramolecular Lewis acid–base interaction; an example is given in Fig. 3.24. It is possible to further subdivide this categorisation by considering *diastereoselective* and *enantioselective* processes. An example of a diastereoselective reaction is the formation of the *trans*-dibromide in Fig. 3.23b.

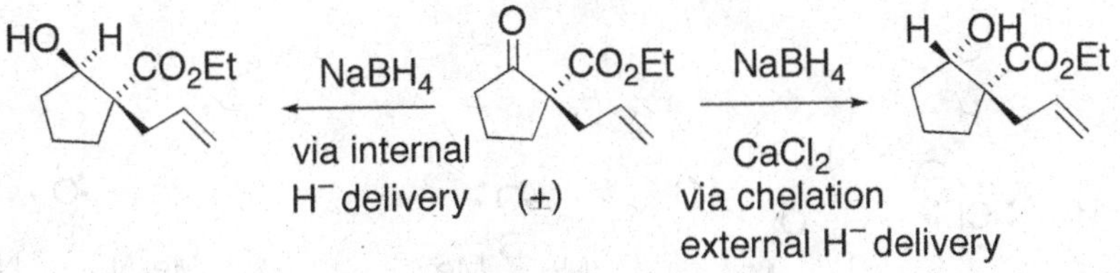

Figure 3.24 Example of a chelation-controlled process.

3.7 Solvents in organic chemistry

Solvents are liquids that are used to dissolve a solute. Although commonly assumed to fulfil a mundane role, solvents can in fact be of fundamental importance to the success of a reaction, since they can profoundly affect solubility (and therefore concentration), acidity/basicity of reagents, solvate charged intermediates and also act as heat sinks in exothermic processes. Solvents are characterised by several properties:

1. *Protic* solvents are those which exhibit proton-donor activity, having large dipole moment and hydrogen-bond-donor activity, for example, H_2O, ROH, RC(O)OH and RNH_2. *Aprotic* solvents cannot act as a proton donor, for example, Et_2O, DCM and hexane.
2. *Polar* solvents have a high dielectric constant ($\varepsilon > 15$) and *apolar* solvents a low dielectric constant ($\varepsilon < 15$).
3. Solvents which are *Lewis bases* are lone-pair donors, for example, Et_2O, THF and CH_3OH. Solvents which are not Lewis bases and are therefore unable to donate a lone pair include pentane, benzene and chloroform.

The dissolution of an ionic solute requires separation of the charges in its lattice and effective solvation of the ions thus generated. This is best achieved when the dielectric constant is large; that is, a solvent is polar. The solvation of ions in solution is aided by hydrogen bonding and Lewis basicity, which coordinate to the ions and assist in the separation of charge. This is exemplified by the solvation of chloride anions by methanol, sodium cations by dimethyl sulfoxide (DMSO) and lithium cations by hexamethylphosphoramide (HMPA) (Figs. 3.25a, 3.25b and 3.25c respectively). For an organic solute, the rule that 'like dissolves like' is often useful; thus, non-polar compounds tend to be highly soluble in non-polar solvents, but highly polar compounds require polar solvents for effective solvation. Dipolar aprotic solvents are important, since they are both polar and good Lewis bases but do not possess acidic protons, for example, Me_2SO, RCN, RCONHR' and RCOR'. Table 3.5 gives an indication of their approximate order of polarity of common solvents. (see section 3.2.3).

More recently, however, a new class of solvents has been developed, called 'ionic liquids'. These compounds are unusual in that they are salts, are frequently composed of organic cations and inorganic anions, are liquids at or slightly above room temperature, are excellent solvents with extremely low vapour pressures and product is easily recovered from the solvent. Because they can be immiscible with both organic and aqueous solvent systems, they can be readily recycled. Some typical structural components are given in Fig. 3.25; cations are usually based on a 1,3-disubstituted imidazolium system (Fig. 3.25a), and anions are

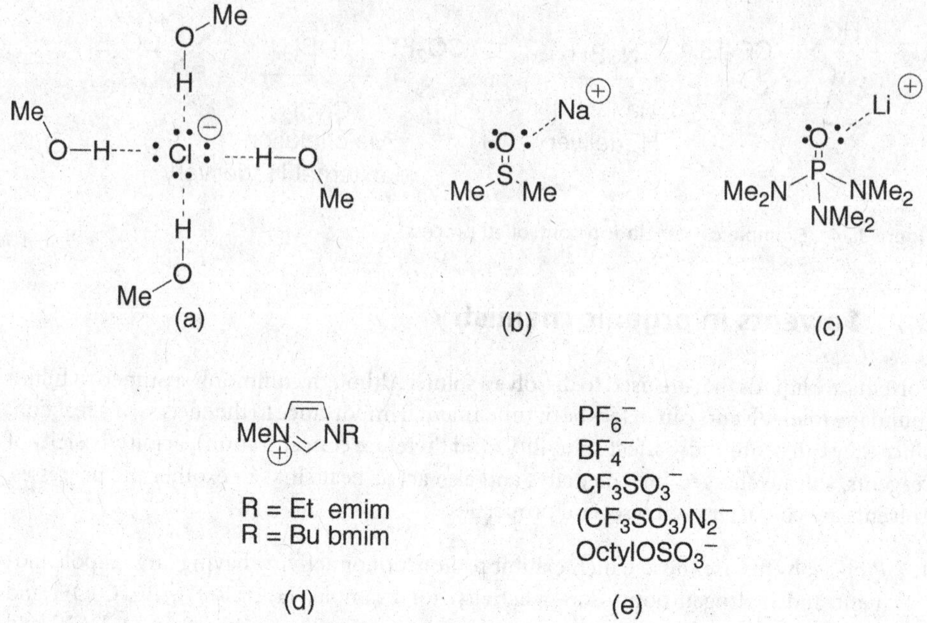

Figure 3.25 Solvation of (a) anions and (b, c) cations, and (d) cationic and (e) anionic components of ionic liquids.

typically inorganic but may contain organic groups too as shown by octylsulfate (Fig. 3.25b). These solvents have excited considerable interest as a result of their potential application in 'green' or at least low-environmental impact synthesis, in which solvent release to the atmosphere is minimised and solvent recovery at the end of a synthetic sequence allows effective recycling. By appropriate choice of a range of cations and anions, a diversity of possible ionic liquids can be readily accessed, and the preparation of chiral solvents is readily achieved by using one or other component in enantiopure form.

3.8   Redox reactions in organic chemistry

We have considered a number of reaction types thus far, which proceed with or without the formation of intermediates, but for which it is possible to describe a coherent mechanism that allows us to understand the course of a reaction. There is, however, another way of thinking about reactions, and this is by considering whether it leads to a change in the state of oxidation or reduction of the products relative to the reactants. We consider that an atom is considered to be *reduced* if it gains electrons and to be *oxidised* if it loses electrons. In order to apply this concept to organic compounds, we ignore the fact that its bonds are covalent and consider how the sharing of electrons, which is never equal, changes in the course of a reaction; this is easily done by a consideration of electronegativity. In order to decide if carbon has been reduced or oxidised in any reaction, the following rules are important:

Rule 1. Elemental carbon has an oxidation number of 0.

Rule 2. The oxidation state of any chemically bonded carbon is assigned as -1 for each more electropositive atom, which is attached, and $+1$ for each more electronegative atom. In practice, this means:

-1 for H, B, Na, Li, Mg
$+1$ for O, N, S, Cl, Br, I
  0 for C

Rule 3. In compounds with multiple bonds, each atom is counted as often as the multiple bond dictates, that is, twice for a double bond and thrice for a triple bond.

$$\text{>C=O} \quad \equiv \quad \text{C(O)(O)} $$

$$\text{—C}\equiv\text{N} \quad \equiv \quad \text{C(N)(N)(N)}$$

Rule 4. A formal positive charge on carbon changes the oxidation number by $+1$ and a formal negative charge by -1, but an odd electron (i.e. a radical) does not change the oxidation number.

The application of these rules is best understood by a consideration of some examples (Fig. 3.26) for some common compounds. Note that the oxidation number of any constituent atom in a molecule must be computed separately, and these are indicated. Using this approach, it is possible to compile a table of oxidation numbers of the carbon atoms of common functional groups, and these are given in Table 3.7. Note that functional groups in the same column are at the same oxidation level; this is not always immediately obvious, such as a 1,1-dihaloalkane and an aldehyde or a tertiary alcohol, or a trihaloalkane and a carboxylic acid.

We can use this concept in a classification of oxidation and reduction reactions. Using a traditional definition, reduction has been considered to be any reaction which leads to an increased hydrogen content or a lowered oxygen content, and oxidation is the reverse, that is, a decreased hydrogen content or an increased oxygen content. The use of the concept of oxidation number, however, allows this definition to be extended to a reaction which does not change the hydrogen or oxygen content of a molecule. In this approach, *oxidation is any reaction that leads to an increase in oxidation number, while reduction is any reaction that leads to a decrease in oxidation number.* In practice, oxidation occurs when a bond between a carbon and an atom that is less electronegative than carbon is replaced by one that is more electronegative than carbon; this is indicated in Fig. 3.27. A functional group interconversion which moves from left to right is an oxidation and that from right to left is a reduction, but one from top to bottom or bottom to top is neither a reduction nor an oxidation. It is instructive to consider some examples (Fig. 3.28). For example, both carbons of ethane have oxidation numbers of -3, and removal of hydrogen increases this to -2; that is, it corresponds to an oxidation. Similarly, removal of hydrogen from ethene gives ethyne, oxidation number -1, another oxidation (Fig. 3.28a). The three carbons of propene have different oxidation numbers (-3, -2 and -1, total -6), and reaction

Table 3.7 Oxidation numbers for common functional groups

Compound class	Oxidation number								
	-4	-3	-2	-1	0	$+1$	$+2$	$+3$	$+4$
Hydrocarbons (saturated)	CH_4	RCH_3	R_2CH_2	R_3CH	R_4C				
Hydrocarbons (unsaturated)			$CH_2{=}CH_2$	$HC{\equiv}CH$					
Alkyl halides			CH_3X	RCH_2X	CH_2X_2	$RCHX_2$	CHX_3	RCX_3	CX_4
Alcohols			CH_3OH	RCH_2OH	R_2CHOH H_2CO $CH_2(OR')_2$	R_3COH	HCO_2H		CO_2
					$RCHO$ $RCH(OR')_2$			RCO_2H	
Carbonyl compounds							R_2CO $R_2C(OR')_2$		

$(R = $ alkyl or aryl$)$.

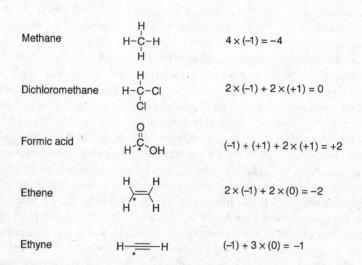

Figure 3.26 Oxidation number of simple organic compounds.

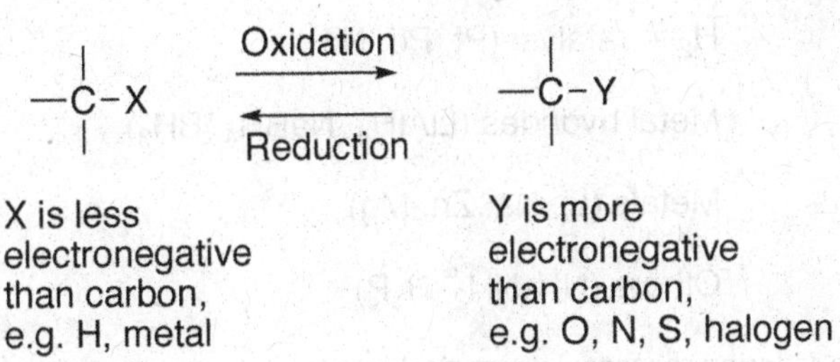

$$\underset{|}{\overset{|}{-C}-X} \quad \underset{\text{Reduction}}{\overset{\text{Oxidation}}{\rightleftarrows}} \quad \underset{|}{\overset{|}{-C}-Y}$$

X is less
electronegative
than carbon,
e.g. H, metal

Y is more
electronegative
than carbon,
e.g. O, N, S, halogen

Figure 3.27 Redox transformations in organic compounds.

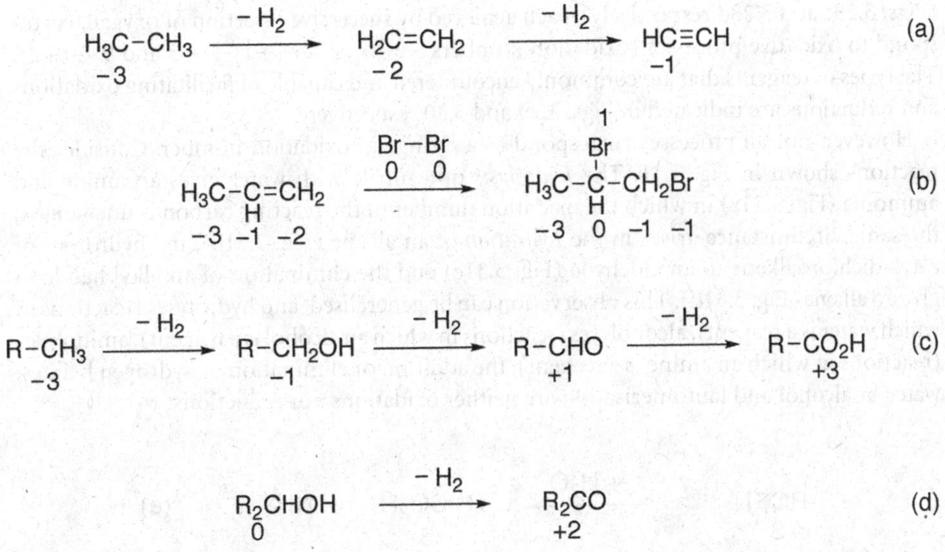

$$\underset{-3}{H_3C-CH_3} \xrightarrow{-H_2} \underset{-2}{H_2C=CH_2} \xrightarrow{-H_2} \underset{-1}{HC\equiv CH} \qquad (a)$$

$$\underset{-3\ -1\ -2}{H_3C-\underset{H}{C}=CH_2} \xrightarrow[0\quad 0]{Br-Br} \underset{-3\ \ 0\ \ -1\ -1}{H_3C-\overset{\overset{1}{Br}}{\underset{H}{C}}-CH_2Br} \qquad (b)$$

$$\underset{-3}{R-CH_3} \xrightarrow{-H_2} \underset{-1}{R-CH_2OH} \xrightarrow{-H_2} \underset{+1}{R-CHO} \xrightarrow{-H_2} \underset{+3}{R-CO_2H} \qquad (c)$$

$$\underset{0}{R_2CHOH} \xrightarrow{-H_2} \underset{+2}{R_2CO} \qquad (d)$$

Figure 3.28 Examples of redox reactions.

O_2	HNO_3	Ag_2O	MnO_4
O_3	X_2	HgO	OsO_4
$HOOH$	R_3N-O	$Hg(OAc)_2$	IO_4^-
$RC(O)OH$		$Pb(OAc)_4$	$Cr_2O_7^{2-}$

Figure 3.29 Some oxidising reagents in organic chemistry.

H$_2$ + catalyst (Pt, Pd, Ni)

Metal hydrides (LiAlH$_4$, NaBH$_4$, BH$_3$)

Metals (Li, Na, Zn, Mg)

Others (NH$_2$NH$_2$, R$_3$P)

Figure 3.30 Some reducing reagents in organic chemistry.

with bromine gives 1,2-dibromopropane, with oxidation numbers of -3, 0 and -1 (total -4), and therefore the process corresponds to an oxidation (Fig. 3.28b). The interconversions of alkanes, alcohols, aldehydes and carboxylic acids and of alcohols and ketones (Figs. 3.28c and 3.28d respectively), each achieved by successive insertion of oxygen, correspond to oxidative processes (oxidation numbers $-3 \rightarrow -1 \rightarrow +1 \rightarrow +3$ and $0 \rightarrow +2$). The types of reagents that are commonly encountered and capable of facilitating oxidations and reductions are indicated in Figs. 3.29 and 3.30 respectively.

However, not all processes correspond to a change in oxidation number. Consider the reactions shown in Fig. 3.31. The *hydrolysis* of a nitrile with water gives an amide and ammonia (Fig. 3.31a) in which the oxidation number of the reacting carbon is unchanged; the same circumstance arises in the *hydration* of an alkene (Fig. 3.31b), the hydrolysis of a 1,1-dichloroalkane to an aldehyde (Fig. 3.31c) and the elimination of an alkyl halide to give an alkene (Fig. 3.31d). This observation can be generalised, and hydrolyses (reactions in which water is a reagent), alcoholyses (reactions in which an alcohol is a reagent), aminolyses (reactions in which an amine is a reagent), the addition or elimination of hydrogen halides, water or alcohol and tautomerisation are neither oxidations nor reductions.

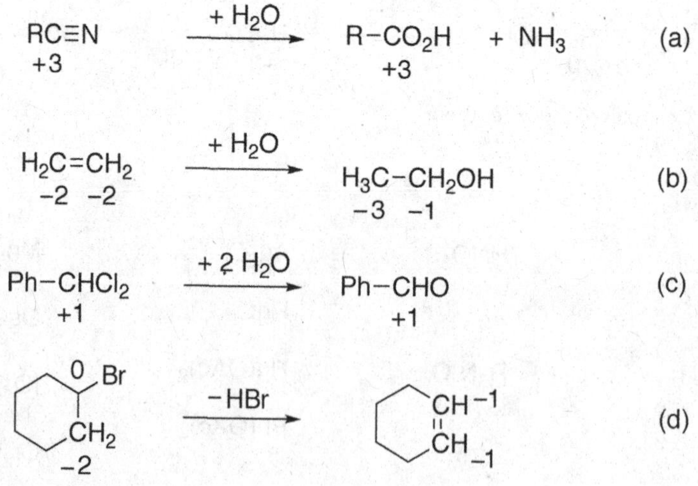

$$RC{\equiv}N \xrightarrow{\ +\ H_2O\ } R{-}CO_2H \ + \ NH_3 \qquad (a)$$
$$+3 \qquad\qquad\qquad +3$$

$$H_2C{=}CH_2 \xrightarrow{\ +\ H_2O\ } H_3C{-}CH_2OH \qquad (b)$$
$$-2\ \ -2 \qquad\qquad\qquad -3\ \ -1$$

$$Ph{-}CHCl_2 \xrightarrow{\ +\ 2\ H_2O\ } Ph{-}CHO \qquad (c)$$
$$+1 \qquad\qquad\qquad +1$$

(d)

Figure 3.31 Hydrolysis and elimination reactions in organic chemistry.

Chapter 4
Intermediates

So far we have considered in some detail the structure of organic compounds and how energy changes determine the course of a reaction. In particular, we have seen that it is possible for reactive intermediates to be formed en route from reactants to products; these are species which occur at a relative energy minimum on a reaction pathway. In this chapter, we consider the structure, stability and formation of reactive intermediates.

Carbon typically exhibits tetravalency as its universally preferred bonding arrangement. Reactive intermediates, however, have lower valency than this and as a result are not stable. There are two main types of reactive intermediates: trivalent and divalent carbon species. Included among the former are carbocations, radicals and carbanions (Fig. 4.1), each with three bonds and six, seven or eight valence electrons respectively. Carbocations, being two electrons short of their valence octet, are electron deficient, whereas carbanions, with their full complement of electrons, are electron rich. Carbenes are divalent, and since they too have only six electrons, they are electron deficient. The stability of these intermediates (I) depends on the magnitude of the activation energy which is required for them to return to reactants (Fig. 4.1, $\Delta G_r^{\ddagger}$) or to go on to products (Fig. 4.1, $\Delta G_p^{\ddagger}$); since these energies are typically small, it is difficult (but not impossible) to isolate intermediates. As we will see later, factors which help to stabilise intermediates (i.e. increase the magnitude of $\Delta G_r^{\ddagger}$ and $\Delta G_p^{\ddagger}$) can be very important in determining the course of a reaction.

4.1 Carbocations

4.1.1 Structure

A carbocation is a trivalent carbon and with only six electrons in its valence shell it is positively charged, since the carbon has one fewer electron than elemental carbon ($4 - 6/3 = +1$). Such intermediates are not particularly stable and must be carefully contrasted with the species shown in Fig. 4.2, which are also positively charged but have filled octets; they occur very commonly in organic chemistry. Carbocations are planar species with 120° bond angles, and the carbon is sp^2 hybridised; they are called *trigonal* (Fig. 4.3). The remaining 2p orbital is empty and hence positively charged; thus, a carbocation is isoelectronic with BF_3.

4.1.2 Factors stabilising carbocations

Although carbocations are not intrinsically stable, occurring as they do at the relative energy minimum indicated in Fig. 4.1, several factors are known to stabilise them, and some can do this very significantly.

1. *Inductive effects*: Alkyl substituents stabilise a carbocation because they release electrons to the electron-deficient centre, thereby reducing the positive charge on that carbon.

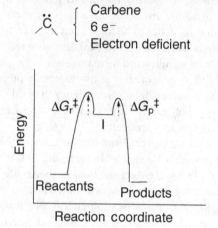

Trivalent reactive intermediates

$-\overset{|}{\underset{|}{C}}\oplus$ $\left\{\begin{array}{l}\text{Carbocation}\\ 6\ e^-\\ \text{Electron deficient}\end{array}\right.$ $-\overset{|}{\underset{|}{C}}\cdot$ $\left\{\begin{array}{l}\text{Radical}\\ 7\ e^-\\ \text{Neutral}\end{array}\right.$ $-\overset{|}{\underset{|}{C}}\ominus$ $\left\{\begin{array}{l}\text{Carbanion}\\ 8\ e^-\\ \text{Electron rich}\end{array}\right.$

Divalent reactive intermediates

$\overset{\cdot\cdot}{C}\diagdown$ $\left\{\begin{array}{l}\text{Carbene}\\ 6\ e^-\\ \text{Electron deficient}\end{array}\right.$

Figure 4.1 Characteristics of reactive intermediates.

$(CH_3)_4N^+$ $(CH_3)_3O^+$ H_3O^+

Tetramethylammonium ion Trimethyloxonium ion   Hydroxonium ion

Figure 4.2 Some stable positively charged species.

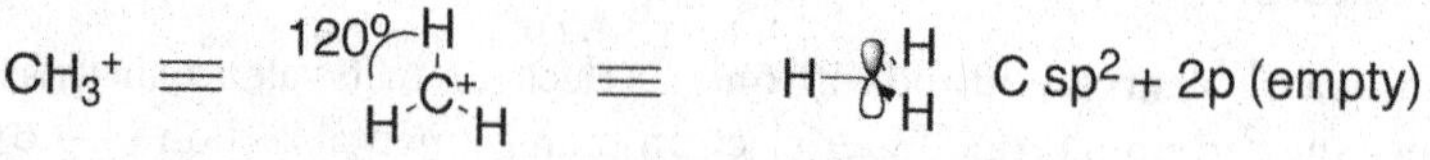

CH_3^+ $\equiv$ $\underset{H^{\diagup}\overset{\displaystyle 120^\circ\diagdown H}{C^+}\diagdown H}{}$ $\equiv$ $H-\overset{H}{\underset{H}{C}}$ $C\ sp^2 + 2p\ (\text{empty})$

Figure 4.3 Structure of a methyl carbocation.

Not surprisingly, the more the alkyl groups that can do this, the more the stabilisation. The order of carbocation stability is shown in Fig. 4.4; the magnitude of this effect, as illustrated by the bond dissociation energies (kJ mol^{-1}), for the reaction leading to the carbocation (R–H → R$^+$ + H$^-$) is also shown. This stabilisation is so important that simple carbocations such as methyl cation are so unstable that they cannot be observed under normal conditions.

2. *Resonance effects*: The delocalisation of π-electrons into the empty 2p orbital of a carbocation leads to significant stabilisation. There are three important cases:
   - *Allyl carbocations*: A positively charged carbon next to a carbon–carbon double bond experiences strong stabilisation, since the adjacent electron density can enter the empty 2p orbital, spreading the positive charge over the molecule; this is represented

Order of stability　　$R_3C^+ > R_2HC^+ > RH_2C^+ > H_3C^+$

	3°	2°	1°	Me
Bond dissociation energy (kJ mol^{-1})	987	1050	1150	1310

Figure 4.4　Stability of carbocations.

in Fig. 4.5a, in which the two possible hybrids allow positive charge to be located on the ends of the three-carbon system.

- *Benzyl carbocations*: A positively charged carbon next to the carbon–carbon double bonds which comprise an aromatic ring experiences very strong stabilisation, since similar stabilisation to that for the allyl system is possible, except that the positive charge can now reside on a total of four carbons, as shown in Fig. 4.5b. The stabilisation afforded by a phenyl group is approximately the same as that of a *t*-butyl group, and the corresponding bond dissociation energy leading to the formation of a benzyl carbocation PhCH$_2^+$ is 1003 kJ mol^{-1}. However, this effect is not necessarily fully additive, and the stabilisation of Ph$_3$C$^+$ ('trityl cation') is not three times that of the benzyl carbocation. This is because steric interactions between the adjacent ortho hydrogens force the aromatic rings out of planarity,

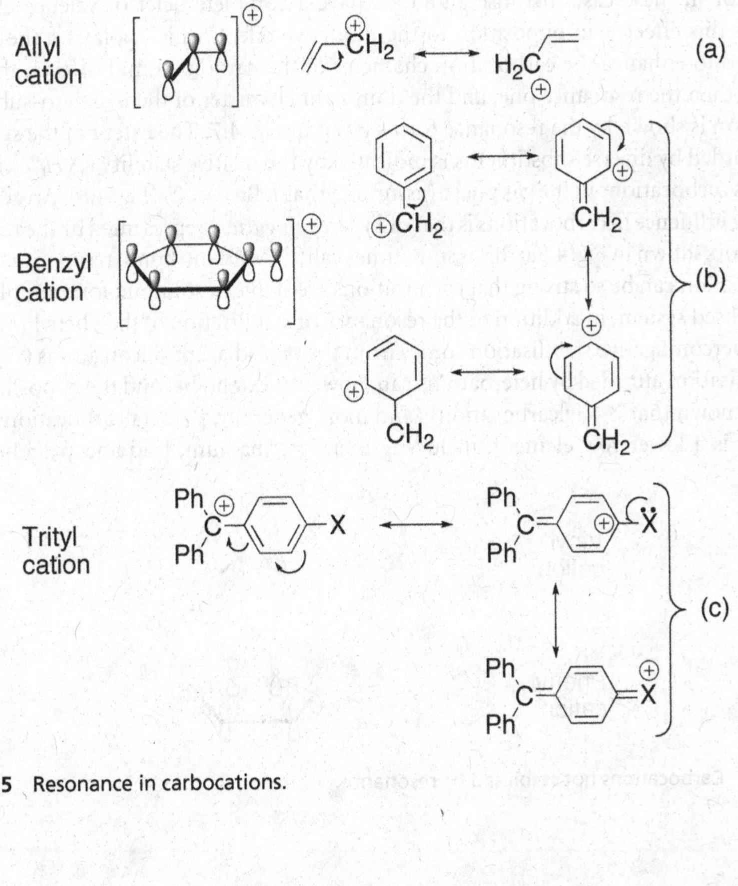

Figure 4.5　Resonance in carbocations.

Table 4.1 Relative stability of substituted trityl cations (Fig. 4.5c)

X	NO$_2$	H	Me	OH	NMe$_2$
kJ mol^{-1}	55	0	−17	−50	−100

reducing the stabilisation afforded by the resonance interactions. The stabilisation of an aromatic unit is even more marked if there are electron-releasing groups (X) on the rings (Fig. 4.5c), but conversely, the carbocation can be destabilised if there are electron-withdrawing groups (X) attached. The data in Table 4.1 indicate the magnitude of the (de)stabilisation, relative to the trityl (X = H) cation. These especially stabilised cases should not be confused with vinyl and phenyl carbocations, which are particularly destabilised because resonance between the empty 2p orbital and the π-system is not possible by virtue of their orthogonal spatial relationship (Fig. 4.6).

- α-*Heteroatom carbocations*: A positively charged carbon next to a heteroatom which has a non-bonded lone pair can be significantly stabilised by delocalisation of that lone pair. Examples include protonated acetone (Fig. 4.7a), the acetylium ion (Fig. 4.7b), the iminium ion (Fig. 4.7c) and the sulfonium ion (Fig. 4.7d).

The canonical structure with the positively charged heteroatom is the more important contributor in these cases because all atoms have a complete octet of valence electrons. Note that this effect is in opposition to the inductive effect, which polarises the C–O or C–N bond to enhance the carbocation character of the carbon atom; however, this effect is weaker than the resonance one, and the dominant character of the α-hetero-substituted carbocation is shown by the resonance forms given in Fig. 4.7. The extent of the stabilising effect afforded by diverse substituents is indicated by the relative stability (ΔH_f°, kJ mol^{-1}) of various carbocations in the gas phase (assuming that t-Bu$^+$ = 0) (Fig. 4.8). An additional stabilising influence in carbocations is that of *hyperconjugation*, represented by the resonance contributors shown in Fig. 4.9a; this is sometimes called 'bond–no bond' resonance. This hyperconjugation can be so strong that carbocations are stable in solution; for example, in the silyl-stabilised system, in addition to the resonance of stabilisation of the phenyl ring, additional hyperconjugative stabilisation comes from the two adjacent silicon atoms (Fig. 4.9b). The stabilisation afforded by heteroatoms can, however, extend beyond the α-position, and it is well known that β-silylcarbocations (and more generally, β-metalcarbocations, where the metal is a lower row element, including mercury, thallium, lead and palladium) are

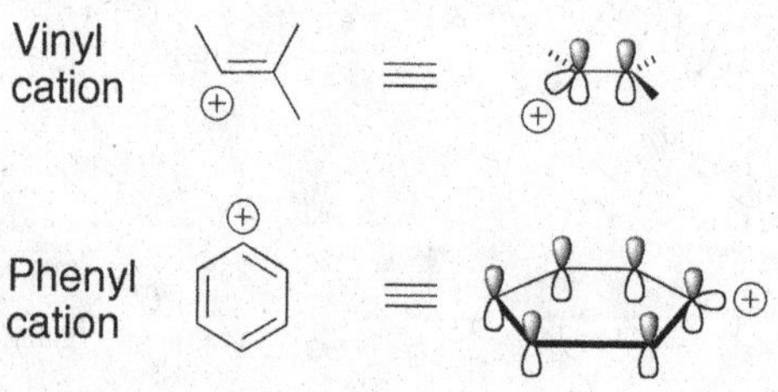

Vinyl cation

Phenyl cation

Figure 4.6 Carbocations not stabilised by resonance.

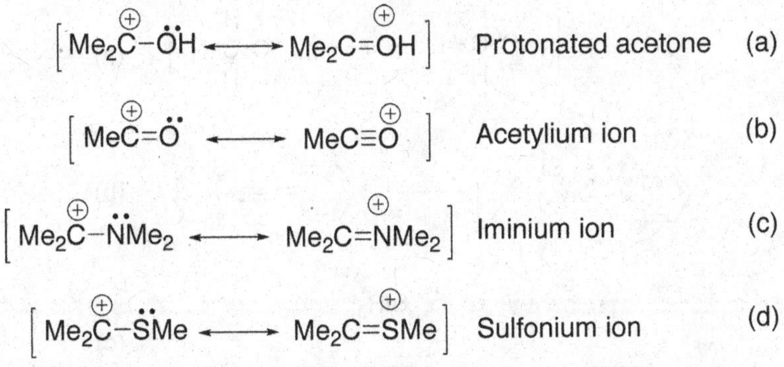

$$\left[\overset{\oplus}{Me_2\overset{\cdot\cdot}{C}}-\overset{\cdot\cdot}{O}H \longleftrightarrow Me_2C=\overset{\oplus}{O}H \right] \quad \text{Protonated acetone} \quad \text{(a)}$$

$$\left[Me\overset{\cdot\cdot}{C}=\overset{\cdot\cdot}{O} \longleftrightarrow MeC\equiv\overset{\oplus}{O} \right] \quad \text{Acetylium ion} \quad \text{(b)}$$

$$\left[\overset{\oplus}{Me_2C}-\overset{\cdot\cdot}{N}Me_2 \longleftrightarrow Me_2C=\overset{\oplus}{N}Me_2 \right] \quad \text{Iminium ion} \quad \text{(c)}$$

$$\left[\overset{\oplus}{Me_2C}-\overset{\cdot\cdot}{S}Me \longleftrightarrow Me_2C=\overset{\oplus}{S}Me \right] \quad \text{Sulfonium ion} \quad \text{(d)}$$

Figure 4.7 Carbocations stabilised by resonance with an adjacent heteroatom.

favourable species on account of overlap of the empty p orbital of the carbocation with the carbon–metal σ-bond (Fig. 4.9c).

4.1.3 Generation of carbocations

Carbocations are very reactive electrophiles as well as good proton donors (i.e. Lowry–Brønsted acids), since both these processes allow the positively charged carbon to regain a full valence octet, typically reacting either with nucleophiles or with bases as indicated in Figs. 4.10a and 4.10b. To be isolable, carbocations must therefore be generated in the absence of base or nucleophiles. This can be achieved either by the reaction of alkyl halides with strong electrophiles (Fig. 4.10c) or by the reaction of alkenes with acids (Fig. 4.10d). In the former, SbF$_5$ acts as a solvent and Lewis acid; the SbF$_6^-$ which is formed is a poor nucleophile, and so the reverse reaction is not favoured. In the latter, superacids which have non-nucleophilic counteranions (HF or HSO$_3$F) in non-nucleophilic, inert and polar solvents, such as SO$_2$, SO$_2$ClF, SO$_2$F$_2$ and CH$_2$Cl$_2$, effectively protonate alkenes. Simple primary cations have never been observed in solution, whereas tertiary cations are reasonably stable. Secondary cations are midway between the two and can be stable, for example, Me$_2$CH$^+$ and the cyclohexyl cation C$_6$H$_{11}^+$.

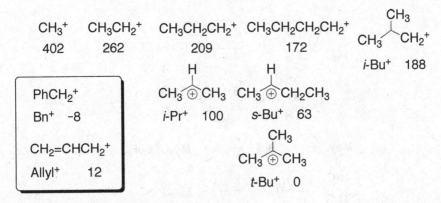

Figure 4.8 Relative stability of substituted carbocations.

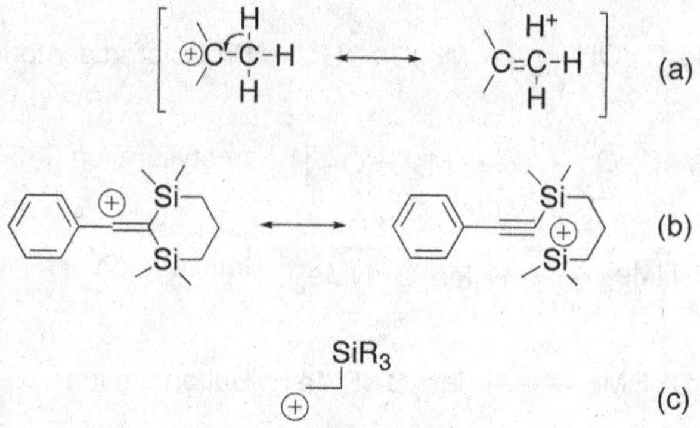

Figure 4.9 Hyperconjugation which stabilises a carbocation.

4.1.4 Rearrangements of carbocations

Carbocations can rearrange by migration of hydrogen, alkyl or phenyl groups; these are called Wagner–Meerwein rearrangements. This is an important pathway, with a low $\Delta G^{\ddagger}$, especially if it gives a thermodynamically more stable carbocation. Particularly favourable are rearrangements which convert a primary or secondary to tertiary carbocation (Figs. 4.11a and 4.11b) or which generate a benzylic carbocation (Fig. 4.11c). Note that carbocations, after rearrangement, may react either with a nucleophile or with a base to give the product of substitution or elimination respectively (Figs. 4.11d and 4.11e).

4.2 Carbanions

4.2.1 Structure

Carbanions possess a trivalent carbon, are negatively charged ($4 - (6/2 + 2) = -1$) and, although having a filled valence shell, are nonetheless *reactive* because carbon has a low electronegativity, enabling the lone pair to be easily shared with or donated to other reactive entities. Formally, at least, a carbanion can be obtained by deprotonation of an alkane

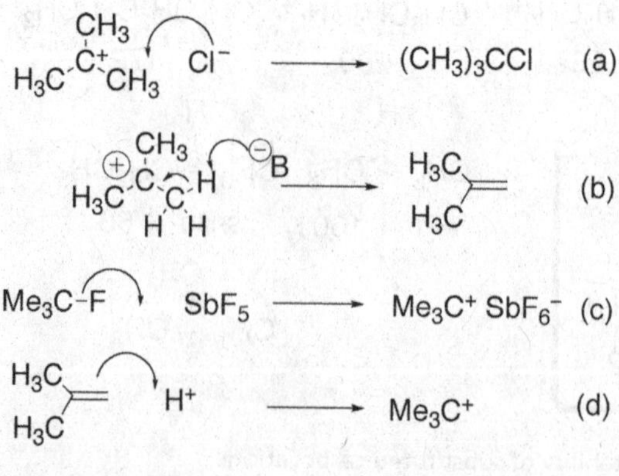

Figure 4.10 Generation of carbocations.

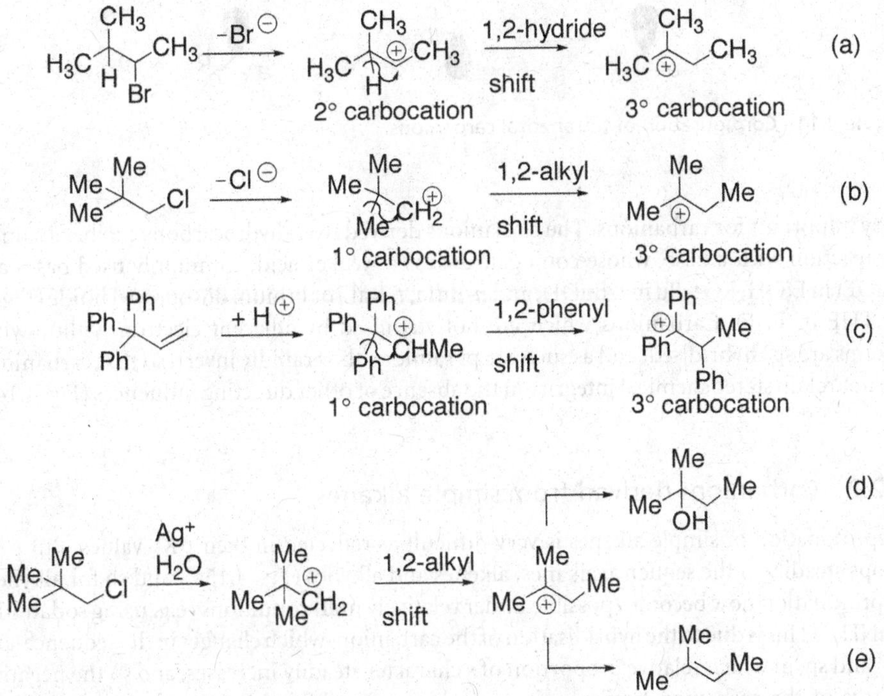

Figure 4.11 Rearrangements of carbocations.

$$CH_4 \longrightarrow CH_3^- + H^+$$

Figure 4.12 Generation of carbanions.

$$\text{Order of stability} \quad R_3C^- < R_2HC^- < RH_2C^- < H_3C^-$$
$$3° \qquad 2° \qquad 1° \qquad Me$$

Figure 4.13 Stability of carbanions.

(Fig. 4.12) but the use of such a mode to generate carbanions requires a solvent which is of lower acidity than the reacting hydrocarbon; ethers such as diethyl ether and tetrahydrofuran (THF) are ideal for this purpose. However, this mode of dissociation for a simple alkane is highly unfavourable, as reflected in the extremely high pK_a value for alkanes (for methane, approximately 60), and is due to the localisation of negative charge on the carbon without any stabilising influences; these species are therefore highly reactive. For carbanions, and in contrast to carbocations, substitution with more alkyl groups leads to a *reduction* in stability, because their electron-releasing effect increases the electron density of the negatively charged carbon atom. The stability of carbanions follows the order shown in Fig. 4.13. In general, carbanions are less stable than nitrogen or oxygen anions, since carbon has a lower electronegativity than these heteroatoms, and this can be seen by inspection of the pK_a values as shown in Table 5.1 (see chapter 5); thus, electron-withdrawing stabilising substituents are

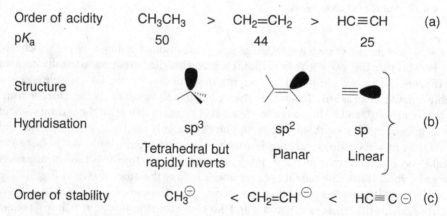

Figure 4.14   Configuration of tetrahedral carbanions.

very important for carbanions. The carbanions derived from hydrocarbons can be obtained by reaction with a base, whose conjugate acid is a weaker acid; commonly used bases are NaOEt in EtOH, KOt-Bu in t-BuOH, and n-BuLi, t-BuLi or lithium diisopropylamide (LDA) in THF or Et$_2$O. Carbanions which are not stabilised by adjacent electron-withdrawing groups are sp^3 hybridised, and as such are pyramidal; they rapidly invert, so that carbanions do not retain stereochemical integrity in the absence of other directing influences (Fig. 4.14).

4.2.2   Carbanions derived from simple alkanes

Deprotonation of simple alkanes is very difficult as reflected in their pK_a values. But pK_a drops steadily in the sequence alkanes, alkenes and alkynes (Fig. 4.15a), and so for alkynes, deprotonation now becomes possible under relatively mild conditions (e.g. using sodamide, NaNH$_2$). This is due to the hybridisation of the carbanion, which changes in the sequence sp^3, sp^2 and sp, in which relative proportion of s character steadily increases, and so the negative charge of the carbanion becomes successively more stabilised by virtue of its increasing proximity to the nucleus (Fig. 4.15b). The stability of carbanions therefore steadily increases in the series sp, sp^2 and sp^3 (Fig. 4.15c). Thus, the acetylide anions are easily formed and are very useful nucleophiles, but the formation of ethyl anions is much more difficult.

The result is that some types of carbanions are difficult to generate, for example, vinylic carbanions. One way by which such carbanions can be prepared is to make use of the powerful leaving-group ability of nitrogen gas; thus, treatment of tosylhydrazones with an excess of strong base readily extrudes nitrogen and generates the vinyl anion in a process called the Shapiro reaction (Fig. 4.16).

Order of acidity	CH$_3$CH$_3$	>	CH$_2$=CH$_2$	>	HC≡CH	(a)
pK_a	50		44		25	

Structure			
Hydridisation	sp^3	sp^2	sp (b)
	Tetrahedral but rapidly inverts	Planar	Linear

Order of stability	CH$_3^{\ominus}$	< CH$_2$=CH$^{\ominus}$	< HC≡C$^{\ominus}$ (c)

Figure 4.15   Types of carbanions.

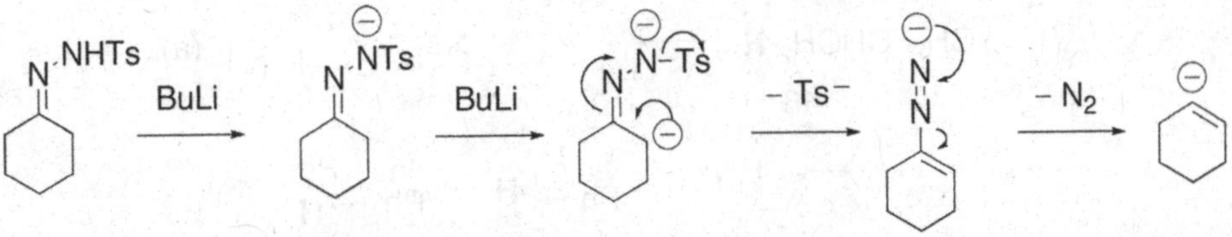

Figure 4.16 The Shapiro reaction.

4.2.3 Factors stabilising carbanions

1. *Inductive effects*: The withdrawal of electron density by adjacent electronegative groups can lead to stabilisation. This can be especially important, and multiple heteroatoms give significant stabilisation (Fig. 4.17).
2. *Resonance effects*: The delocalisation of negative charge through a π-system leads to substantial stabilisation. This can occur where the carbanion is adjacent to a carbon–carbon, carbon–oxygen or carbon–nitrogen double bond, or an aromatic system. In these cases, the negative charge is located in an sp^3 orbital which is capable of overlap with the adjacent π-network. For example, in allylic anions, significant resonance stabilisation arises due to the presence of two equivalent ('degenerate') canonical structures (Fig. 4.18a). Similar stabilisation is possible for benzylic anions (Fig. 4.18b), in which four contributing resonance structures are possible (Fig. 4.18c). The Ph_3C^- ('trityl') anion is not so stable that it is unreactive; it is in fact an excellent base (Fig. 4.19) and will remove a proton from any substrate more acidic (i.e. with a lower pK_a) than itself, including acetylenes and cyclopentadiene. These especially stabilised cases should not be confused with vinyl and phenyl carbanions, which are particularly destabilised, since the negative charge is located in sp^2 hybrid orbitals which are orthogonal to the π-system, and so there is no benefit from resonance stabilisation (Fig. 4.20). In contrast to carbocations, carbanions do not usually rearrange, but it can occur in some cases (Fig. 4.21). The most common rearrangement is a 1,2-phenyl shift, in which the migrating phenyl group is capable of forming a bridged intermediate.

One very important class of stabilised carbanions is that in which the negative charge is adjacent to a carbonyl group. In this case, the charge benefits from stabilising inductive withdrawal from the highly electronegative oxygen atom, as well as resonance withdrawal of electron density in a manner similar to the allylic system described above (Fig. 4.22a). Significantly, the negative charge in one of the resonance contributors resides on oxygen. This effect is even greater for two carbonyl groups (Fig. 4.22b), and the effect of such stabilisation can be easily seen by comparing pK_a data for a series of carbonyl compounds

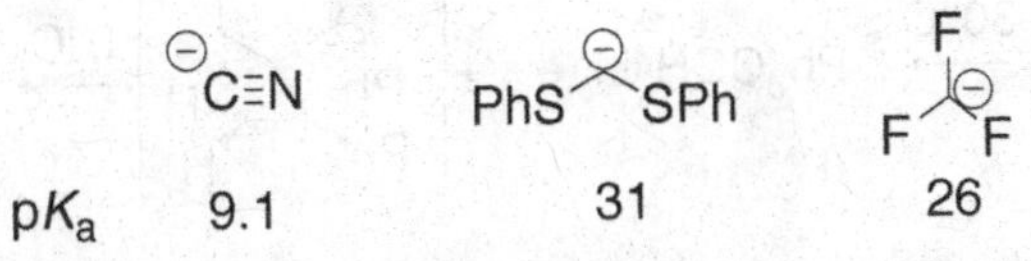

	$^{\ominus}C{\equiv}N$	$PhS{\overset{\ominus}{\frown}}SPh$	$F{\overset{F}{\underset{F}{\overset{\ominus}{\frown}}}}F$
pK_a	9.1	31	26

Figure 4.17 Heteroatom-stabilised carbanions.

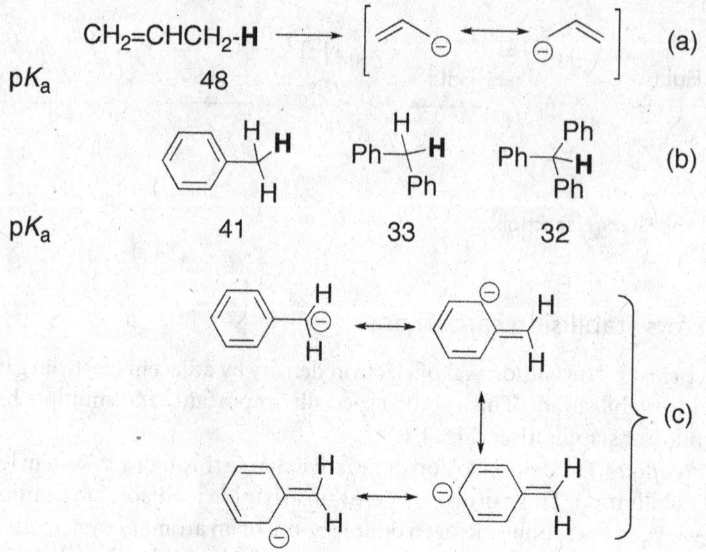

$$CH_2=CHCH_2\text{-}\mathbf{H} \longrightarrow \left[\text{⊖} \longleftrightarrow \text{⊖} \right] \quad \text{(a)}$$

pK_a 48

pK_a 41   33   32   (b)

(c)

Figure 4.18 Resonance-stabilised carbanions.

$$R\text{—}\!\!\equiv\!\!\text{—H} \xrightarrow{Ph_3C^\ominus} R\text{—}\!\!\equiv\!\!\text{⊖} + Ph_3CH$$

pK_a 25

pK_a 15

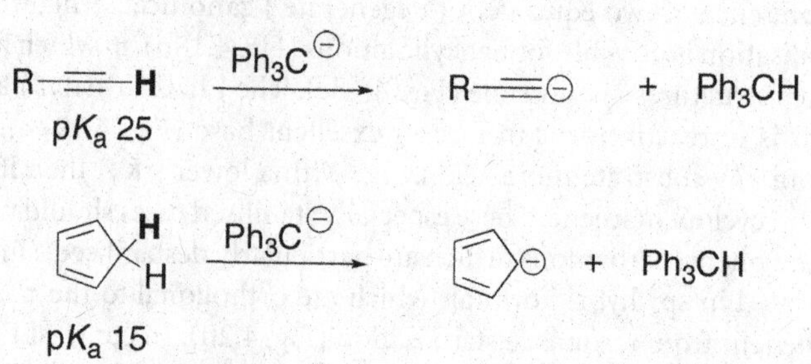

Figure 4.19 Stabilised carbanions and their reaction with acidic hydrocarbons.

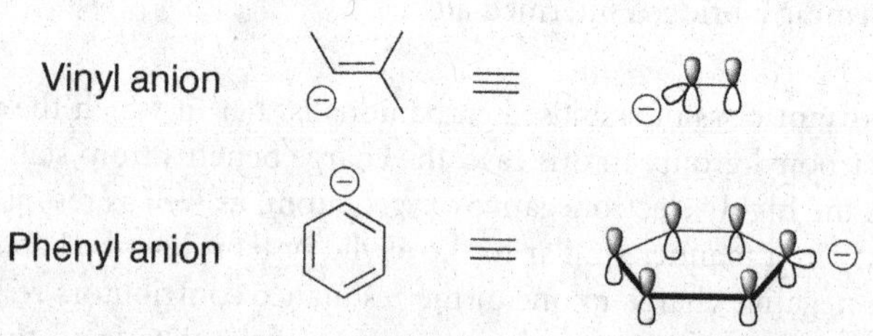

Vinyl anion

Phenyl anion

Figure 4.20 Carbanions not stabilised by resonance.

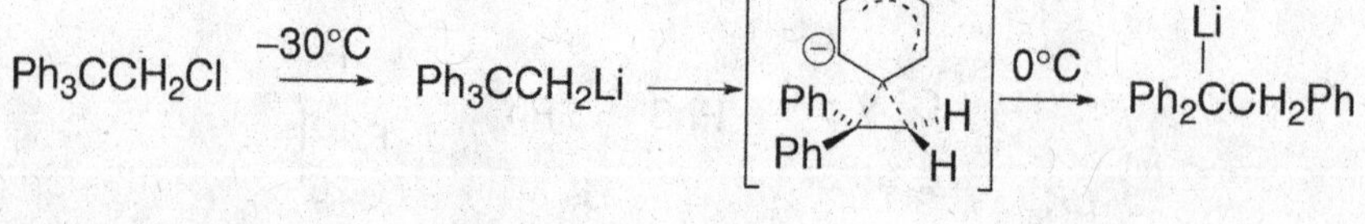

$$Ph_3CCH_2Cl \xrightarrow{-30°C} Ph_3CCH_2Li \longrightarrow \left[\text{...} \right] \xrightarrow{0°C} Ph_2CCH_2Ph$$

Figure 4.21 Rearrangements of carbanions.

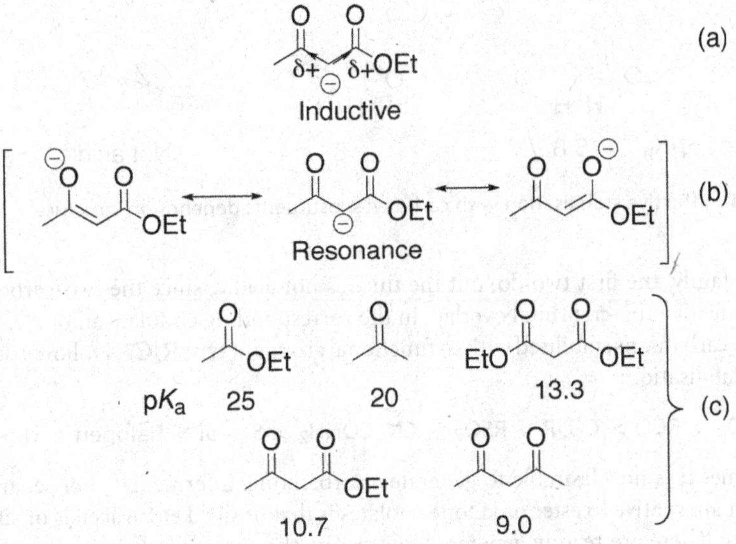

Figure 4.22 Carbanions stabilised by carbonyl substituents.

(Fig. 4.22c) in which significant enhancement of acidity relative to acetone can be readily achieved: Increases in acidity of up to 10 pK_a units relative to acetone are easily achieved.

Other groups which behave in a similar way include the cyano, nitro and sulfoxide groups (Fig. 4.23), since they possess π-unsaturation. Sulfur is also capable of stabilising adjacent (α) negative charge, since the sulfur, being a second-row element, atom has low-lying d orbitals which can accept the additional electron density of the carbanion; a similar situation arises in organosilanes. Anion stabilisation can be cumulative, and the acidity of the α-proton in dithioacetals, adjacent as it is to two sulfur atoms, is high enough that it can be conveniently removed with an amide base. Species in which there is adjacent positive and negative charges, such as for triphenylphosphonium- and sulfonium-derived carbanions, are called *ylides*.

However, it is important to remember that this stabilisation is possible only if a key requirement for resonance stabilisation is met: The sp^3 orbital containing the negative charge must be coplanar with the interacting π-system. If this does not occur, no stabilisation is possible. This may seem obvious in theory, but it may not seem so in practice. For example, the three dicarbonyl compounds shown in Fig. 4.24 might be expected to have similar pK_a

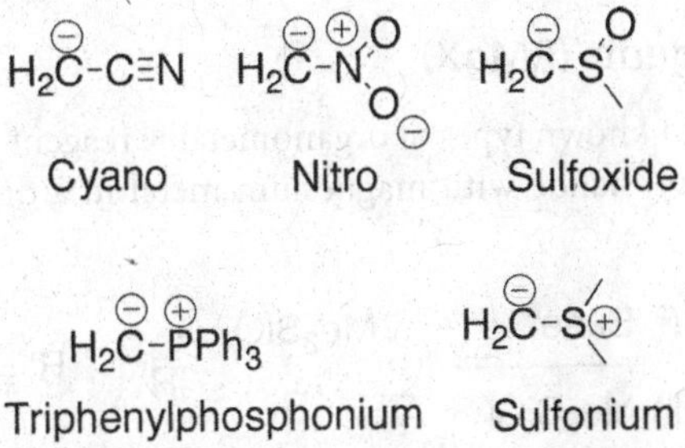

Figure 4.23 Carbanions stabilised by substituents.

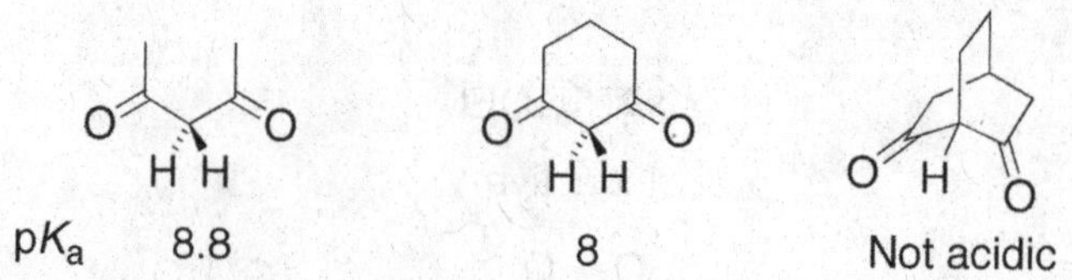

Figure 4.24 Effective stabilisation with carbonyl substituents depends on structure.

values; certainly, the first two do, but the third is not acidic, since the two-carbon bridge prevents effective sp^3–π-orbital overlap, in the corresponding endolate anion.

Overall, carbanions stabilised with α-functional groups of type R_2C^-X follow a decreasing order of stabilisation:

$$X = NO_2 > RCO > CO_2R > RSO_2 > CN, CONH_2 > S > Si > halogen > H > alkyl$$

Sometimes it is not desirable to generate a carbanionic intermediate per se, and a very convenient alternative to ester or ketone enolates is that of silyl ketene acetals or silyl ethers respectively. These are readily generated by treating the corresponding enolate with a trialkylsilyl halide and have the advantage that they can be isolated and purified prior to onward reaction (Fig. 4.25), but readily react as nucleophiles; they do, however, tend to be quite moisture- and acid-sensitive and therefore need careful handling.

4.3 Carbanions with covalent character

In addition to the stabilisation which comes from appropriate substitution as discussed in Section 4.2.3, carbanions are easily stabilised by forming partial covalent bonds with metals; such *organometallic compounds*, often represented as 'R–M', are very important in organic chemistry. Only the most electropositive of metals are capable of forming ionic bonds with carbon (e.g. Na, K, Rb and Cs), and such derivatives are highly reactive, non-volatile solids insoluble in benzene or other organic solvents. These compounds are strongly basic, and strictly anhydrous solvents must be used for their preparation and reactions. However, many other metals form covalent carbon–metal bonds much more readily, which possess covalent character, for example, Li, Mg, Zn, Sn, Hg, Cu, Ni, Fe, Pb and Si. These species are very important in organic chemistry, since they provide a convenient source of a carbanion equivalent 'R$^-$'.

4.3.1 Grignard reagents (RMgX)

These are one of the oldest known types of organometallic reagent and are readily obtained by treating an alkyl or aryl halide with magnesium metal in a dry ether solvent at room

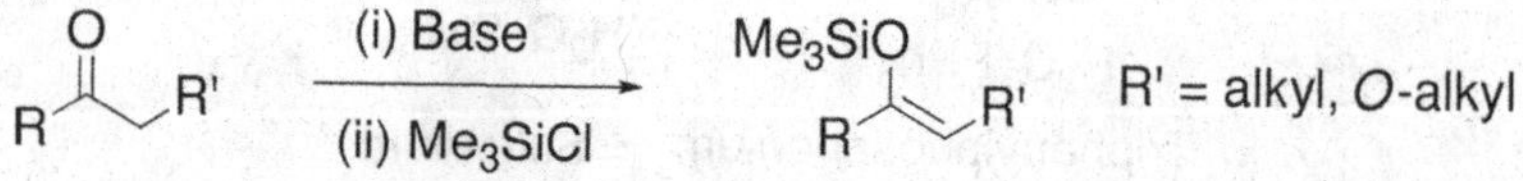

Figure 4.25 Formation of silyl enol ethers.

$$RX \ + \ Mg \ \longrightarrow \ RMgX \quad (a)$$

$$RX \ + \ i\text{-PrMgCl} \longrightarrow \ RMgX \quad (b)$$

$$2\,RMgX \ \rightleftharpoons \ R_2Mg \ + \ MgX_2 \quad (c)$$

Figure 4.26 Formation and structure of Grignard reagents.

temperature; conversion to the organometallic reagent is often spontaneous and exothermic (Fig. 4.26a). The structure of Grignard reagents in solution is complicated by the existence of the Schlenk equilibrium, which leads to the formation of dimers, as well as the fact that the solvent can act as a ligand for the metal (Figs. 4.26c–4.26e). More recently, the Knochel modification has provided an alternative approach for the formation of organomagnesium species (Fig. 4.26b) and is particularly valuable for those substrates which would not survive the classical reaction approach.

4.3.2 Organolithium reagents (RLi)

The reaction of an alkyl halide with lithium metal leads to the formation of an organo-lithium reagent, as shown in Fig. 4.27. Organolithium species readily aggregate in solution or coordinate with solvent, giving complex mixtures. They are much more reactive than Grignard reagents, particularly with moisture, and so need to be handled carefully under an inert gas atmosphere such as nitrogen or argon. It should be noted that although these reagents possess a covalent carbon–lithium bond, they nonetheless react like a carbanion. In fact, many organolithium reagents are now marketed commercially, and alkyllithiums can be handled industrially on many kilo scales despite their reactivity.

4.3.3 Organocadmium reagents

These are obtained by treating Grignard reagents with $CdCl_2$ (Fig. 4.28a) and are among the least reactive of all organometallic reagents, but this can be useful as high selectivity can be

$$RCl \ + \ 2\,Li \ \longrightarrow \ RLi \ + \ LiCl \quad (a)$$

$$Me_3C\text{-Br} \ \xrightarrow[Et_2O,\,-30°C]{2\,Li} \ Me_3C\,Li \ + \ LiBr \quad (b)$$

Figure 4.27 Organolithium reagents as carbanion equivalents.

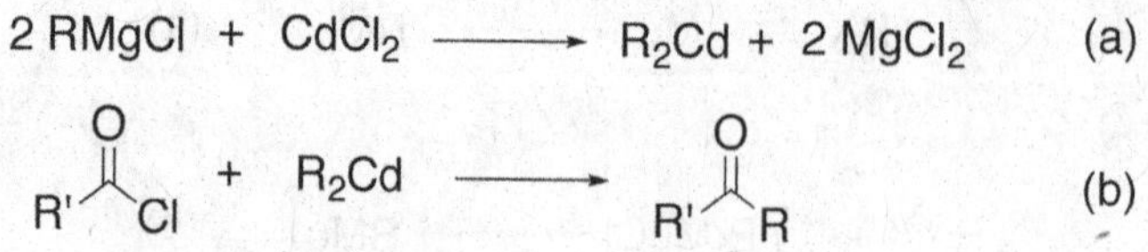

$$2\ RMgCl\ +\ CdCl_2\ \longrightarrow\ R_2Cd\ +\ 2\ MgCl_2 \qquad (a)$$

Figure 4.28　Organocadmium reagents as carbanion equivalents.

achieved in their reactions with acid halides to give the corresponding ketones (Fig. 4.28b). However, organocadmium reagents are highly toxic and are rarely used now.

4.4 Radicals

Carbocations and carbanions result due to the heterolytic cleavage of a carbon–hydrogen bond; formal removal of a H^- or a H^+ from an alkane leads to a carbocation or carbanion respectively. A third possibility is homolytic cleavage of a carbon–hydrogen bond, which gives rise to free-radical intermediates, in which the carbon has one unpaired electron but is uncharged since it has its usual complement of four electrons (Fig. 4.29).

4.4.1 Structure

Radicals derived from simple alkanes are planar or near planar in structure, consistent with sp^2+p hybridisation, with the unpaired electron located in the p orbital (Fig. 4.30a). This hybridisation pattern means that any stereochemical information at the radical centre is lost. In this regard, they could be considered to be similar to carbocations. However, radicals can also be accommodated in sp^2 orbitals of vinylic or phenyl substrates (Fig. 4.30b).

4.4.2 Factors stabilising radicals

Radicals tend to be of fleeting existence unless there is some suitable structural stabilisation or extremes of conditions used for their preparation. (For example, radicals can be trapped in crystal lattices at low temperature, although even under these conditions they are not very stable: The methyl radical has a $t_{0.5}$ of 10–15 min in MeOH at 77 K.)
　　Radicals can be stabilised by appropriate substitution as for carbocations and carbanions; however, such radicals are generally still too unstable to isolate. They can be stabilised by the following:

1. *Inductive effects*: For alkyl-substituted systems, inductive release of adjacent alkyl residues helps to stabilise the incomplete octet at the radical centre, and so the order of stability of alkyl radicals is $3° > 2° > 1°$.
2. *Resonance effects*: The delocalisation of electron density leads to high levels of stabilisation, and this is possible for allylic and benzylic systems, as shown in Figs. 4.31a and

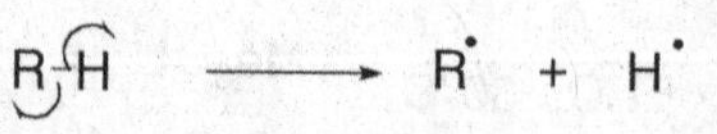

Figure 4.29　Homolysis leading to formation of radicals.

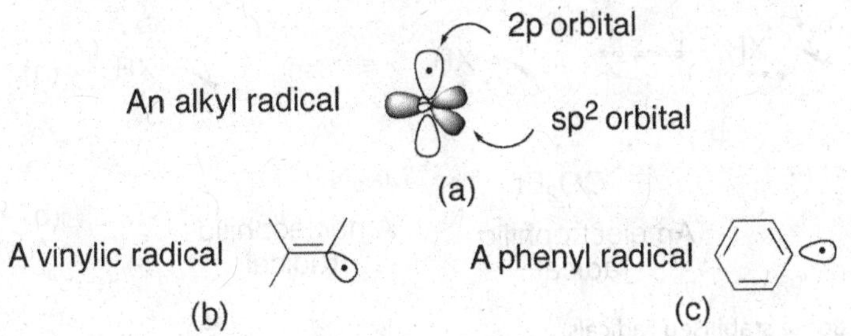

Figure 4.30   Orbital representation of radicals.

4.31b respectively, in a manner similar to the corresponding carbocations (see Fig. 4.5) and carbanions (see Fig. 4.18). Two or three phenyl groups are especially stabilising, giving the benzhydryl and trityl systems respectively (Fig. 4.31c). Similarly, the penta-cyclopentadienyl radical is highly stabilised by resonance interactions over the entire molecule. Both electron-withdrawing and electron-releasing substituents stabilise radicals very significantly, giving electrophilic and nucleophilic radicals respectively (Fig. 4.32). Radicals may in fact be simultaneously stabilised by both of these types of substituents, in which case the system is referred to as captodatively stabilised. It is also claimed that hyperconjugation leads to significant stabilisation, as shown in Fig. 4.33.

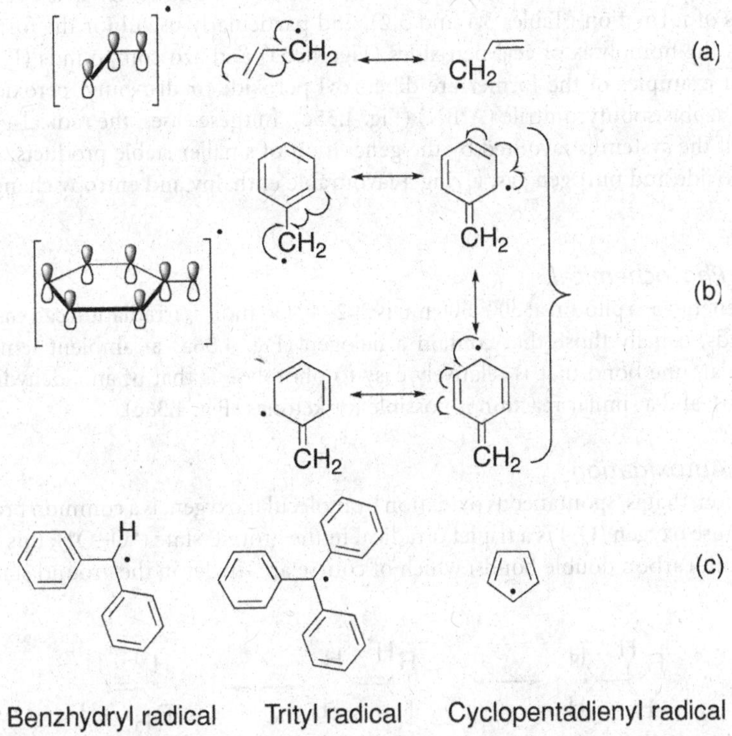

Figure 4.31   Resonance stabilisation of radicals.

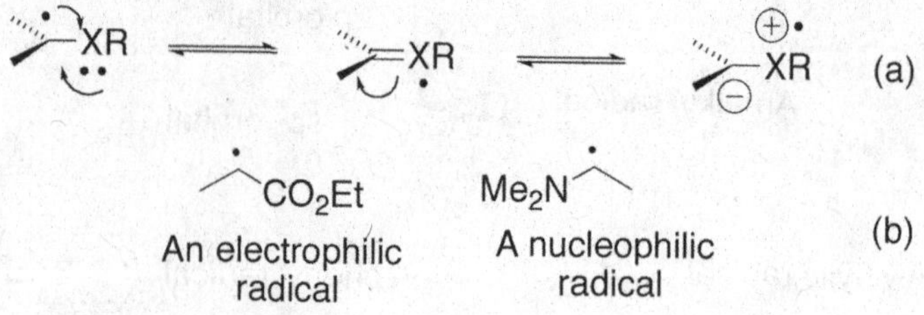

Figure 4.32 Some stabilised radicals.

There are a few examples of radicals so stable that they can be isolated, as shown in Fig. 4.34a, and this stability comes from a combination of inductive and resonance electron withdrawal and steric effects. Although the triphenyl radical is very stabilised, even at room temperature it exists in equilibrium with the dimeric form as shown in Fig. 4.34b.

4.4.3   Generation of radicals

Radicals may be commonly formed under four conditions.

4.4.3.1   Thermal

When a compound is heated, weak bonds can be easily homolysed. Examples of weak bonds are heteroatom–heteroatom bonds (e.g. O–O, N–N, O–N and X–X) as is evident from enthalpies of formation (Tables 3.1 and 3.2), and particularly useful for the formation of radicals is the homolysis of acyl peroxides (Fig. 4.35a) and azo compounds (Fig. 4.35b). Important examples of the former are dibenzoyl peroxide or di-*t*-butyl peroxide and of the latter, azobisisobutyronitrile (AIBN) (Fig. 4.35c). In these cases, the radical-generating reaction of the system is favoured by the generation of smaller stable products, including carbon dioxide and nitrogen gas, giving a favourable enthalpy and entropy change for the process.

4.4.3.2   Photochemical

Since the energy of a photon at 300–600 nm is 202–404 kJ mol^{-1}, irradiation can easily cleave weak bonds, usually those that contain a halogen (Fig. 4.36a) at ambient temperature. In particular, one bond that is relatively easy to photolyse is that of an aldehydic proton (Fig. 4.36b), and a similar reaction is possible for ketones (Fig. 4.36c).

4.4.3.3   Autoxidation

Autoxidation, that is, spontaneous oxidation by molecular oxygen, is a common process and arises because oxygen (O_2) is a triplet diradical in the ground state ($^{\bullet}O{-}O^{\bullet}$); this contrasts with carbon–carbon double bonds, which of course are singlet in the ground state (C=C)

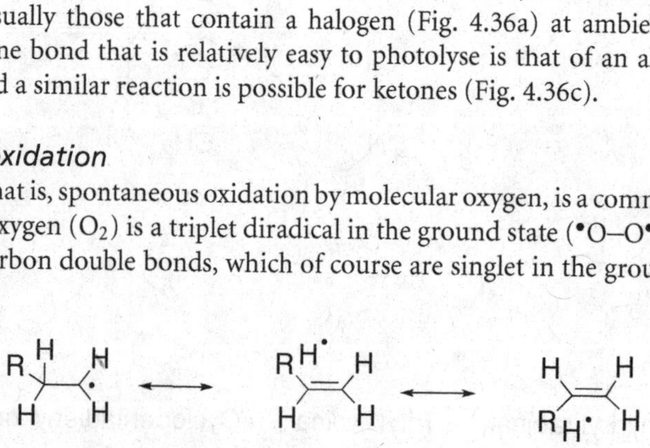

Figure 4.33 Hyperconjugative stabilisation of a radical.

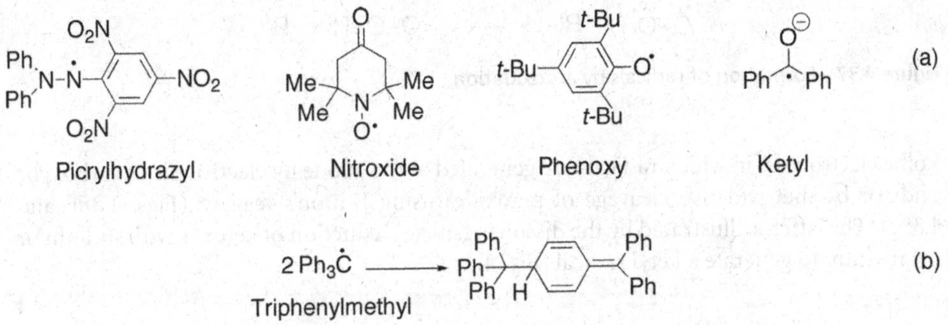

Picrylhydrazyl   Nitroxide   Phenoxy   Ketyl   (a)

$$2\,Ph_3\dot{C} \longrightarrow$$

Triphenylmethyl   (b)

Figure 4.34 Some stable radicals.

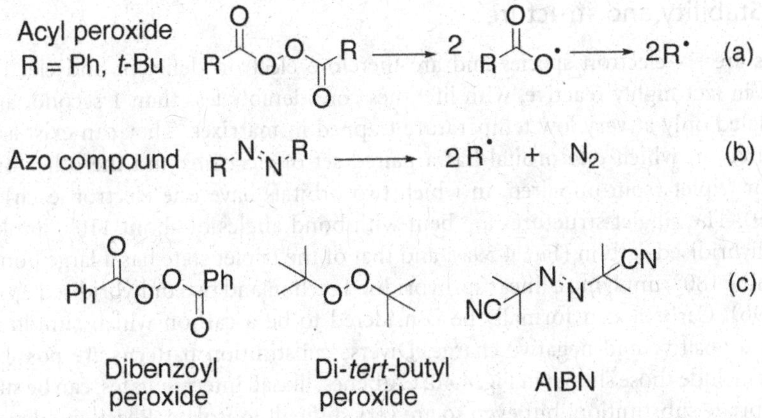

Acyl peroxide
$R = Ph, t\text{-}Bu$

Azo compound   (b)

Dibenzoyl
peroxide   Di-*tert*-butyl
peroxide   AIBN   (c)

Figure 4.35 Thermal formation of radicals.

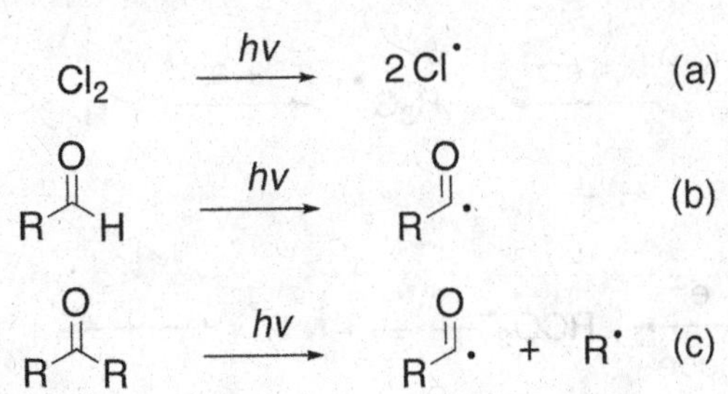

$$Cl_2 \xrightarrow{h\nu} 2\,Cl^{\bullet} \quad (a)$$

(b)

(c)

Figure 4.36 Photolytic formation of radicals.

and triplet in the excited state ($^{\bullet}C{\equiv}C^{\bullet}$). Ground-state oxygen is therefore capable of generating radical intermediates by hydrogen abstraction, as shown in Fig. 4.37.

4.4.3.4 Redox processes
Radicals may easily be generated by the one-electron oxidation of anions or by the one-
electron reduction of cations (Fig. 4.38a). The former may be achieved electrochemically by

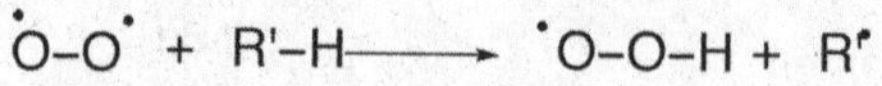

$$\dot{O}-O^{\bullet} + R'-H \longrightarrow {}^{\bullet}O-O-H + R^{\bullet}$$

Figure 4.37 Formation of radicals by autoxidation.

Kolbe electrolysis, in which radicals are generated at the anode by electrolysis of carboxylic acids or by the oxidative cleavage of peroxides using Fenton's reagent (Figs. 4.38b and 4.38c). The latter is illustrated by the dissolving-metal reduction of ketones with sodium or magnesium to generate a ketyl radical (Fig. 4.38d).

4.5 Carbenes

4.5.1 Stability and structure

Carbenes are six-electron species and are therefore electron deficient and electrophilic. They are in fact highly reactive, with lifetimes considerably less than 1 second, and have been isolated only at very low temperature trapped in matrixes. They can exist as singlet (spin-paired, in which one orbital has a paired set of electrons and the other orbital is empty) or triplet (spin-unpaired, in which two orbitals have one electron each) species (Fig. 4.39). The singlet structures are bent with bond angles of about 110°, implying an $sp^2 + 2p$-hybridised carbon (Fig. 4.39a), and that of the triplet state has a large bond angle approaching 180°, implying a linear sp-hybridised carbon and two unhybridised 2p orbitals (Fig. 4.39b). Carbene can formally be considered to be a carbon which simultaneously possesses a positive and negative charge. Diverse substitution patterns are possible, and examples include those shown in Fig. 4.40. Carbenes, like all intermediates, can be stabilised by appropriate substitution, but even so are very difficult to isolate. Reactivity depends on substituents, but both electron-withdrawing and electron-releasing substituents will lead to

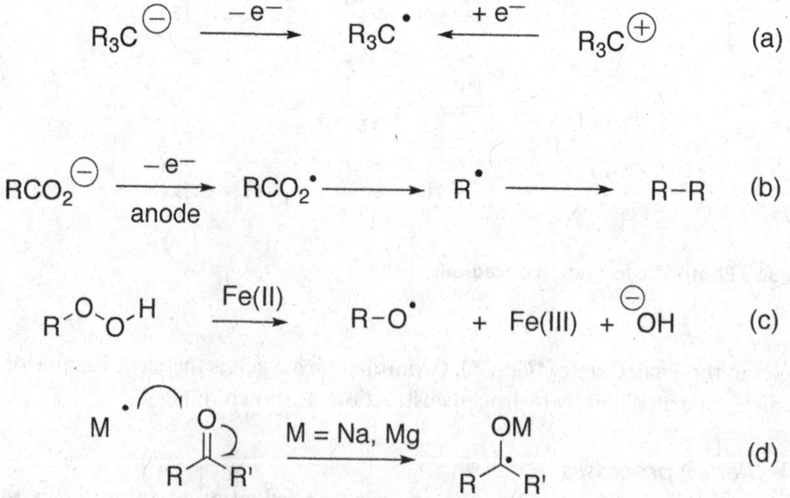

$$R_3C^{\ominus} \xrightarrow{-e^-} R_3C^{\bullet} \xleftarrow{+e^-} R_3C^{\oplus} \qquad (a)$$

$$RCO_2^{\ominus} \xrightarrow[\text{anode}]{-e^-} RCO_2^{\bullet} \longrightarrow R^{\bullet} \longrightarrow R-R \qquad (b)$$

$$R^{\diagdown O}{\diagup}O^{\diagdown}H \xrightarrow{Fe(II)} R-O^{\bullet} + Fe(III) + {}^{\ominus}OH \qquad (c)$$

$$M^{\bullet} \quad \underset{R \quad R'}{\overset{O}{\parallel}} \quad \xrightarrow{M = Na, Mg} \quad \underset{R \quad R'}{\overset{OM}{\underset{\bullet}{\mid}}} \qquad (d)$$

Figure 4.38 Redox generation of radicals.

Figure 4.39 Orbital representation of (a) singlet and (b) triplet carbenes.

Figure 4.40 Examples of some carbenes.

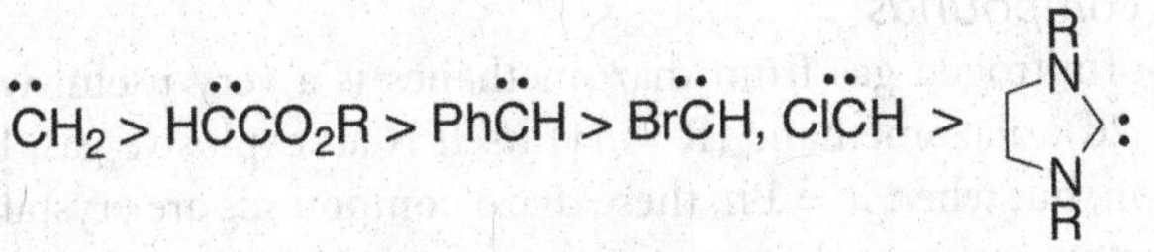

Figure 4.41 Carbenes in decreasing order of reactivity.

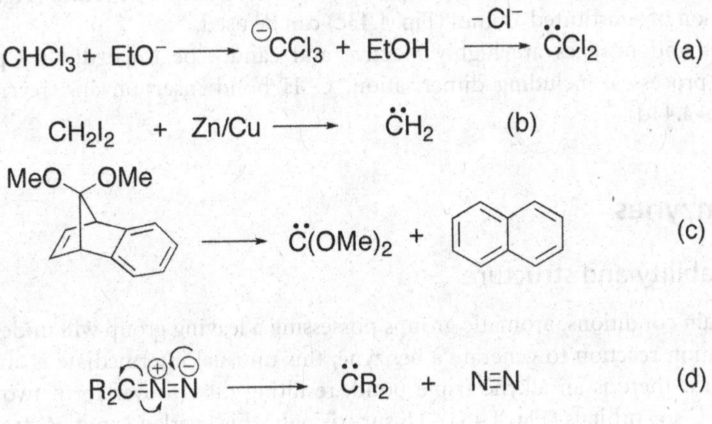

Figure 4.42 Formation of carbenes.

stabilisation (Fig. 4.41): Methyl carbene is the most reactive, and substituted analogues are of lower reactivity.

4.5.2 Generation of carbenes

4.5.2.1 α-Elimination

The elimination of HX from the same carbon leads to direct formation of the corresponding carbene. This is particularly valuable for haloalkanes, and chloroform, for example, is easily converted to the highly reactive dichlorocarbene by treatment with base (Fig. 4.42a). A similar reaction can be achieved using zinc/copper couple with diiodomethane, giving the so-called Simmons–Smith reagent (Fig. 4.42b). Other types of eliminations from cyclic structures can give dimethoxycarbene (Fig. 4.42c).

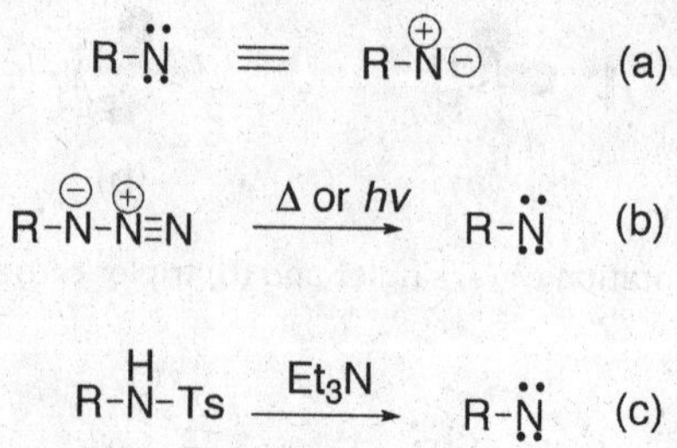

$$R-\ddot{N} \quad \equiv \quad R-\overset{\oplus}{N}{\ominus} \qquad \text{(a)}$$

$$R-\overset{\ominus}{N}-\overset{\oplus}{N}{\equiv}N \xrightarrow{\Delta \text{ or } h\nu} R-\ddot{N} \qquad \text{(b)}$$

$$R-\overset{H}{N}-Ts \xrightarrow{Et_3N} R-\ddot{N} \qquad \text{(c)}$$

Figure 4.43 Formation of nitrenes.

4.5.2.2 Diazo compounds

The elimination of nitrogen gas from diazomethanes is a very useful way of generating carbenes (Fig. 4.42d); diazomethane (R = H) itself is an explosive gas, but can be easily handled in solution, but when R = Ph, these diazo compounds are crystalline solids.

Species analogous to carbenes also possessing six valence electrons, but this time involving a nitrogen atom, are called nitrenes (Fig. 4.43a) and can be formed and behave in much the same way as carbenes. For example, photolysis or thermolysis of azides (Fig. 4.43b) or α-elimination of substituted amines (Fig. 4.43c) can be used.

Carbenes and nitrenes are highly reactive and cannot be isolated; they participate in diverse processes, including dimerisation, C—H bond insertion and rearrangement (Figs. 4.44a–4.44d).

4.6 Benzynes

4.6.1 Stability and structure

Under certain conditions, aromatic groups possessing a leaving group will undergo a *syn*-1,2-elimination reaction to generate a benzyne; this unusual intermediate is an aromatic ring in which there is an alkyne triple bond, resulting due to overlap of two electrons in adjacent C sp^2 orbitals (Fig. 4.45a). Unsurprisingly, this overlap is not as strong as in a

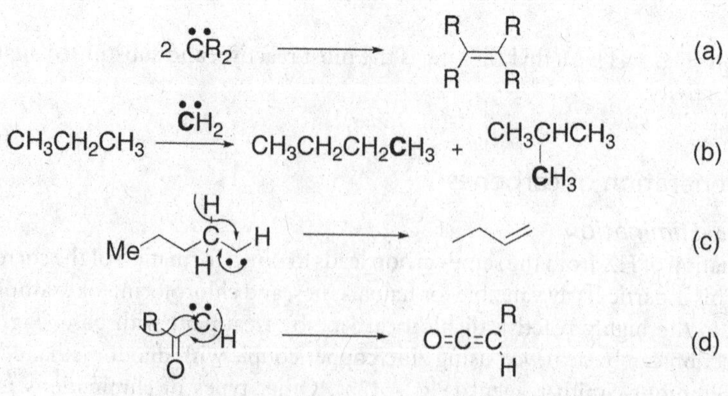

$$2 \ \ddot{C}R_2 \longrightarrow \overset{R}{\underset{R}{}}C=C\overset{R}{\underset{R}{}} \qquad \text{(a)}$$

$$CH_3CH_2CH_3 \xrightarrow{\ddot{C}H_2} CH_3CH_2CH_2CH_3 + \underset{CH_3}{\overset{CH_3CHCH_3}{\mid}} \qquad \text{(b)}$$

$$Me \overset{\overset{H}{|}}{\underset{H}{\overset{C}{\underset{\bullet}{\cdot}}}}H \longrightarrow \diagup\!\!\!\diagdown\!\!=\!\!\diagup \qquad \text{(c)}$$

$$R\overset{\ddot{C}}{\underset{O}{\diagdown}}H \longrightarrow O=C=C\overset{R}{\underset{H}{\diagdown}} \qquad \text{(d)}$$

Figure 4.44 Some reactions of carbenes.

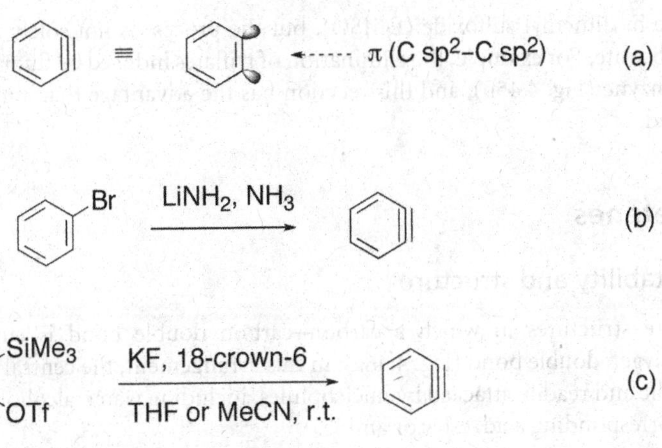

Figure 4.45 Formation of benzynes.

π(p–p)-orbital, since the sp^2 orbitals do not overlap so favourably, and benzynes are therefore highly reactive.

4.6.2 Generation of benzynes

Benzyne intermediates are characteristically generated by 1,2-elimination from alkyl halides using lithium or sodium amide as base in ammonia as solvent (Fig. 4.45a) or even

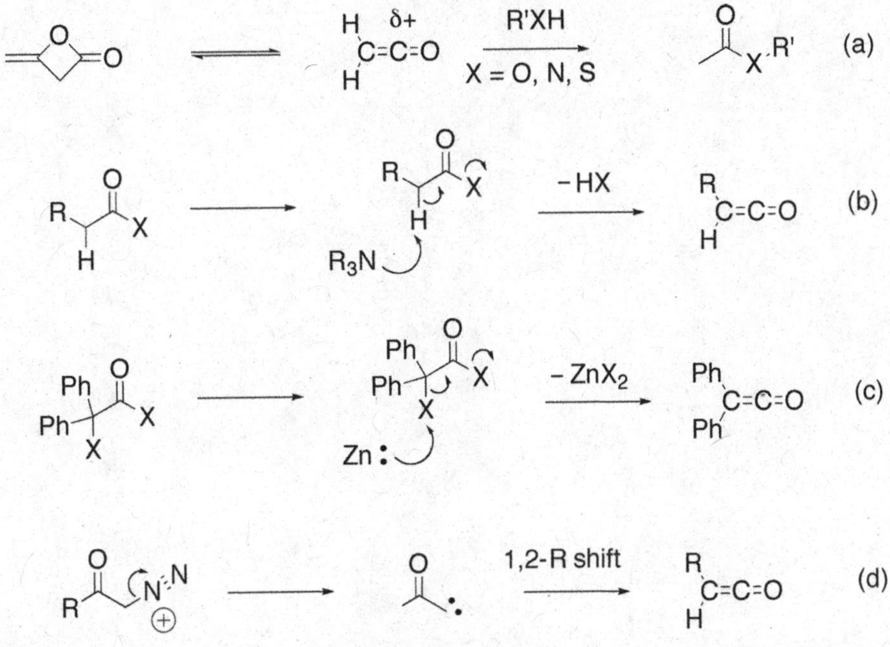

Figure 4.46 Ketenes and their formation.

t-butoxide in dimethyl sulfoxide (DMSO), but the process is not solely restricted to this type of substrate. For example, the elimination of triflates induced by fluoride will also generate a benzyne (Fig. 4.45b), and this reaction has the advantage that ambient conditions can be used.

4.7 Ketenes

4.7.1 Stability and structure

Ketenes are structures in which a carbon–carbon double bond is cumulated with a carbon–oxygen double bond (Fig. 4.46a); in this arrangement, the central carbon is highly electrophilic and readily attacked by nucleophiles, including water, alcohols and amines, to give the corresponding acid, ester or amide.

4.7.2 Generation of ketenes

Ketene itself is most conveniently prepared by thermolysis of diketene (Fig. 4.46a), but substituted ketenes are typically generated by the reaction of acid chlorides with a tertiary amine, by elimination of HX to generate the new double bond (Fig. 4.46b). Reductive elimination of α-haloacid chlorides using zinc gives an analogous elimination process (Fig. 4.46c). Alternatively, they can be formed by Wolff rearrangement of α-diazoketones (Fig. 4.46d).

Chapter 5
Acidity and Basicity

Thus far we have considered the nature of structure and bonding of organic compounds, but an additional value of this study is that it allows us to understand the *reactivity* of compounds as well; we will see how this is possible by considering reactions involving organic acids and bases.

5.1 Lowry–Brønsted Acid–Base theory

Lowry–Brønsted acid–base theory considers that an *acid* is a compound capable of *donating* a proton and that a *base* is capable of *accepting* a proton; a generalised reaction can be written as shown in Fig. 5.1. After an acid (HA) donates a proton to a base (B), the acid is converted to its *conjugate base* (A^-) and the base is converted to its *conjugate acid* (BH^+). Significantly, this process is an equilibrium reaction, whose equilibrium constant K is given by the indicated formula. If these reactions are conducted in aqueous solution as is commonly the case, the base is water, and the reaction is given by Fig. 5.2a; in this case, the expression for the equilibrium constant simplifies to K_a and the value of pK_a ($= -\log K_a$) is commonly quoted to define the acidity of the acid HA. The stronger the acidity of a compound, the higher will be the value of $[A^-]$ and $[H_3O^+]$ and the smaller the value of $[HA]$, so that the absolute value of pK_a will be larger, and pK_a will have a larger negative value. Conversely, for weaker acids, pK_a will have a high (positive) value. The typically encountered range of pK_a values is from about −12 for the most acidic compounds to +70 for the least acidic compounds. Because the acidity scale is a logarithmic one, a difference of x pK_a units between two acids in fact corresponds to a difference in proton concentration of 10^x; thus, apparently small changes in pK_a can in fact reflect considerable changes in acidity. Examples of this process include the ionisation of acetic acid and the protonation of benzylamine in water (Figs. 5.2b and 5.2c respectively).

Because acidity in aqueous systems is an equilibrium phenomenon, the acid–base reaction is therefore governed by thermodynamic factors. In particular, the equilibrium position is influenced by solvent and temperature; although oddly, these two parameters are usually ignored because acid–base reactions are so commonly done in water and at ambient temperature. But not all acids possess the same ability to donate protons; for example, HCl is fully ionised in aqueous solution ($pK_a = -7$), but acetic acid (CH_3CO_2H, $pK_a = 4.7$) is incompletely ionised. Using many of the principles discussed in the previous chapters, it is possible to understand what makes an organic compound acidic and why some are more acidic than others; in other words, why in some cases the equilibrium lies further to the right than in others.

$$H-A \quad + \quad B \quad \rightleftharpoons \quad A^- \quad + \quad BH^+$$

Acid   Base   Conjugate   Conjugate $K = \dfrac{[A^-][BH^+]}{[HA][B]}$
                             base   acid

Figure 5.1 A general acid–base equilibrium.

$$H-A \quad + \quad H_2O \quad \rightleftharpoons \quad A^- \quad + \quad H_3O^+$$

$$K_a = \dfrac{[A^-][H_3O^+]}{[HA]} \qquad pK_a = -\log K_a \qquad (a)$$

$$CH_3CO_2H \quad + \quad H_2O \quad \rightleftharpoons \quad CH_3CO_2^- \quad + \quad H_3O^+ \qquad (b)$$

$$PhCH_2NH_2 \quad + \quad H_2O \quad \rightleftharpoons \quad PhCH_2NH_3^+ \quad + \quad HO^- \quad (c)$$

Figure 5.2 Acid–base equilibria in aqueous solution.

5.2 Organic acidity

Ignoring the fact that the position of many equilibria can be significantly affected by the solvent and the temperature, the acidity of organic compounds is most importantly determined by the structure of the acid HA: For any given acid, the strength of the H—A bond, the electronegativity of A and the stability of the conjugate base A⁻ are of considerable importance. Of obvious relevance is the strength of the H—A bond; anything which weakens this bond by polarising it in the sense indicated in Fig. 5.3 will increase the acidity of H—A, since this effectively leads to partial breaking of the bond. This is best achieved if the H—A bond is one where A is an electronegative atom, especially halogen and oxygen and to a lesser extent, nitrogen; if it is carbon, the bond polarisation is usually weak. Once the H—A bond has broken, the position of equilibrium is significantly affected by the stability of A⁻. Therefore, the intrinsic electronegativity of A, and any inductive and resonance effects which stabilise A⁻, will also affect the acidity of H—A. Thus, inductive and resonance electron-withdrawing effects which disperse electron density will enhance acidity, with the latter being the more important of the two. Also of relevance is the hybridisation of the orbital in which the negative charge resides and any stabilisation which may arise due to the presence of empty adjacent π or empty d orbitals that are capable of accepting electron density. However, the nature of the solvent can be very important, since this has a direct influence on the stabilisation of charged species in solution. Water, in particular, is a highly effective solvent because it both stabilises charged species due to its high dielectric constant and solvates charged species either by hydrogen bonding or by sharing of the lone pair of its oxygen atom.

$$\overset{\delta+}{H}-\overset{\delta-}{A} \quad \rightleftharpoons \quad H^+ \quad + \quad A^-$$

Figure 5.3 Acid–base equilibrium for a generalised acid HA.

Table 5.1 Some typical pK_a values for common functional groups

Acid	pK_a	Acid	pK_a
HX	-7	$ROC(O)CH_2R$	24
RCO_2H	4	$HC\equiv CH$	26
HCN	9	$CH_3C\equiv N$	31
$RC(O)CH_2C(O)R$	9	Ph_3CH	32
ArOH	10	NH_3	33
RCH_2NO_2	10	$CH_3S(O)CH_3$	35
H_2O	15.7	C_6H_6	43
ROH	18	$H_2C=CH_2$	45
$CH_3C(O)CH_3$	20	CH_4	50

Note: Acidic protons are shown in bold; X = halide anion.

5.2.1 Organic acids

The relative ease with which common functional groups may be deprotonated is indicated by the pK_a data in Table 5.1; as usual, the smaller the value of pK_a, the stronger the acid. The relative strength of protonated functions as acids is shown in Table 5.2; these data give an indication of the relative difficulty of protonating different functional groups, with most easily protonated functions possessing larger pK_a values.

It is instructive to consider what factors determine the acidity of some common functional groups and that leads to the ordering in Tables 5.1 and 5.2.

1. *Aliphatic carboxylic acids*: Carboxylic acids are good acids, indicated by the equilibrium reaction shown in Fig. 5.4, because firstly, the O—H bond of the carboxylic acid is highly polarised and weakened by virtue of the electronegativity difference between oxygen and hydrogen. Secondly, the carboxylate anion is stabilised because the oxygen bearing

Table 5.2 Some typical pK_a values for common protonated functional groups

Acid	Base	pK_a
$RC\equiv N$ $-H^+$	$RC\equiv N$	-12
$ArOH_2^+$	ArOH	-7
$H_{\backslash}O^{\oplus}$, $R\overset{\parallel}{C}R$	O, $R\overset{\parallel}{C}R$	-5
R_2OH^+	R_2O	-3
$ROH_2{}^+$	ROH	-3
$H_{\backslash}O^{\oplus}$, $R\overset{\parallel}{C}NHR$	O, $R\overset{\parallel}{C}NHR$	-2
$ArNH_3{}^+$	$ArNH_2$	-2
$NH_4{}^+$	NH_3	9
$RNH_3{}^+$	RNH_3	11

Note: Acidic protons are shown in bold.

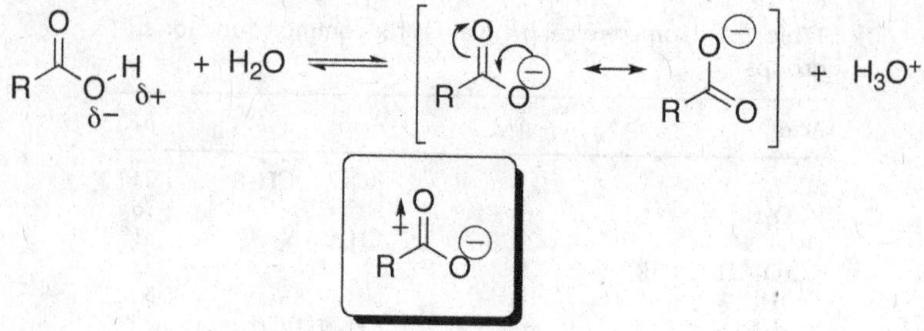

Figure 5.4   Acidity of carboxylic acids.

the negative charge is stable, and there is a stabilising inductive effect from the adjacent oxygen atom. Resonance effects strongly stabilise the carboxylate anion, which allow localisation of the negative charge on either of the two oxygen atoms.

It is the combination of these three factors, given in Table 5.2, which is responsible for the observed acidity, and if any of them are enhanced, the resulting acidity will increase. For example, if the R group is substituted with electron-withdrawing groups, both the polarisation of the O—H bond and the inductive stabilisation of the carboxylate are improved, and it is for this reason that trifluoroacetic acid ($pK_a = 0.23$) is a much stronger acid than acetic acid ($pK_a = 4.7$). In fact, the effect is approximately additive, increasing with the number of halogens. (pK_a values for CH_2ClCO_2H, $CHCl_2CO_2H$ and CCl_3CO_2H are 2.87, 1.35 and 0.66.) However, this effect falls away rapidly with distance. (pK_a value for $ClCH_2CH_2CO_2H$ is 3.98.) Other inductive electron-withdrawing groups exert this influence, including nitro, nitrile, methoxy, carbonyl and thio groups (for RCH_2CO_2H, with $R = NO_2$, NC, MeO, $CH_3C(O)$, and for MeS, pK_a values are 1.7, 2.5, 3.5, 3.6 and 3.7 respectively), with the order of pK_a values reflecting the approximate electronegativity of the adjacent functional groups.

Other factors can also be important: For example, if the carboxylate anion is stabilised by hydrogen bonding, enhancement of acidity results, as shown by the relative pK_a values for the first ionisation of maleic acid relative to fumaric acid (Fig. 5.5); however, for the second ionisation, removal of the proton from maleate anion is more difficult due to the same hydrogen bonding, but that from fumarate is easier, leading to a dianion which

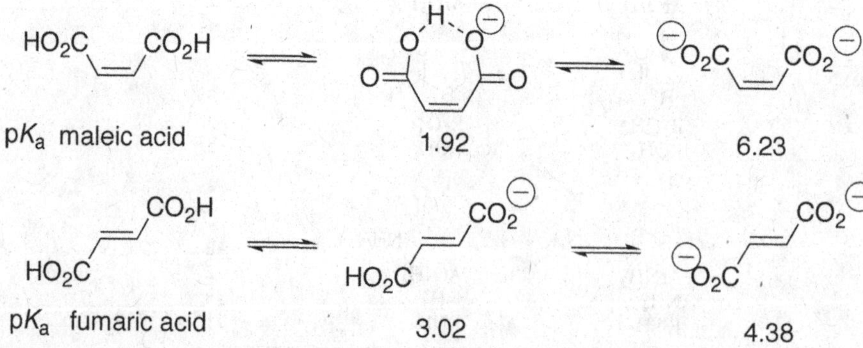

Figure 5.5   Acidity of dicarboxylic acids.

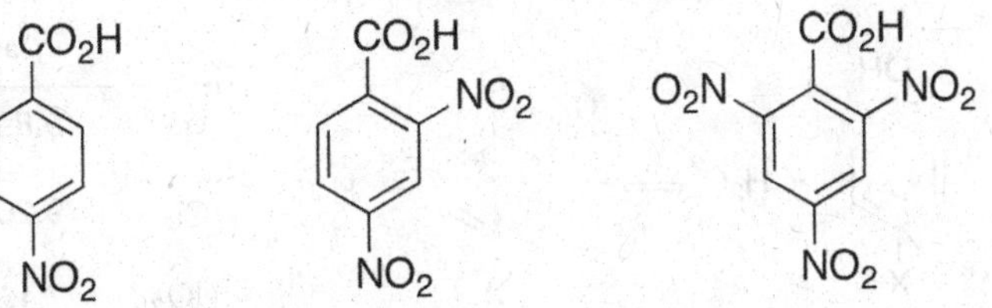

pK_a 3.43 1.43 0.65

Figure 5.6 Acidity of aromatic carboxylic acids.

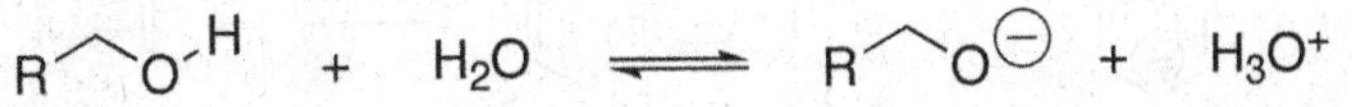

H_2O > MeOH > EtOH > Me_2CHOH > Me_3COH

pK_a 15.7 16 18 18 19

Figure 5.7 Acidity of alcohols.

places the negative charges as far apart as possible. Similar phenomena are observed in the case of aromatic carboxylic acids: Increasing substitution of the ring by electron-withdrawing groups leads to enhanced acidity, as shown in Fig. 5.6.

2. *Alcohols and phenols*: Aliphatic alcohols are only weak acids, in contrast to carboxylic acids discussed in point (1) above; the relevant equilibrium reaction is shown in Fig. 5.7. Although the O–H bond is polarised because of the electronegativity difference of those atoms, and the alkoxide anion also reasonably stabilised due to the electronegativity of oxygen, there is no resonance stabilisation. However, like carboxylic acids, it is possible to improve acidity by substituting the R group with electron-withdrawing groups, such as fluorine; thus, trifluoroethanol ($pK_a = 12.4$) is a much stronger acid than ethanol ($pK_a = 15.5$). Increasing alkyl substitution reduces the acidity of the alcohol, since in this case the alkoxide anion is less readily solvated due to the presence of the adjacent bulky groups; the trend is shown in Fig. 5.7. Alkoxides are therefore important bases, and sodium ethoxide (NaOEt) and potassium-*t*-butoxide (KO*t*-Bu) are often used synthetically to deprotonate other more acidic compounds.

In contrast, phenols are much stronger acids than simple aliphatic alcohols, since they benefit from an additional resonance stabilisation as shown in Fig. 5.8, in which

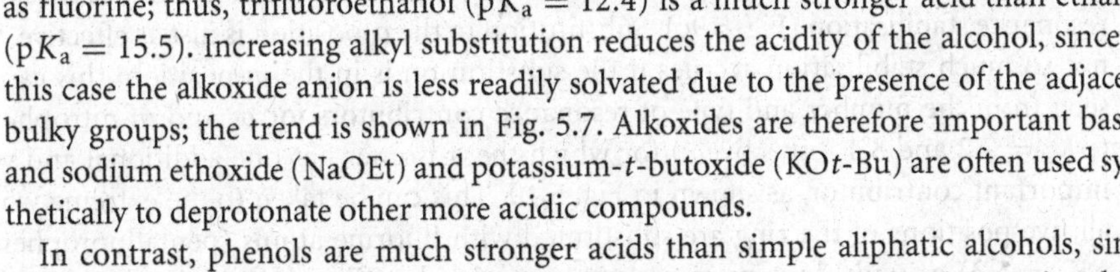

Figure 5.8 Acidity of phenols.

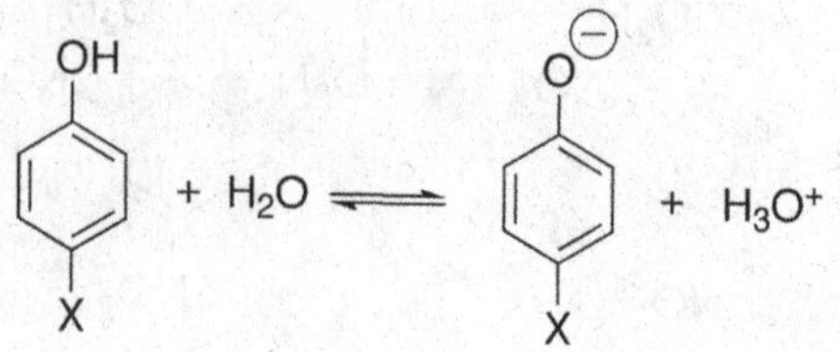

Figure 5.9 Resonance effects in the acidity of phenols.

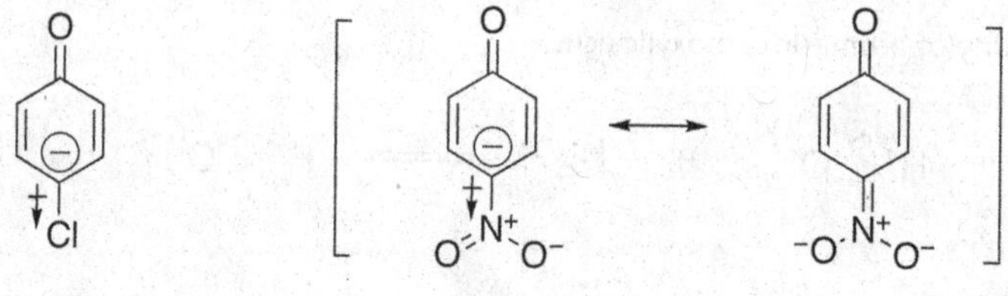

negative charge is delocalised onto the aromatic ring; their typical pK_a value is 6 orders of magnitude lower than an alcohol, that is, about a million times more acidic. However, phenols are less acidic than carboxylic acids since the negative charge can reside on carbon atoms for the former, but only oxygen atoms for the latter. This stabilisation can be improved still further by substituting the ring with electron-withdrawing groups, such as *p*-chloro and *p*-nitro groups (Fig. 5.9). In this case, in addition to the usual resonance structures that can be drawn in which negative charge is delocalised around the ring as shown in Fig. 5.8, there are additional contributors in which the negative charge is located immediately adjacent to the substituent and therefore benefits from the full inductive stabilisation of the electronegative chlorine atom (Fig. 5.9b), or even better for the nitro substituent, from both the inductive stabilisation and the additional resonance stabilisation (Fig. 5.9c). Substitution in the *o*-position is just as effective, but not so much stabilisation accrues if the substitution is in the *m*-position; this can be seen from the number and type of resonance contributors for *o*- and *m*-nitrophenol ($pK_a = 7.2$ and 8.4 respectively) for which the *o*-isomer has one additional and very important contributor, as shown in Fig. 5.10. This can be taken to the extreme where all five positions of the ring are substituted with fluorine atoms (pentafluorophenol, $pK_a = 5.3$) or with three nitro groups (picric acid, $pK_a = 0.4$), giving highly acidic phenols.

3. *Amines*: Alkylamines on the whole are very weak acids ($pK_a \approx 35$), since the resultant amide anion (as it is called, not to be confused with the amide functional group ($-C(O)NH-$)) is relatively unstabilised, at least compared to alkoxides; the equilibrium shown in Fig. 5.11a therefore lies heavily to the left. However, in non-aqueous solvents (such as tetrahydrofuran (THF) or ether), such amide anions can be formed, but will readily deprotonate any available acids. Two in particular, sodamide ($NaNH_2$) and lithium diisopropylamide (LDA, $LiN(i\text{-}Pr)_2$), are often used to deprotonate organic acids. The acidity, however, of arylamines is greater, because of the possibility of resonance in the conjugate base; the acidity of aniline ($pK_a = 28$) can be further

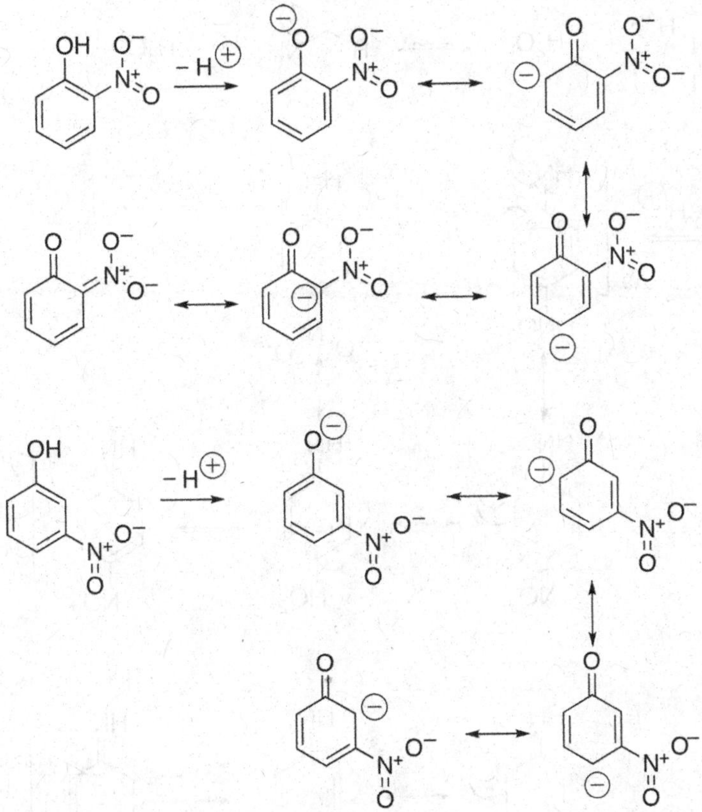

Figure 5.10 Stabilisation of phenolate anions.

enhanced by the inclusion of electron-withdrawing substituents such as nitro or cyano, which permit additional resonance stabilisation of the anilide anion ($pK_a = 18$ and 24, respectively, Figs. 5.11b and 5.11c). Similarly, amides are more acidic ($pK_a \approx 15$) than simple amines because of resonance stabilisation in the conjugate base (Fig. 5.11d). Conversely, amines are good bases, as illustrated by the pK_a value for RNH_3^+ (typically 9–10), and this will be discussed in more detail in Section 5.3.1.

4. *Carbonyl compounds*: The α-proton of carbonyl compounds is acidic, with pK_a values typically of 20–25 for the equilibrium reaction shown in Fig. 5.12. In this case, although the C–H bond is not polarised by virtue of a large electronegativity difference between carbon and hydrogen, it is in fact polarised and weakened by an inductive effect from the adjacent oxygen atom, and a similar inductive effect also stabilises the enolate anion once formed. More important though is the resonance stabilisation that operates, which places the negative charge on the most electronegative (oxygen) atom. Using similar principles, which we have seen already, it is possible to enhance acidity further in these systems. Thus, by placing additional electron-withdrawing groups proximal to the enolate, significant additional stabilisation is also possible. For example, a second carbonyl, ester, nitrile or nitro function (Fig. 5.13) can provide additional resonance structures that significantly enhance acidity, as shown in the pK_a values included in

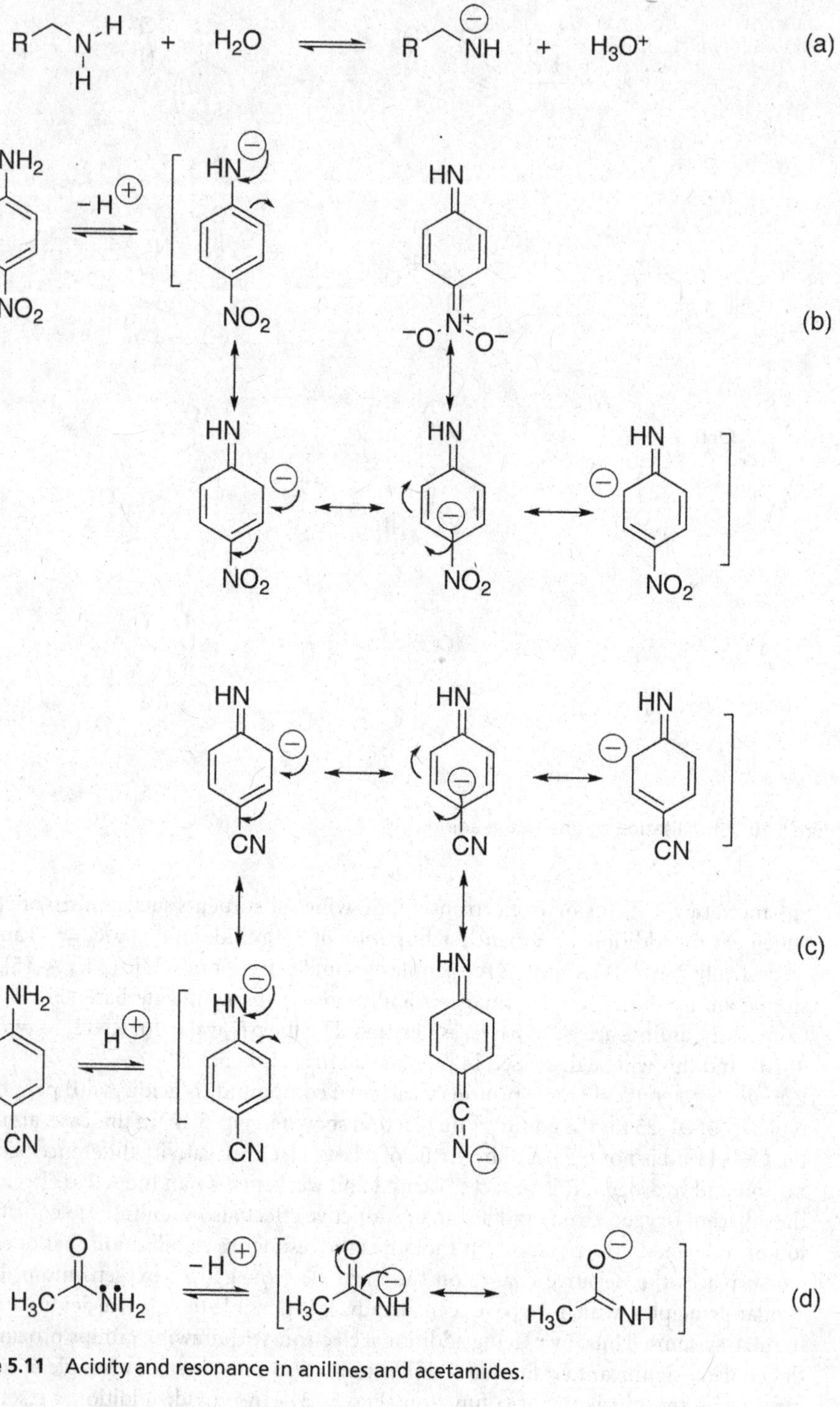

Figure 5.11 Acidity and resonance in anilines and acetamides.

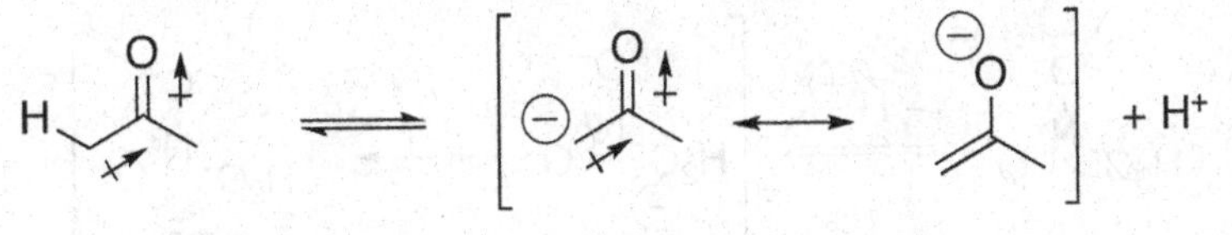

Figure 5.12 Acidity of carbonyl compounds.

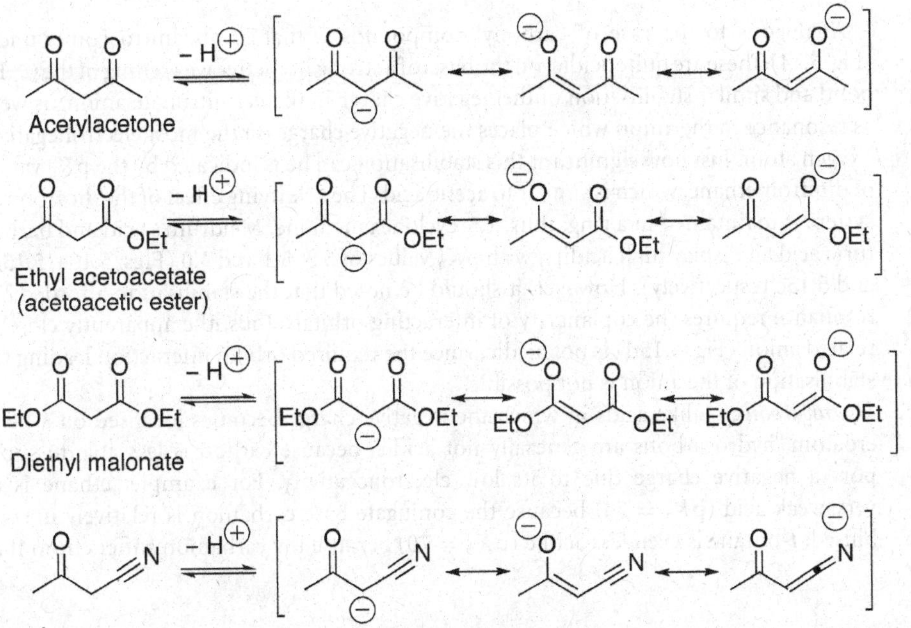

Figure 5.13 Acidity of active methylene compounds.

Table 5.3. For each of these cases, those canonical structures which place negative charge on the heteroatoms are particularly stabilising, placing negative charge on the most electronegative atom in the molecule. However, for diethyl malonate, delocalisation of charge onto the ester also takes away additional resonance from the alkoxy oxygen atom, and so this does not give as large a contribution to stabilisation as might be expected.

Table 5.3 pK_a values for some carbonyl compounds

$O_2NCH_2NO_2$	4
$Me(O)CCH_2C(O)Me$	9
$Me(O)CCH_2C(O)OEt$	11
$NCCH_2CN$	12
$EtO(O)CCH_2C(O)OEt$	13
$H_3CC(O)CH_3$	20
$H_3CC(O)OEt$	25

Note: Acidic protons are shown in bold.

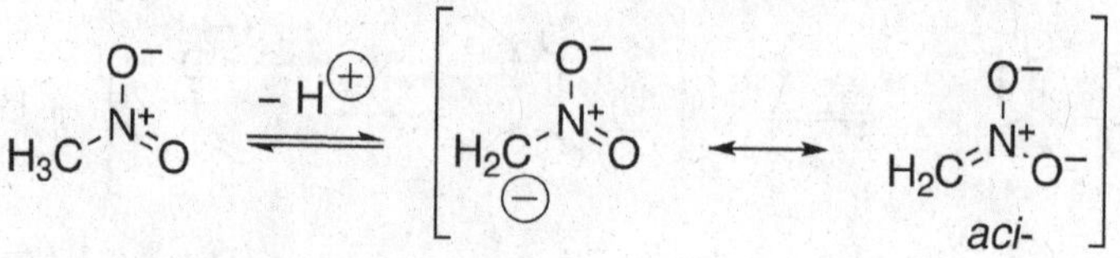

Figure 5.14   Acidity of alkyl nitro compounds.

Analogous to the case of carbonyl compounds is that of alkylnitro compounds (Fig. 5.14); these are quite acidic, on the basis of a strong inductive weakening of the C–H bond and similar stabilisation of the negative charge in the *aci*-nitronate anion, as well as resonance in the anion which places the negative charge on the most electronegative oxygen atom. Just how significant this stabilisation can be is indicated by the pK_a value of dinitromethane, which is similar to acetic acid. The acidifying effect of the dicarbonyl system is maintained in a ring; thus, 1,3-cyclohexanedione, Meldrum's acid and barbituric acid all display high acidity, with pK_a values of 5.3, 5.1 and 4.0 (Figs. 5.15a, 5.15b and 5.15c respectively). However, it should be noted that the stabilisation afforded by resonance requires the coplanarity of interacting orbitals; thus, the apparently closely related anion (Fig. 5.15d) is not acidic, since the required orbital interaction leading to stabilisation of the anion is not possible.

5. *Hydrocarbons*: Unlike acids in which the negative charge becomes localised on a heteroatom, hydrocarbons are generally not acidic, because carbon is less able to support a negative charge due to its low electronegativity. For example, ethane is a very weak acid ($pK_a = 50$) because the conjugate base carbanion is relatively unstabilised; *t*-butane is even less acidic ($pK_a = 70$) because the carbanion suffers from the

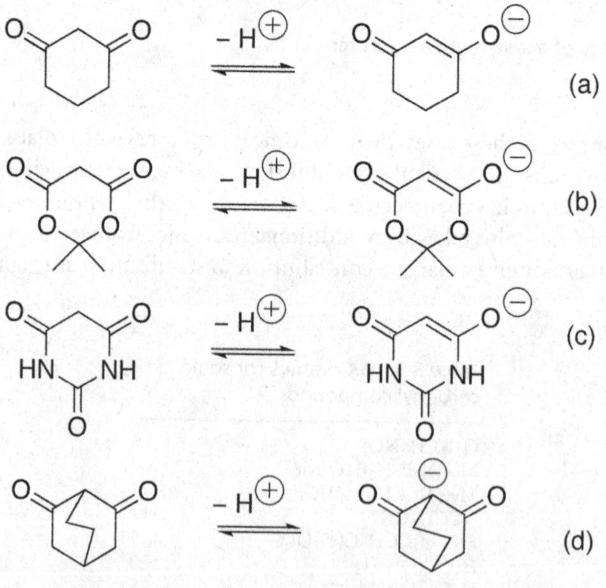

Figure 5.15   Acidity of cyclic dicarbonyl compounds.

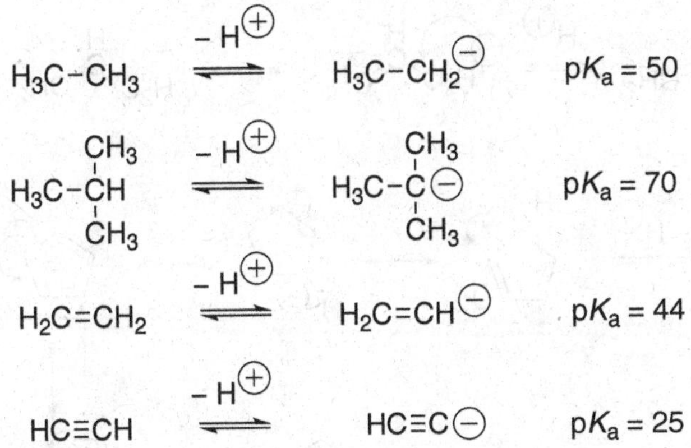

$$H_3C-CH_3 \xrightleftharpoons{-H^{\oplus}} H_3C-CH_2^{\ominus} \qquad pK_a = 50$$

$$\underset{CH_3}{\overset{CH_3}{H_3C-\underset{|}{\overset{|}{CH}}}} \xrightleftharpoons{-H^{\oplus}} \underset{CH_3}{\overset{CH_3}{H_3C-\underset{|}{\overset{|}{C}}^{\ominus}}} \qquad pK_a = 70$$

$$H_2C=CH_2 \xrightleftharpoons{-H^{\oplus}} H_2C=CH^{\ominus} \qquad pK_a = 44$$

$$HC{\equiv}CH \xrightleftharpoons{-H^{\oplus}} HC{\equiv}C^{\ominus} \qquad pK_a = 25$$

Figure 5.16 Acidity of hydrocarbons.

presence of three methyl inductively releasing groups which destabilise the negative charge (Fig. 5.16). However, alkenes and alkynes are more acidic than alkanes, because the conjugate base has negative charge located in an sp^2- or sp-hybridised orbital rather than an sp^3 one, and this means that the orbital carrying the negative charge has a higher s contribution, allowing the negative charge to be held nearer to the nucleus, that is, to be more stabilised. This stabilisation can be significant; for example, ethene and ethyne have pK_a values of 44 and 25, representing considerable enhancement of acidity over ethane.

However, this acidification is modest in comparison to that which results if the carbanion can be stabilised by inductive effects. For example, cyanide anion, which benefits from having a proximal electronegative nitrogen atom giving inductive stabilisation, as well as negative charge located in an sp hybrid orbital, has a pK_a value of 9 (Fig. 5.17). Even better, however, is resonance stabilisation, and two important examples illustrate this situation. Propene has a pK_a value of 35, and this results from the resonance stabilisation of the conjugate base, as shown in Fig. 5.18a. Cyclopentadiene is even more acidic ($pK_a = 16$), since not only does it benefit from an additional double bond allowing for more resonance possibilities (Fig. 5.18b), but the carbanion has six π-electrons, and this is a particularly stable arrangement similar to that of benzene: The anion is said to be *aromatic*. Similarly, triphenylmethane is an unusually strong carbon acid ($pK_a = 31$) (Fig. 5.18c) since the trityl anion is strongly stabilised by resonance delocalisation into the three aromatic rings. However, the incorporation of resonance groups does not afford unlimited stabilisation. In the series $PhCH_3$, Ph_2CH_2 and Ph_3CH, the additional stabilisation afforded by successive increases in phenyl substituents decreases; this is because the steric interaction of the adjacent ortho substituents on the phenyl rings eventually forces the phenyl rings out of planarity, impeding

$$N{\equiv}C-H \xrightleftharpoons{-H^{\oplus}} N{\equiv}C^{\ominus} \qquad pK_a = 9$$

Figure 5.17 Acidity in compounds with adjacent heteroatoms.

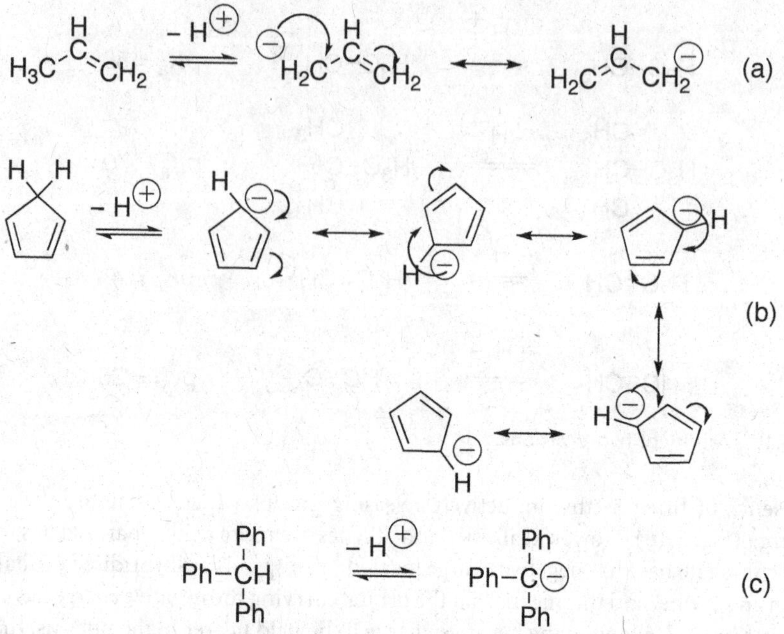

Figure 5.18 Acidity in resonance-activated hydrocarbons.

the stabilising resonance interaction. Thus, the incremental acidifying effect diminishes in the series PhCH$_3$, Ph$_2$CH$_2$ and Ph$_3$CH (pK_a values of 41, 33 and 31 respectively). Because these carbanions are derived from hydrocarbons with pK_a values larger than that of almost any other functional group, carbanions can be used for the reliable deprotonation of most organic compounds. In this regard, the most commonly used are butyllithium (BuLi), s-butyllithium (CH$_3$CH$_2$CH(Li)CH$_3$) and t-butyllithium ((CH$_3$)$_3$CLi), since they are strong bases and the resulting hydrocarbon products are inert and easily removed.

Heteroatoms have the capacity to stabilise adjacent negative charge by electron-withdrawing inductive or chelation effects. For example, in dithioacetals (also called dithianes), a hydrogen atom α to both of the sulfur atoms is acidified (pK_a = 31) and can be readily removed by a suitably strong base such as an organolithium (Fig. 5.19a); this arises because the anion is stabilised by inductively withdrawing effects by each sulfur. Alternatively, in aromatic systems which possess heteroatom-containing substituents, it has been found that the spatially proximal ortho protons can often be readily removed by strong base, usually s-butyllithium, in a process called ortho-lithiation (Fig. 5.19b). This reaction is very general, and a wide variety of aromatic ring substituents are accommodated, including those containing O, N and S; it proceeds because the formation of a five- or six-membered chelated structure with the lithium countercation is very favourable. The resulting carbanions are nucleophilic in character and are valuable synthetic intermediates, providing access to substituted aromatic compounds which would not otherwise be very easily available. This effect has been exploited in other systems, for example, t-butyloxyamines and epoxides, where the combination of inductive and chelating stabilisation permits ready deprotonation at the α-position (Figs. 5.19c and 5.19d).

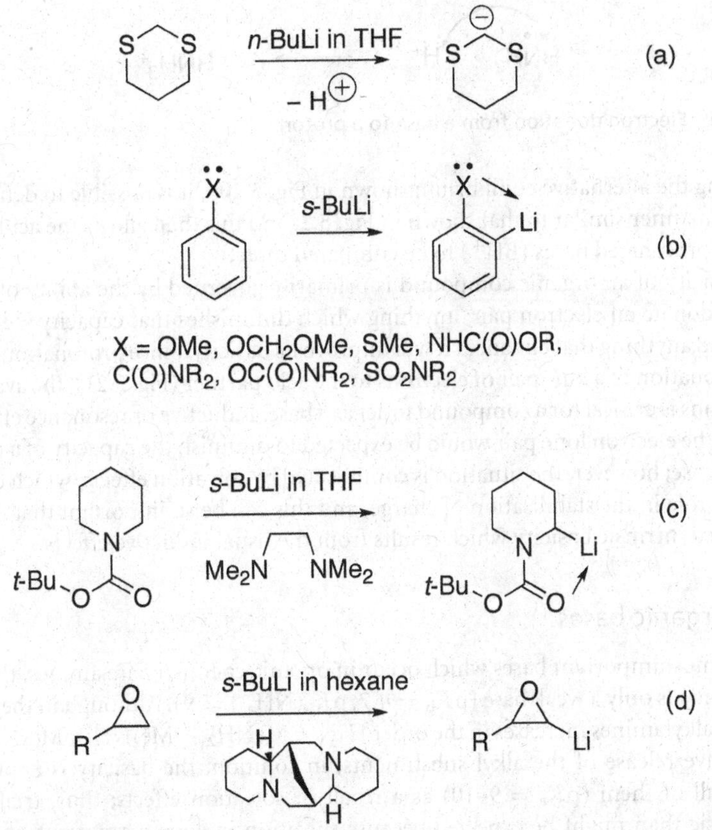

Figure 5.19 Acidity in heteroatom-activated hydrocarbons: (a) dithianes, (b) ortho-lithiation, (c) amines and (d) epoxides.

5.3 Organic basicity

We have seen that acidity corresponds to the ability of a compound to donate a proton; *basicity*, on the other hand, corresponds to the ability of a compound to *accept* a proton, given by the equilibrium shown in Fig. 5.20a for a generalised base B. K_b and pK_b are defined as indicated. Thus, a strong base will have a large value for K_b and a small one for pK_b, while a weak base will have a small value for K_b and a large one for pK_b. However, by

$$B + H_2O \rightleftharpoons BH^+ + HO^- \qquad K_b = \frac{[BH^+][HO^-]}{[B]} \qquad pK_b = -\log K_b \quad \text{(a)}$$

$$BH^+ + H_2O \rightleftharpoons B + H_3O^+ \qquad K_a = \frac{[B][H_3O^+]}{[BH^+]} \qquad pK_a = -\log K_a \quad \text{(b)}$$

Figure 5.20 Acid–base equilibria in aqueous solution.

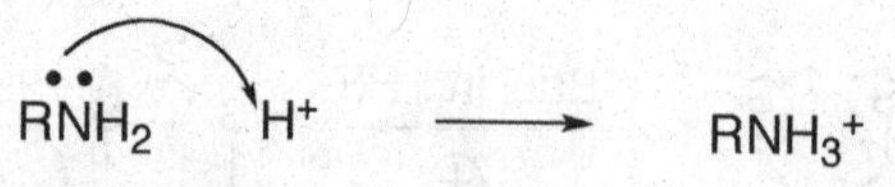

Figure 5.21 Electron donation from a base to a proton.

considering the alternative equilibrium shown in Fig. 5.20b, it is possible to define the pK_a value in a manner similar to that shown in Fig. 5.2, and this then allows the acidity of acids (HA) and protonated bases (BH$^+$) to be compared directly.

The basicity of an organic compound is primarily governed by the ability of that compound to donate an electron pair; anything which diminishes that capacity will reduce its basicity and anything that enhances it will improve its basicity. Since protonation of the base involves donation of a lone pair of electrons to the acid partner (Fig. 5.21), the availability of these protons is critical for a compound to act as a base. Inductive or resonance effects which delocalise the electron lone pair would be expected to diminish the capacity of a compound to act as a base; however, the situation is complicated by solvation effects, which can play an important role in the stabilisation of charge, and this can be so important that it overrides the expected intrinsic basicity which results from the usual inductive effects.

5.3.1 Organic bases

By far the most important bases which occur in organic chemistry are amines; the simplest one ammonia is only a weak base ($pK_b = 4.7$, $pK_a(NH_4^+) = 9$). Although in the gas phase, basicity of alkylamines increases in the order $H_3N < MeNH_2 < Me_2NH < Me_3N$, reflecting the inductive release of the alkyl substituents, in solution, the basicity is approximately equal for all of them ($pK_a = 9$–10) as a result of solvation effects; thus, trialkylamines are less basic than might be expected because solvation in aqueous systems of R_3NH^+ is impeded by the adjacent bulk of the alkyl groups. Resonance effects, however, are much more significant: this can either diminish or enhance basicity, depending on the circumstances. For example, acetamide is effectively non-basic ($pK_a(CH_3CONH_2^+) = -2$) because firstly, resonance delocalisation of the nitrogen lone pair diminishes its ability to donate a lone pair and secondly if protonation does occur, there is a loss of this delocalisation energy, which raises the energy of the product relative to starting materials. Overall, then, the equilibrium lies heavily to the left (Fig. 5.22).

This situation is even more important in anilines, in which the nitrogen lone pair is adjacent to an aromatic ring (Fig. 5.23); here, a number of resonance structures can be drawn which indicate the delocalisation of negative charge onto the aromatic ring, the sum total of which is diminution of the availability of the nitrogen lone pair for donation to a proton. Protonation, if it does occur, gives the anilinium cation, but in this case no resonance

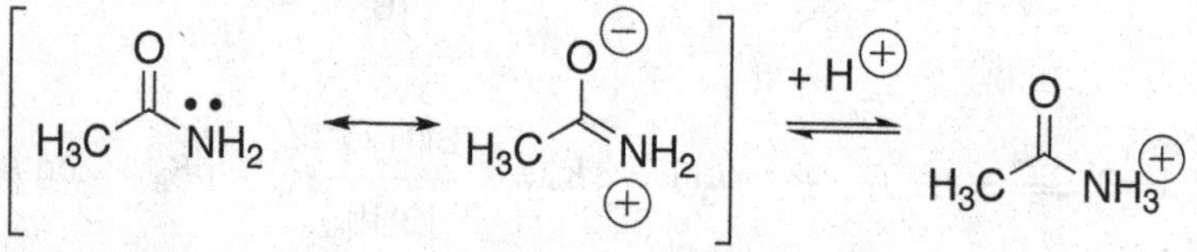

Figure 5.22 Resonance stabilisation and protonation in acetamide.

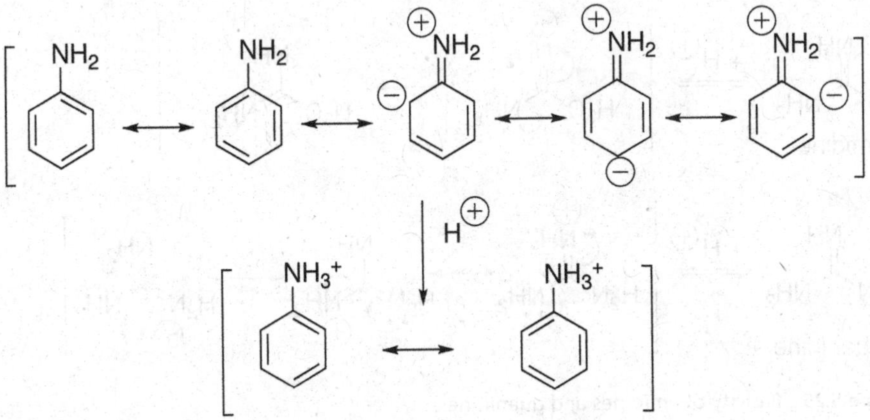

Figure 5.23 Resonance stabilisation in aniline.

stabilisation involving the nitrogen atom is possible, meaning that the protonated product is raised in energy relative to reagents. Aniline, therefore, is a relatively weak base, with a pK_b of 9.4. However, their basicity can be modified by substituents (Fig. 5.24a). Thus, electron-withdrawing groups, such as nitro, cyano and halo, increase the value of pK_b and therefore diminish basicity relative to aniline, and electron-releasing groups, such as methyl, methoxy and amino, decrease the value of pK_b and enhance basicity relative to aniline. This is because electron-withdrawing groups, such as nitro or cyano, have important contributors

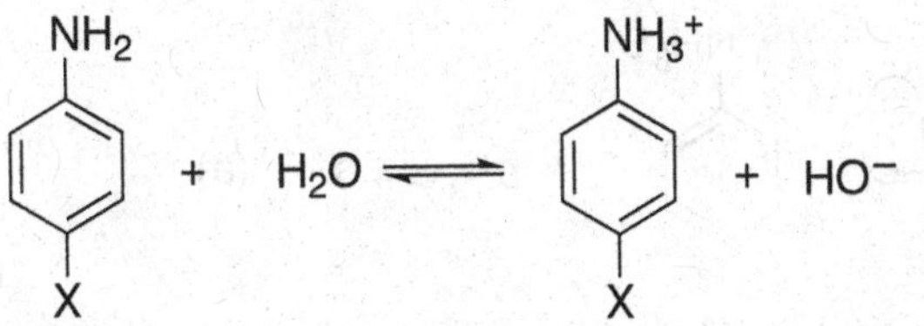

X	pK_b	
NH_2	7.85	
OCH_3	8.66	
CH_3	8.92	
H	9.37	(a)
Cl	10.02	
Br	10.14	
CN	12.26	
NO_2	13.0	

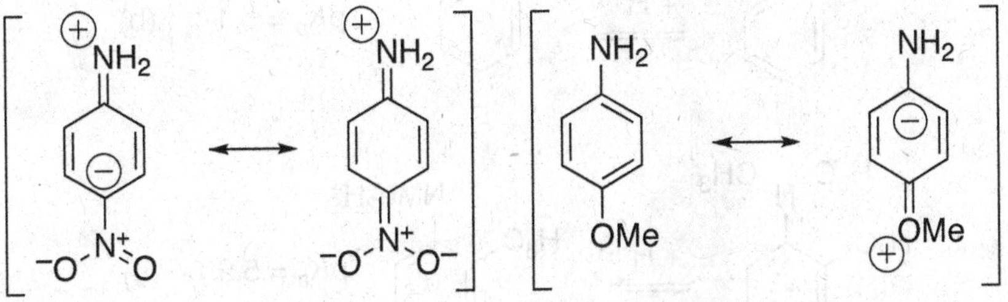

Figure 5.24 Basicity and resonance stabilisation in substituted anilines.

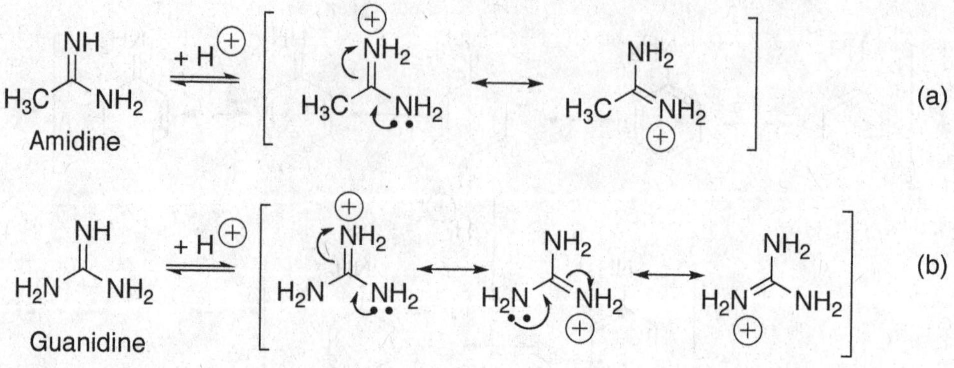

Figure 5.25 Basicity of amidines and guanidines.

(Fig. 5.24b), in which the lone pair is delocalised away from the nitrogen atom, and in the case of electron-releasing groups, such as methoxy, because adjacent negative charge to the nitrogen lone pair enhances its donating ability on grounds of electrostatic repulsion (Fig. 5.24c). If this substitution pattern is extended to 2,4,6-trinitroaniline (picramide), the amine is not basic at all and is in fact a strong acid ($pK_a = -9$).

Amidine and guanidine bases (Figs. 5.25a and 5.25b) are examples of cases in which protonation allows for the possibility of resonance in the product, making for a very stable outcome. Guanidinium ion is particularly stabilised, because it has three identical resonance structures contributing equally to the overall structure; such *degenerate* canonical forms are very energetically favourable. These compounds as a result are very strong organic bases ($pK_a(RNH_2^+) = 12.4$ and 13.6 respectively).

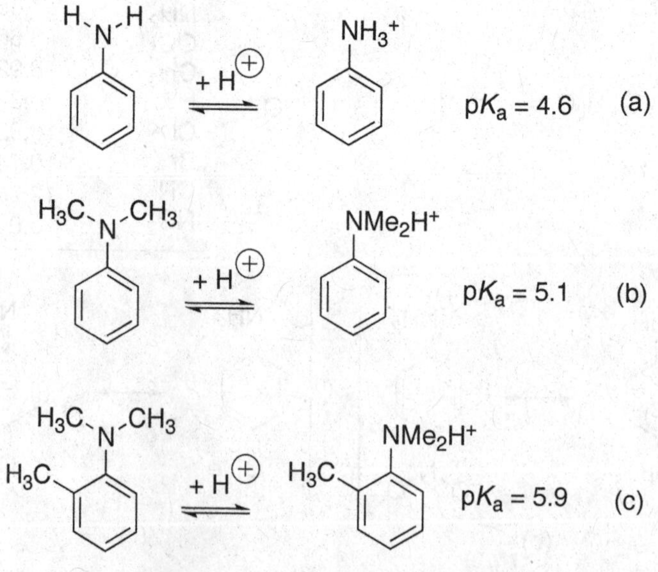

Figure 5.26 Steric interactions in substituted anilines.

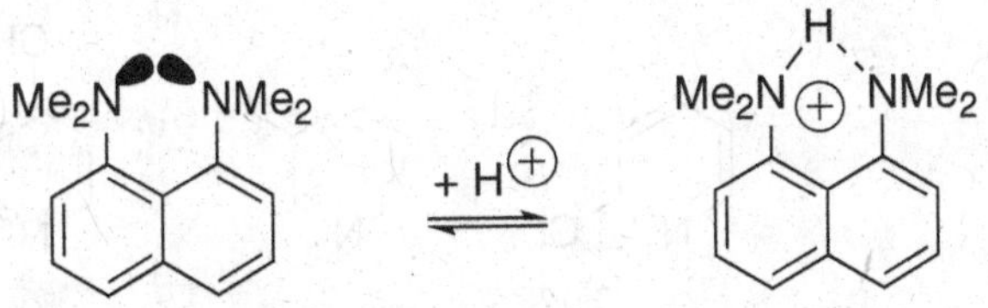

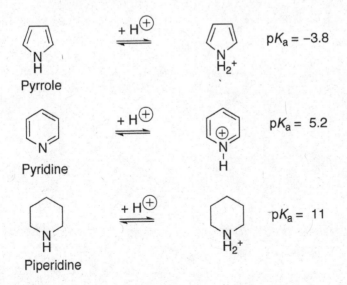

Figure 5.27 Enhanced basicity of diaminonaphthalenes.

It is worth noting, however, that steric effects too can be important. Aniline and dimethylaniline have approximately the same basicity (Figs. 5.26a and 5.26b respectively), but *o*-dimethylaminotoluene (Fig. 5.26c) is a stronger base (weaker acid) than the latter by nearly 1 pK_a unit, since steric interactions of the methyl substituent impedes resonance delocalisation of the nitrogen lone pair into the aromatic ring. On the other hand, for 1,8-diaminotetramethylnaphthalene (Fig. 5.27), the proximity of the two nitrogen lone pairs is particularly destabilising, and the resulting electrostatic repulsion can be readily removed by protonation of one of the nitrogen atoms; in this case, the proton is stabilised by an additional hydrogen bond with the second amine nitrogen, and these naphthalene derivatives are surprisingly strong bases.

Heterocyclic bases, in which the nitrogen is incorporated into a carbocyclic ring which may or may not be aromatic, also exhibit basic character (Fig. 5.28). The strength of this effect critically depends on whether the nitrogen lone pair is required for the aromatic system, as for pyrrole, or not, as for pyridine; as a result, the former is a much weaker base than the latter. For pyrrole, the nitrogen lone pair resides in a p-orbital capable of overlapping with the π-system, and hence it is much less available for protonation, weakening pyrrole as a base. On the other hand, the nitrogen lone pair of pyridine resides in an sp^2 hybrid orbital, which is orthogonal to the π-system of the heterocyclic ring, precluding resonance interaction. The lone pair is therefore easily able to interact with a proton, and pyridine

Figure 5.28 Basicity of nitrogen heterocycles.

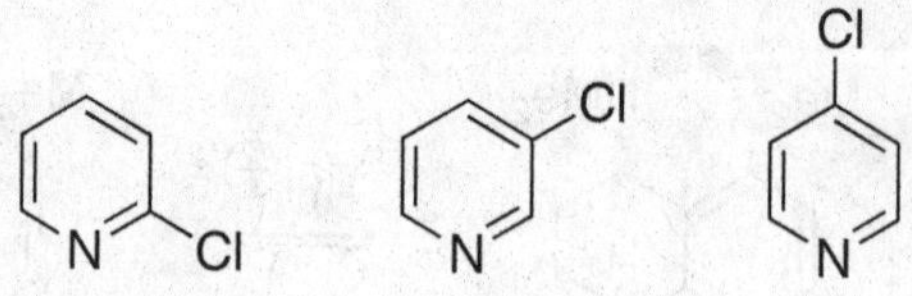

pKa (py·HCl)   0.5   2.8   3.8

Figure 5.29   Effect of substitution on the basicity of pyridine.

exhibits basic behaviour $(pK_a(pyH^+) = 5.2)$, although it is significantly weaker than the related amine, piperidine $(pK_a(RNH_3^+) = 11)$.

The same effects as discussed above can diminish basicity. For example, substitution of pyridine with electron-withdrawing atoms can lead to a significant reduction in basicity, depending on the location on the ring (Fig. 5.29); thus, substitution with electron-withdrawing groups closest to the nitrogen has the largest effect on basicity, but this falls off rapidly with distance.

Chapter 6
Nucleophilic Substitution

Since many other elements (Y) are more electronegative than carbon (C), a C—Y bond will be polarised as shown in Fig. 6.1, and this means that such a carbon is readily attacked by nucleophiles (Nu⁻), leading to replacement of Y⁻ by the nucleophile; this mode of attack is particularly common in organic chemistry, and reactions of this type are designated S_N (substitution, nucleophilic). In principle, this nucleophilic attack could happen in three ways, and these are illustrated in Figs. 6.2a–6.2c, using the reaction of hydroxide (a good nucleophile) with methyl iodide (the electrophile):

1. The nucleophile (HO⁻) adds before the C—I bond is broken – but this never happens as carbon is unable to expand its stable octet of electrons, as formation of the pentavalent intermediate would require (Fig. 6.2a).
2. The C—I bond breaks first, by departure of the leaving group I⁻, leaving a carbocation intermediate, and the nucleophile (HO⁻) then adds – in this case, the reaction is called an S_N1 reaction (Fig. 6.2b).
3. Simultaneous formation of the C—Nu and cleavage of the C—I bonds – in this case, the reaction is called an S_N2 reaction (Fig. 6.2c).

The S_N1 and S_N2 reactions occur very widely in organic chemistry and will be considered in more detail below.

6.1 The S_N1 reaction

The S_N1 (substitution, nucleophilic, unimolecular) reaction involves ionisation of an organic substrate by initial slow departure of the leaving group X, to generate a carbocation intermediate, followed by the fast entry of a suitable nucleophile Y, to give the product (Fig. 6.3a); the energetic course of this reaction is shown in Fig. 6.4. The initial step is often called ionisation, or less commonly solvolysis, since it occurs spontaneously on dissolution of the substrate in the solvent, without the interaction of any other reagents. Examples of S_N1 reactions include the reaction of t-butyl chloride in water to give t-butanol (Fig. 6.3b) and the reaction of diphenylchloromethane in aqueous acetic acid to give solvolysis products in which water or the carboxylate anion has been trapped by the intermediate carbocation (Fig. 6.3c).

For the generalised reaction shown in Fig. 6.3a, with rate constants for the first step of k_1 and for the second step of k_2 and with $k_2 > k_1$ (i.e. the first step being the slower of the two), the rate of the reaction, as measured by the disappearance of reactant RX, follows the time course of Fig. 6.5; that is, the rate of the reaction at any given time is directly proportional to the concentration of RX. Notice that the rate is dependent only on the concentration of the starting material, since the concentration of no other reactive species appears in the rate equation, and the reaction is therefore called unimolecular.

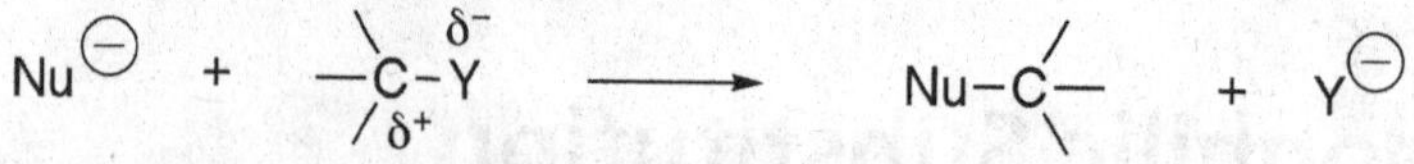

Figure 6.1 The general nucleophilic substitution reaction.

6.1.1 Factors affecting the S_N1 reaction

The efficiency of the S_N1 reaction depends on several factors, and for the generalised reaction shown in Fig. 6.3a, these are:

(a) *The nature of the substrate R*: The more stable the intermediate carbocation R^+, the more feasible it is for an S_N1 reaction to occur, since the activation energy ($\Delta G^{\ddagger}$) leading to the formation of the carbocation intermediate is lowered, and this is one of the most important factors which determines the likelihood of a reaction. We have seen in Chapter 5 that the stability of a carbocation R^+ depends vitally on its structure, with tertiary carbocations being the most stable and primary the least; each substitution of a hydrogen atom by a methyl group improves the stability of the corresponding carbocation by about 55 kJ mol^{-1}. Therefore, the ease of S_N1 reaction follows the same order (Fig. 6.6).

Significant stabilisation of the carbocation intermediate occurs with adjacent aryl substituents by virtue of resonance interactions between the empty 2p orbital of the carbocationic centre and the nearby π-orbitals of the aromatic ring; the rates of solvolysis of alkyl chlorides in EtOH at 25°C reflect this (Figs. 6.7a and 6.7b). Similar stabilisation is possible in allylic carbocations. In a similar way, α-heteroatoms will readily stabilise an adjacent carbocation by a resonance interaction involving the heteroatom lone pair (Fig. 6.8a), and solvolysis rates reflect this too (Fig. 6.8b).

Any steric or electronic influence that stabilises the intermediate carbocation will enhance the S_N1 reaction, and conversely, any steric or electronic influence that destabilises the cation will impede the reaction. For this reason, S_N1 reactions of phenyl halides and vinyl halides are not favoured, since the corresponding cation is significantly destabilised. Importantly, if a carbocation cannot be planar, it is very destabilised.

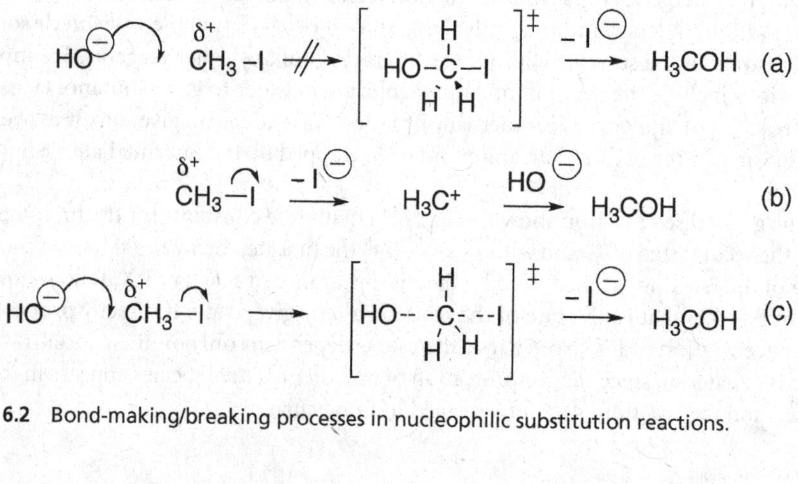

Figure 6.2 Bond-making/breaking processes in nucleophilic substitution reactions.

$$R-X \xrightarrow[{-X^{\ominus}}]{\text{Slow}\,(k_1)} R^{\oplus} + X^{\ominus} \xrightarrow[{+Y^{\ominus}}]{\text{Fast}\,(k_2)} R-Y \qquad \text{(a)}$$

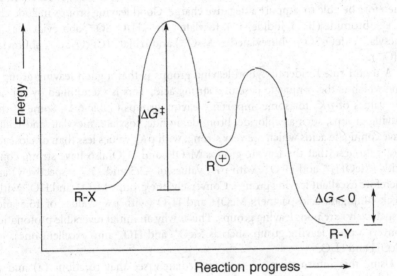

Figure 6.3 Examples of S_N1 reactions.

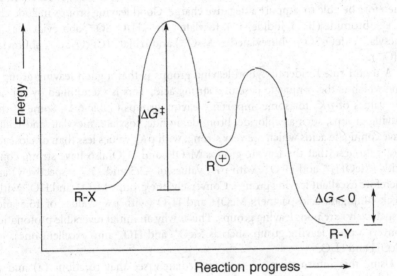

Figure 6.4 Energy profile for the course of an S_N1 reaction.

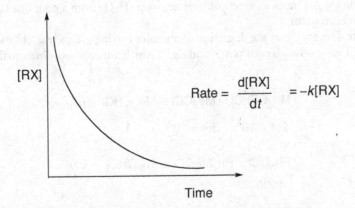

$$\text{Rate} = \frac{d[RX]}{dt} = -k[RX]$$

Figure 6.5 Change of [RX] during the course of an S_N1 reaction.

$$R_3CX > R_2HCX > RH_2CX > H_3CX$$

$$3° \qquad 2° \qquad 1° \qquad Me$$

Figure 6.6 Substrate reactivity in the S_N1 reaction.

This is indicated by the relative rates of hydrolysis of bridged alkyl halides, for which the carbocation cannot adopt a planar structure (Fig. 6.9).

Also of importance is the facility with which carbocations will rearrange, if this leads to a more stable carbocation (see Chapter 4). In particular, 1° carbocations will rearrange to 2° or 3°, and 2° carbocations will rearrange to 3° by 1,2-hydride, carbon or phenyl shifts, if this is possible.

(b) *The nature of the leaving group X:* The leaving group X is of considerable importance, since the more easily it is able to depart, the easier is the S_N1 reaction; it must therefore be able to support a negative charge. Good leaving groups include chloride (Cl^-), bromide (Br^-), iodide (I^-), tosylate ($MeC_6H_4O_2SO^-$, abbreviated as TsO^-), mesylate (MeO_2SO^-, abbreviated as MsO^-) and triflate ($CF_3O_2SO^-$, abbreviated as TfO^-).

A useful rule to identify good leaving groups is that a good leaving group (X) is one which is the conjugate base of a strong acid. This is exemplified by considering the values of pK_a for some important leaving groups (Table 6.1). Some of the best leaving groups, such as chloride, bromide, iodide, tosylate, mesylate and triflate, all have conjugate acids which are very strong, with pK_a values less than or close to zero. Note, though, that the leaving groups MeOH and H_2O also have strong conjugate acids ($MeOH_2^+$ and H_3O^+, with pK_a values of -2.5 and -1.7 respectively) and are therefore excellent leaving groups. Conversely, the groups MeO^- and HO^- with their weak conjugate acid partners MeOH and H_2O (with pK_a values of 15.5 and 15.7 respectively) are poor leaving groups. This is why an initial reversible protonation can convert a poor leaving group, such as MeO^- and HO^- into excellent ones, such as MeOH and H_2O.

Using this information, we can immediately see that reactions (a) and (b) of Fig. 6.10 occur readily, since each liberates a good leaving group, but that (c) and (d) are not favoured at all. On the other hand, initial protonation of the alcohol in reaction (e) generates an excellent leaving group (H_2O) from a poor one (HO^-), and reaction can occur.

Note, however, that the departure of a poorer leaving group (e.g. chloride) can be assisted by Lewis acids such as Ag^+ and Hg^{2+}, which not only coordinate to the halogen

$$Me_2CPhCl > Me_3CCl > Me_2CHCl \quad \text{(a)}$$

$$2.4 \times 10^8 \qquad 5.4 \times 10^4 \qquad 1$$

$$Ph_3CCl > Ph_2MeCCl > Ph_2CHCl \quad \text{(b)}$$

$$1000 \qquad 35 \qquad 1$$

Figure 6.7 Relative reaction rates for different substrates in the S_N1 reaction.

Table 6.1 pK_a values for some possible leaving groups

Leaving group	Conjugate acid	pK_a
I$^-$	HI	-10
Br$^-$	HBr	-8
Cl$^-$	HCl	-7
CH$_3$OH	CH$_3$OH$_2$$^+$	-2.5
H$_2$O	H$_3$O$^+$	-1.7
CF$_3$SO$_2$O$^-$	CF$_3$SO$_2$OH	0.3
p-MeC$_6$H$_4$SO$_2$O$^-$	p-MeC$_6$H$_4$SO$_2$OH	1
F$^-$	HF	3.2
AcO$^-$	AcOH	4.8
CH$_3$O$^-$	CH$_3$OH	15.5
HO$^-$	H$_2$O	15.7
NH$_2$$^-$	NH$_3$	35
H$^-$	H$_2$	36

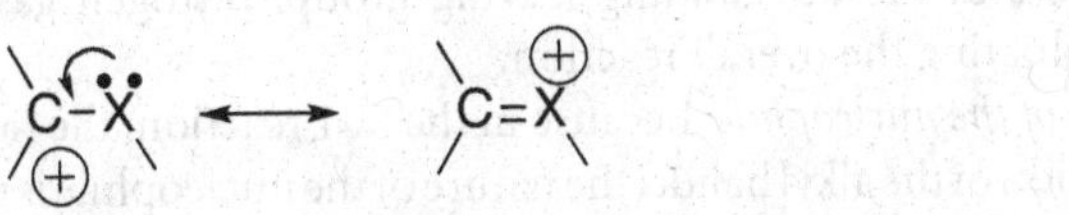

(a)

$$EtOCH_2Cl > CH_3CH_2CH_2CH_2Cl > CH_3CH_2OCH_2CH_2Cl \quad \text{(b)}$$
$$10^9 \qquad\qquad 1 \qquad\qquad\qquad 0.2$$

Figure 6.8 Relative reaction rates for α-substituted substrates in the S$_N$1 reaction.

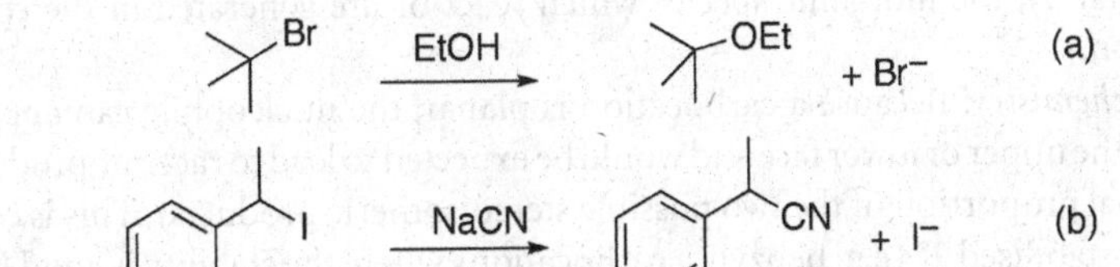

Relative rate of
S$_N$1 reaction in
80% H$_2$O/EtOH at
25°C

1   10^{-3} 10^{-6} 10^{-13}

Figure 6.9 Reactions in sterically hindered substrates.

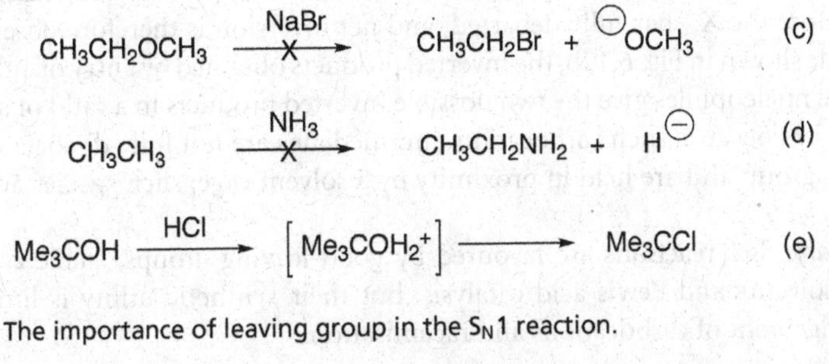

Figure 6.10 The importance of leaving group in the S$_N$1 reaction.

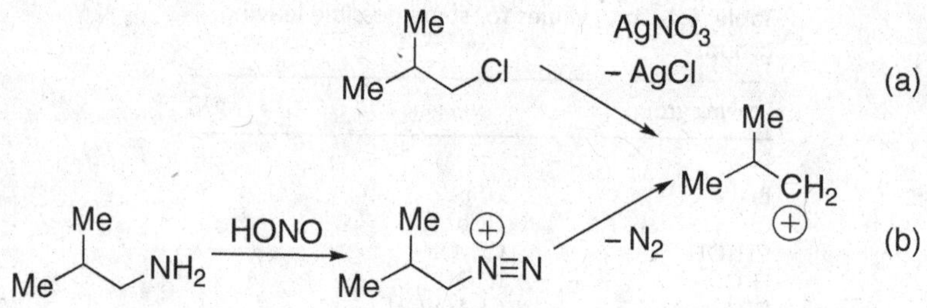

Figure 6.11   Assistance of leaving groups.

leaving group, but in doing so produce an insoluble inorganic salt (AgCl or HgCl$_2$) (Fig. 6.11a). Sometimes it is possible for a poor leaving group to be converted to a good leaving group by a chemical reaction; the conversion of an amine to a diazo group generates the outstanding leaving group, nitrogen gas (Fig. 6.11b), thereby greatly facilitating the overall reaction.

(c)   *The nature of the nucleophile*: Because in the S$_N$1 reaction, the rate depends only on the concentration of the alkyl halide, the nature of the nucleophile is relatively unimportant. However, faced with more than one reactive nucleophile, more stable carbocations (such as Ph$_3$C$^+$) will react faster with more reactive nucleophiles.

(d)   *The nature of the solvent*: The solvent can profoundly influence the course of the reaction, since it must solvate both R$^+$ and X$^-$, but it may also directly partici-pate in the substitution reaction by acting as a nucleophile; if it does so, the reac-tion is called solvolysis. Anions are best solvated by hydrogen bonding to a dipo-lar solvent (e.g. water or alcohol), as exemplified by chloride solvated by methanol, while cations are best solvated by coordination to lone pairs of the solvent (e.g. acetone, dimethyl sulfoxide (DMSO) and alcohols), as exemplified by Na$^+$ coor-dination by DMSO or Li$^+$ by hexamethylphsophoramide (HMPA) (see Fig. 3.25). Therefore, the best solvent mixtures for S$_N$1 reactions are alcohol/water mixtures, since this permits effective solvation of both the carbocation and, probably more importantly, the inorganic species which react or are generated in the course of the reaction.

(e)   *Stereochemistry*: Because a carbocation is planar, the nucleophile can approach from either the upper or lower face and would be expected to lead to racemic products (that is, an equal proportion of the two possible stereoisomeric products). This is common for 3° and stabilised 2° (e.g. benzylic) carbocations where the stability is good (Fig. 6.12a), but for secondary carbocation, which are not so stable, the incoming nucleophile can add before the X$^-$ has fully departed, and net inversion is therefore observed. In the example shown in Fig. 6.12b, the inverted products obtained by entry of either water or ethanol nucleophiles give the two possible inverted products in a ratio of about 85:15. This arises because such carbocation intermediates are not fully dissociated from the leaving group and are held in proximity by a solvent cage; such species are called ion pairs.

(f)   *Summary*: S$_N$1 reactions are favoured by good leaving groups, stable carbocations, polar solvents and Lewis acid catalysis, but their synthetic utility is limited by the rearrangement of carbocations and racemisation.

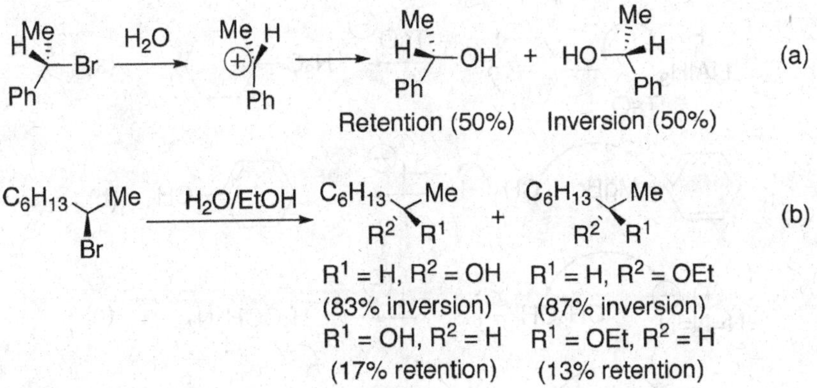

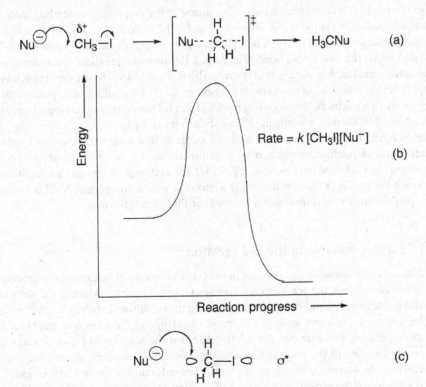

Figure 6.12 illustration (a) and (b):

(a) Retention (50%)   Inversion (50%)

(b) $R^1 = H, R^2 = OH$
(83% inversion)
$R^1 = OH, R^2 = H$
(17% retention)

$R^1 = H, R^2 = OEt$
(87% inversion)
$R^1 = OEt, R^2 = H$
(13% retention)

Figure 6.12 Stereochemical outcomes in S_N1 reactions.

6.2   The S_N2 reaction

Unlike the S_N1 reaction which involves a stepwise reaction pathway, the S_N2 reaction involves simultaneous making and breaking of bonds, via a pentacoordinate transition state (Fig. 6.13a), although the total bond order around carbon never exceeds four (i.e. the octet of electrons). The rate equation is second order, with both reactants appearing in the equation,

Rate = k [CH$_3$I][Nu$^-$]   (b)

Figure 6.13 Aspects of the S_N2 reaction.

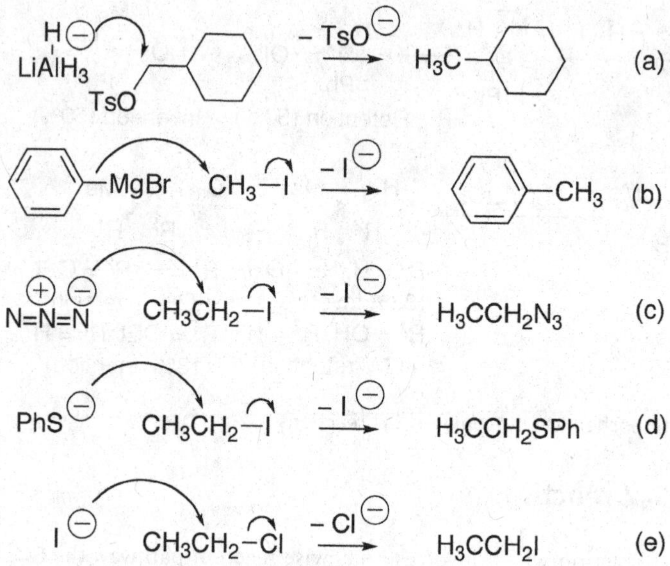

Figure 6.14 Typical S_N2 processes.

and so the reaction is called bimolecular. The course of the reaction from starting materials to products proceeds via the transition state (Fig. 6.13b). Transfer of electron density from the nucleophile to the σ^* of the C–X bond of the substrate simultaneously weakens the C–X bond and forms the new C–Nu bond (Fig. 6.13c). Because the donation of the electrons of the incoming nucleophile occurs into the σ^*-orbital, located 180° to the departing leaving group, strict inversion of stereochemistry is observed. This transition state places both of the electron-rich partners, the incoming nucleophile and the departing leaving group, as far apart as possible, thereby minimising electrostatic interactions.

The S_N2 reaction is very common and will occur with a wide variety of nucleophiles, including hydride, carbon, nitrogen, oxygen, sulfur and halide, and examples of each process are given in Fig. 6.14. The last reaction (Fig. 6.14e), the exchange of a poorer leaving halogen (Cl^-) for a better one (I^-), is performed in acetone in which the product NaCl is insoluble and is preparatively very important; it is called the Finkelstein reaction.

6.2.1 Factors enhancing the S_N2 reaction

(a) *The nature of the substrate*: Because in the transition state of S_N2 reactions the carbon is five coordinate, the reactions are very sensitive to steric crowding in the substrate; this is due to non-bonding interactions among the substituent, the nucleophile and the leaving group, and this leads to a decrease in entropy as the transition state becomes more ordered. For example, the relative rates for the reaction in Fig. 6.15 show that steady increase in the substitution and therefore the steric bulk around the carbon carrying the leaving group leads to a significant reduction in the rate of reaction. This is reflected in the general relative rates of reaction for primary, secondary and tertiary substrates shown in Fig. 6.16.

$$\text{EtO}^{\ominus} \quad \text{RCH}_2\text{—I} \quad \longrightarrow \quad \text{EtOCH}_2\text{R}$$

R	H	CH$_3$	CH$_3$CH$_2$	(CH$_3$)$_2$CH	(CH$_3$)$_3$C
Relative reaction rate	1	6×10^{-2}	1.6×10^{-2}	1.7×10^{-4}	2.4×10^{-7}

Figure 6.15 Relative rates for S$_N$2 processes in different substrates.

(b) *The nature of the leaving group X*: Obviously, the better the leaving group, the easier the reaction, and the same rule applies as for S$_N$1 reactions: A good leaving group (X) is the conjugate base of a strong acid (Table 6.1). The use of solvents which more efficiently solvate leaving groups, for example, dimethylforamide (DMF) or dimethyl sulfoxide (DMSO), can enhance leaving-group ability.

However, it is possible to convert a poor leaving group into a better one, and hydroxyl substituents are particularly effective in this regard. For example, conversion of the alcohol shown in Fig. 6.17 to its corresponding tosylate generates a very effective leaving group, which can be displaced by good (e.g. PhS$^-$) and even weak nucleophiles like acetate (AcO$^-$). Note that the former transformation proceeds with retention of configuration, since it is not a nucleophilic substitution, but the second one proceeds with inversion, since it is an S$_N$2 reaction. Direct acetylation of the opposite enantiomer of the alcohol gives the product of the same stereochemistry.

(c) *The nature of the nucleophile Nu$^-$*: Because the concentration of the nucleophile appears in the rate equation, and because the course of the S$_N$2 reaction crucially depends on the ability of the nucleophile to donate its electrons to the reacting carbon atom, good nucleophiles will enhance the rate of the S$_N$2 reaction, and this is quite different to the S$_N$1 reaction. Nucleophilicity decreases as we go left to right (Fig. 6.18a) and increases as we go top to bottom (Fig. 6.18b) of the periodic table, and also increases with negative charge (Fig. 6.18c), and can be altered by changing the solvent. Examples of good nucleophiles include halides (Cl$^-$, Br$^-$ and I$^-$ but not F$^-$), RO$^-$, NC$^-$, RS$^-$, N$_3^-$, NH$_3$, R$_2$PH and RSH, and some approximate nucleophilicities, relative to water for the standard reaction with methyl iodide, are shown in Fig. 6.19.

It is possible for carbanions to act as nucleophiles, and this can be very effectively used to create new carbon–carbon bonds; these reactions are therefore of substantial importance in synthetic organic chemistry, providing access to more elaborate structures by direct coupling of simpler precursors. Examples of such processes are shown in Fig. 6.20.

(d) *The nature of the solvent*: A bimolecular reaction needs a solvent which will solvate both cations (M$^+$) and anions (Y$^-$); these are usually dipolar aprotic solvents (e.g. acetone (Me$_2$CO), Me$_2$SO, DMF (HC(O)NMe$_2$) and acetonitrile) which have non-bonding

$$\text{R}_3\text{CX} < \text{R}_2\text{HCX} < \text{RH}_2\text{CX} < \text{H}_3\text{CX}$$

$$3° \qquad 2° \qquad 1° \qquad \text{Me}$$

Figure 6.16 Generalised rates of reaction for S$_N$2 processes.

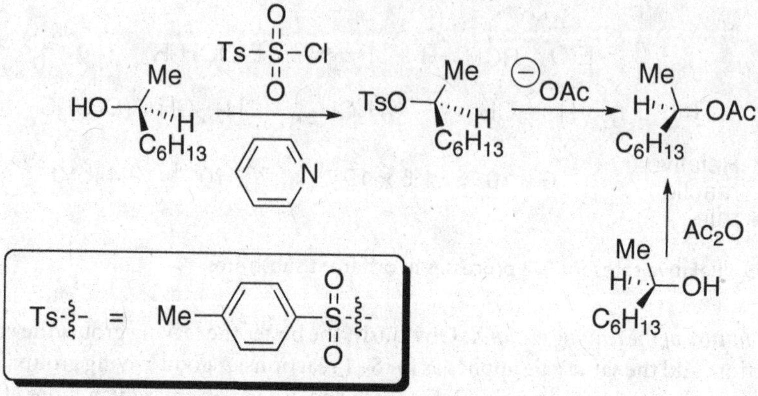

Figure 6.17   Conversion of poor leaving groups into better ones.

$$H_3C^- > H_2N^- > HO^- > F^- \qquad (a)$$

$$H_3COH < H_3CSH < H_3CSeH \qquad (b)$$

$$H_2N^- > NH_3, \quad HO^- > H_2O, \quad H_3CO^- > H_3COH \qquad (c)$$

Figure 6.18   Ranking order of nucleophilicity.

$$HS^-, CN^-, I^- \sim 10^5 > HO^-, N_3^-, Br^-, NH_2C(S)NH_2 \sim 10^4$$
$$> Cl^-, AcO^-, Me_3N, C_5H_5N \sim 10^3 > H_2O \sim 1$$

Figure 6.19   Approximate nucleophilicities relative to water.

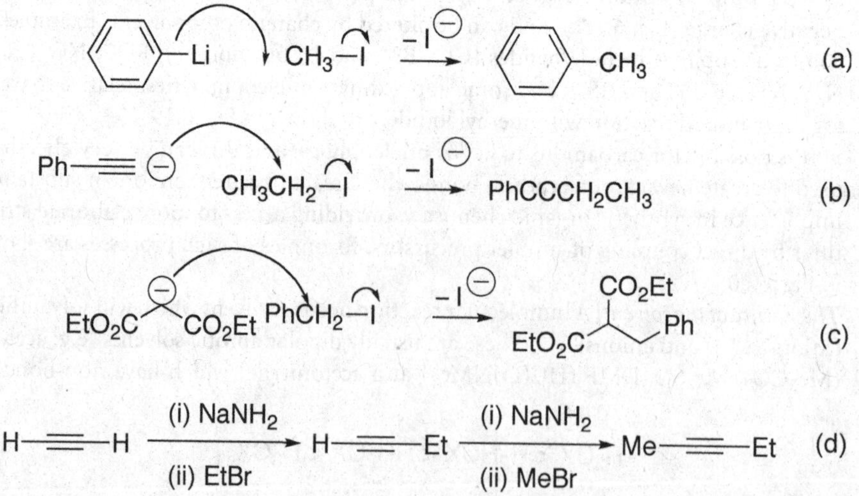

Figure 6.20   Reactions of carbanions as nucleophiles (a–c), and (d) use of acetylide anions as nucleophiles.

$$RS^- > I^- > CN^- > CH_3O^- > Br^- > H_3N > Cl^- > F^- > H_3COH$$

Figure 6.21 Relative nucleophilicity of nucleophiles in methanol.

electrons capable of coordinating with the cation, but which do not form a tight solvation shell around the anion, which would otherwise reduce its nucleophilicity. In fact, the order of nucleophilicity can easily be modified by solvent; for example, for halides in H_2O, for which a solvation shell is most readily formed around the highly polarising and smallest ion (F^-), the order is (from most nucleophilic to least nucleophilic):

$$I^- > Br^- > Cl^- > F^-$$

But for halides in DMSO, in which a solvation shell is not formed, the most nucleophilic halide is the one with the smallest radius, and the order changes to:

$$F^- > Cl^- > Br^- > I^-$$

The relative nucleophilicity of common nucleophiles in methanol solvent is shown in Fig. 6.21.

Since S_N2 reactions do not proceed with significant charge separation, it might be expected that non-polar solvents would promote this bimolecular mechanism; conversely, a more polar solvent would slow the rate of reaction. An example is shown in Fig. 6.22: the reaction is slowed if the acetone reaction medium is diluted with the more polar solvent, water.

Dipolar aprotic solvents lead to significant acceleration of S_N2 reactions, since they very effectively solvate cations but not anions; this lowers the $\Delta G^{\ddagger}$ for the substitution reaction. For example, in the simple displacement of iodomethane with chloride, bromide, azide or cyanide, the rate of reaction in DMF is approximately 10^6 times greater than that of the corresponding reaction in methanol.

(e) *Stereochemistry*: As noted earlier, strict inversion of stereochemistry is observed (Fig. 6.13c). This requirement is so important that substrates which cannot undergo inversion will not react in S_N2 reactions (Figs. 6.23a and 6.23b).

(f) *Competition between the S_N1 and S_N2 reaction*: Because S_N1 and S_N2 reactions are affected in opposite ways by substitution in the reacting alkyl halide, we often see a gradual change in mechanism along the series of substrates Me, 1°, 2° and 3°. However, 1° halides react predominantly by S_N2, but 3° halides react by S_N1; the mechanism for 2° halides depends crucially on the nature of the alkyl group R, the nucleophile, the leaving group X and the solvent, and therefore varies from case to case. In order to assess the likely mechanism, consideration of each of these factors must be made.

(g) *Summary*: S_N2 reactions are favoured by good leaving groups and nucleophiles, dipolar aprotic solvents and non-bulky substrates.

$$HO^{\ominus} + Me_3S^{\oplus} \xrightarrow{\text{Acetone}} MeOH + Me_2S$$

Figure 6.22 The importance of solvent on reactivity.

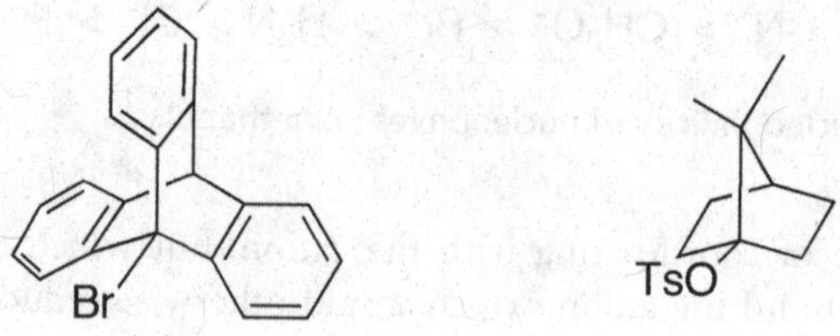

Figure 6.23   Substrates inert to nucleophilic substitution.

6.3   Synthetic applications of nucleophilic substitution reactions

The importance of nucleophilic substitution reactions in synthetic organic chemistry is substantial, and two examples will be used to illustrate this.

6.3.1   Protecting-group chemistry

Frequently in chemical synthesis, it is necessary to block undesirable reactions at certain functional groups, and this can be most conveniently done using protecting groups; such groups normally block reactivity as a result of steric effects or by modification of the electronics of the group that requires protection. The ideal protecting group needs to be easily and efficiently introduced in the first place, stable to a wide range of reaction conditions, and easily and efficiently removed without disruption of the remainder of the molecule. In practice, this turns out to be quite a tall order, and it is rare for protecting groups to be fully effective in all of these categories, and so compromises need to be made.

Because they are so common, carboxylic acids, amines and alcohols frequently require protection, and a wide variety of protecting groups have been developed; many of these rely upon nucleophilic substitution processes both to insert and to remove the protecting group. Carboxylates are weak leaving groups, and this allows carboxylic acids to be protected as esters, from which they can be readily released under conditions dictated by the nature of the ester. For example, a carboxylate may be protected as a *t*-butyl, benzhydryl or trityl ester (Figs. 6.24a, 6.24b and 6.24c respectively); this reaction is achieved under S_N1 reaction conditions and requires excellent leaving groups to generate the sterically hindered products. It is this hindrance which protects the carbonyl group from further nucleophilic attack, since the bulky ester impedes the Burgi–Dunitz nucleophile trajectory. For deprotection, acidic conditions enable protonation of the carbonyl group and then loss of a highly stabilised carbocation (S_N1 reaction). However, by appropriate substitution, it is possible to devise protecting groups of lower lability in acid; an excellent example is the phenylfluorenyl-protected amine (Fig. 6.24d), which is 6000 times less reactive than the corresponding trityl-protected system and requires trifluoroacetic acid for extended reaction time in order for its cleavage. Using this strategy, it is possible to protect different functional groups with differing lability in acid conditions and then to selectively release them by appropriate choice of acid treatment. Alternatively, cleavage of a methyl or ethyl ester (Fig. 6.24e) using a potent nucleophile such as phenylsulfide or phenylselenide allows cleavage, but now as an

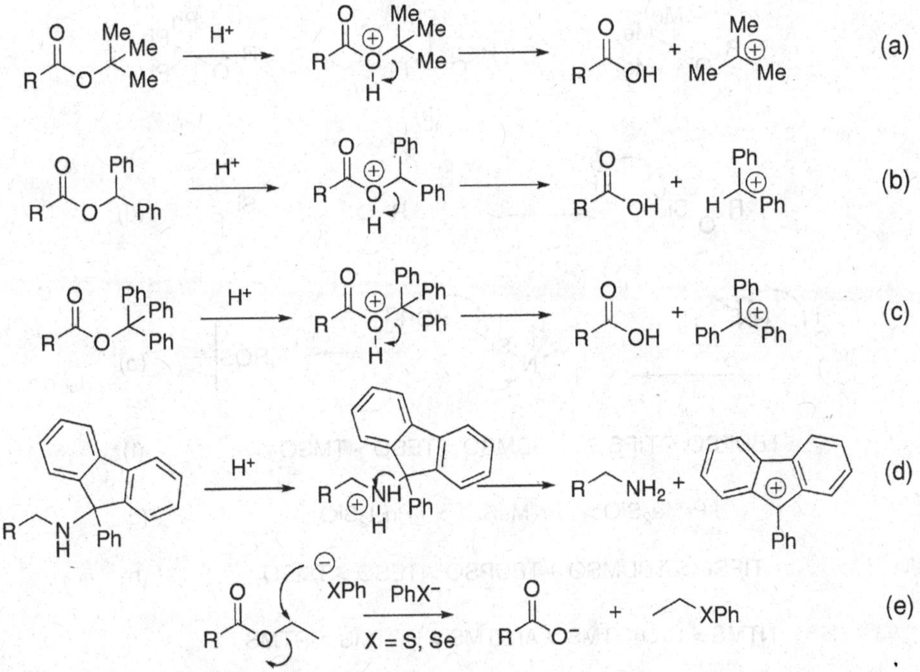

Figure 6.24 Substitution reactions in protecting-group chemistry.

S_N2 reaction. Thus, appropriate choice of a carboxylic acid enables selective deprotection, on the basis of changes in reaction mechanism.

In the same way, protection of alcohols as *t*-butyl, benzhydryl or trityl ethers is also possible (Figs. 6.25a–6.25c), since these can all be cleaved by an analogous mechanism to those shown in Fig. 6.24. However, methyl ethers are not on the whole such useful protecting groups for alcohols, since although they are easily formed, the removal is difficult; one exception is in phenolic ethers, where cleavage with the powerful electrophile BCl_3 is assisted by the better leaving ability of phenol. An alternative protecting group of significant use for alcohols is the trialkylsilyl group (Fig. 6.25d); here, the strong oxygen–silicon bond and bulky groups on the silicon provide for a stable protecting group, which is resistant to diverse chemical processes. These groups are readily introduced by reaction of the alcohol with the relevant silylchloride in the presence of imidazole or triethylamine in a reaction which proceeds by nucleophilic catalysis (Fig. 6.25e). However, the silyl groups are readily released by treatment with acid, alkali or fluoride anion, which leads to an S_N2-like reaction at silicon, and the rate of this process can be easily controlled by appropriate manipulation of the substituents on the silicon atom. The ease of hydrolysis is mainly influenced by bulk of ligands on silicon, with the more bulky ligands decreasing the susceptibility to hydrolysis. A typical order of hydrolysis under acidic conditions is shown in Fig. 6.25f, with *t*-butyldiphenylsilyl being the most robust and successive reductions in steric encumbrance leading to increased lability, and this can be controlled further, for example, by variation of the substitution pattern on isopropylsilyl-protected alcohols (Fig. 6.25g). Trimethylsilyl (TMS) ethers are sufficiently reactive and frequently hydrolysis by water is possible; work-up in these cases need to be

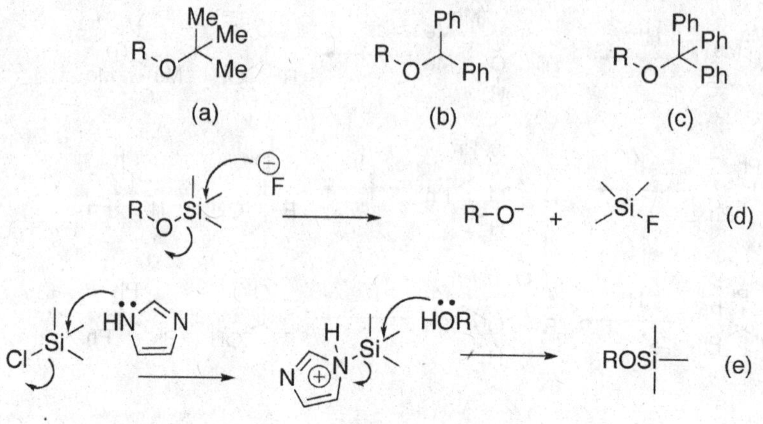

$$\text{TBDPSO} > \text{TIPSO} > \text{TBDMSO} > \text{TESO} > \text{TMSO} \qquad \text{(f)}$$

$$i\text{-PrMe}_2\text{SiO} > i\text{-Pr}_2\text{MeSiO} > i\text{-PrEt}_2\text{SiO} \qquad \text{(g)}$$

$$\text{TIPSO} > \text{TBDMSO} \sim \text{TBDPSO} > \text{TESO} > \text{TMSO} \qquad \text{(h)}$$

$$\text{NTMS} > \text{C(O)OTMS} > \text{ArOTMS} > \text{ROTMS; RSTMS} \qquad \text{(i)}$$

TES = $-\text{SiEt}_3$, TIPS = $-\text{Si}i\text{-Pr}_3$, TBDPS = $-\text{Si}t\text{-BuPh}_2$, TBDMS = $-\text{Si}t\text{-BuMe}_2$, TMS = $-\text{SiMe}_3$

Figure 6.25 Protection of alcohols as ethers.

carefully planned. Under basic conditions, however, the relative ordering changes, since electron-withdrawing groups increase the lability of silyloxy groups, with the relative rates indicated in Fig. 6.25h. Further changes in reactivity can be exerted from electronic effects; for example, phenyl silyl ethers are more reactive than alkyl silyl ethers as a result of their better leaving-group ability under basic conditions, but the latter are more labile under acidic conditions as a result of the lower basicity of the phenyl silyl ether oxygen atom. Furthermore, selective deprotection of different types of heteroatoms is feasible, as shown in Fig. 6.25i, with silylamines being selectively cleavable faster than silyl esters, silyl phenols and lastly silyl ethers or thioethers.

There are several side benefits from the use of silyl protecting groups: One is the concomitant increase in lipophilicity of the protected substrate, which can be particularly beneficial for polar substrates. A further advantage of the use of silyl protection is that deprotection with in situ derivatisation either to a differently protected alcohol or to a different functional group is possible: Direct conversion of silyl ethers to aldehydes, ketones, bromides, acetates and ethers has been established. In addition to acidic and basic hydrolysis, fluoride deprotection is particularly effective as a result of the high Si—F bond strength which provides a potent thermodynamic driving force for the reaction, and this process is predominantly sterically controlled, to liberate the alcohol and generate a highly stable Si—F bond (Fig. 6.25d).

The protection of amines presents a slightly different problem as a result of the great nucleophilicity of nitrogen and the weaker leaving-group ability of an ammonium species.

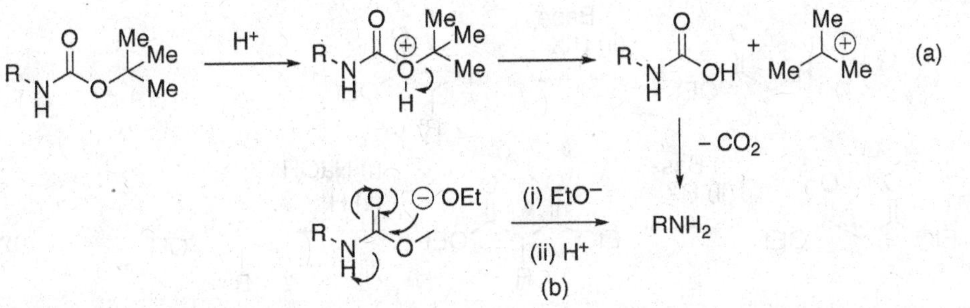

Figure 6.26 Protection of amines as carbamates.

In the case of amines, this difficulty is effectively solved by placing an oxycarbonyl spacer onto the nitrogen and then treating this residue as if it were a carboxylic acid giving the so-called carbamate or urethane group. A particularly important example is the *t*-butylcarbamate- (or BOC-)protecting group, readily removed by acid treatment (Fig. 6.26a) in an S_N1-like process. Conversely, the methyl carbamate is not at all easily cleaved by acid, and generally requires strong alkaline treatment for its removal (Fig. 6.26b), by nucleophilic attack at carbonyl.

6.3.2 Stereocontrolled alkylation reactions

Carbanions are generally highly nucleophilic and participate readily in S_N reactions; some organometallic reagents and enolate anions are especially useful in this regard. For example, the reaction of Grignard reagents (Fig. 6.14b), organolithium reagents and acetylide anions with alkyl halides gives very effective carbon–carbon bond formation by substitution; application of this procedure sequentially to ethyne allows the formation of unsymmetrical alkynes (Fig. 6.21).

The reaction of enols or enolate anions, obtained under mildly or strongly acidic or basic conditions respectively (Fig. 6.27), with alkyl and allyl halides readily generates the corresponding alkylated products very efficiently (Fig. 6.28a). This process is particularly favourable for the substrates diethyl malonate and ethyl acetoacetate (Figs. 6.28b and 6.28c); α-deprotonation is particularly easy, and the resulting enolates are sufficiently stabilised to facilitate their formation but not so stable as to be unreactive. They readily react with a variety of alkyl halides and their equivalents in excellent yield, but significantly, ester hydrolysis and decarboxylation in both cases lead to the corresponding α-substituted acetic

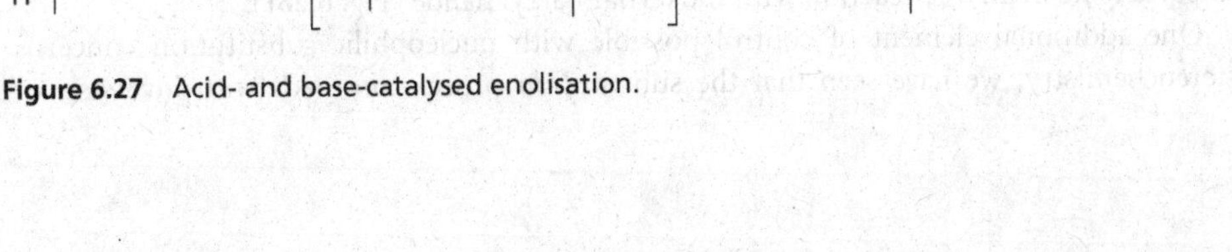

Figure 6.27 Acid- and base-catalysed enolisation.

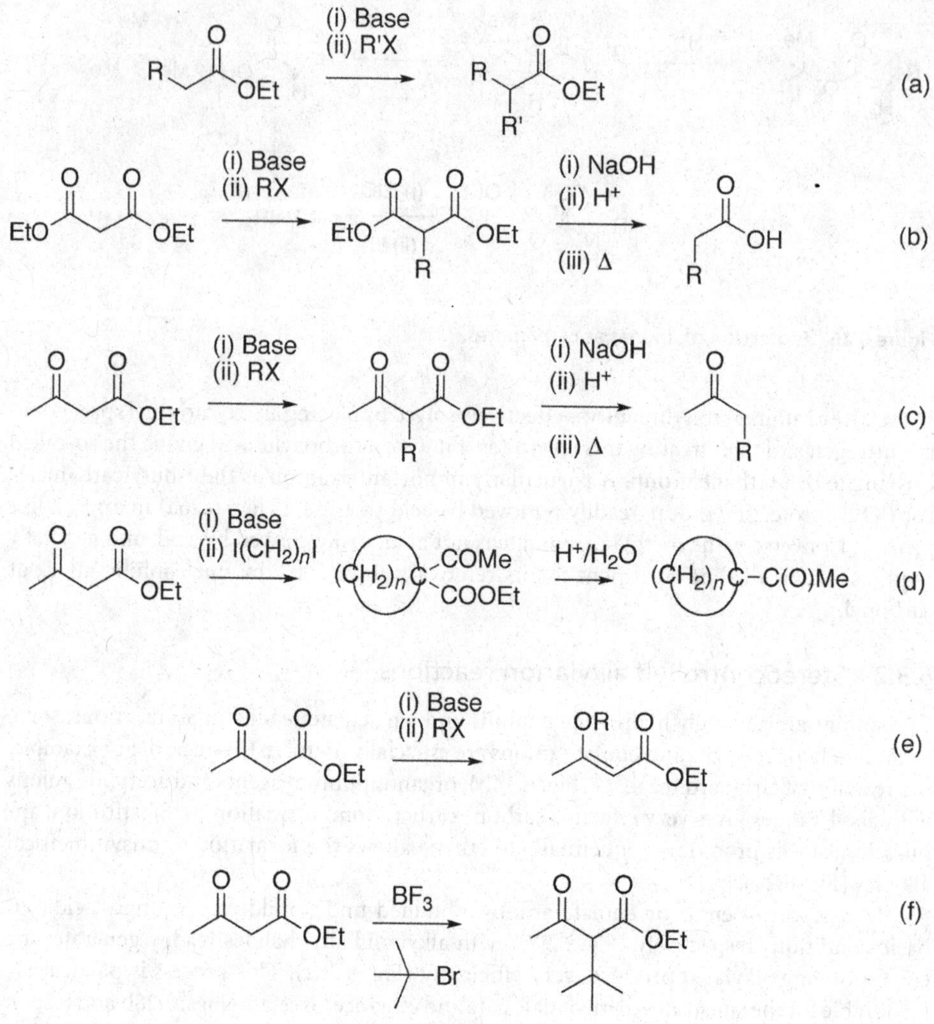

Figure 6.28 Efficient α-alkylation of carbonyl substrates.

acid and acetone respectively (Figs. 6.28b and 6.28c). This reaction can be used with ω-dihalo electrophiles, which will successively alkylate, to generate a cyclic product (Fig. 6.28d). One complication, however, in this sequence is the problem of O- versus C-alkylation, since enolates are ambident nucleophiles (Fig. 6.28e); this can be controlled to some extent by the careful choice of both the base and the solvent. Noteworthy is that alkylation under basic conditions with tertiary alkyl halides is unsuccessful, due to excess steric hindrance in the transition state, which disfavours S_N2 nucleophilic substitution reactions; in this case, it is possible, however, to successfully achieve alkylation under Lewis acid conditions, in which formation of the enol form of the β-dicarbonyl is favoured, and this is nucleophilic enough to participate in an S_N1 reaction with the tertiary alkyl halide (Fig. 6.28f).

One additional element of control possible with nucleophilic substitution concerns stereochemistry; we have seen that the substitution process proceeds with inversion of

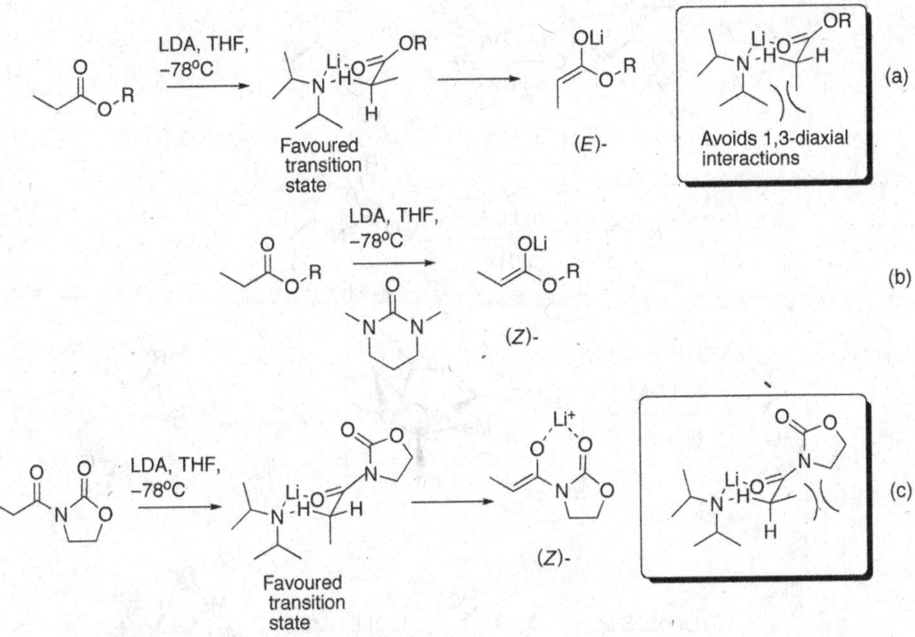

Figure 6.29 Stereocontrolled enolate formation.

stereochemistry, but nucleophiles can also be used to generate a chiral product. Enolates are particularly useful in this regard; firstly, deprotonation of esters and amides can be achieved under conditions which reliably give access to either the *E*- or *Z*-geometric outcomes (Figs. 6.29a–6.29c). For esters, enolate formation is generally kinetically favoured (i.e. faster) for the *E*-enolate, since this avoids 1,3-diaxial interactions in the transition state which would lead to the *Z*-enolate (Fig. 6.29a), but the *Z*-enolate is accessible if the reaction is carried out under equilibrating conditions (i.e. thermodynamic control) (Fig. 6.29b). On the other hand, enolate generation from the amide leads to the *Z*-enolate (Fig. 6.29c). This controlled generation of enolates can be further used to dictate the stereochemical outcome of the reaction using either steric or chelation influences. If the carbonyl compound is chiral, then diastereoselective alkylations become possible; for example, alkylation of the enolate derived from bicyclic systems can be especially effective (Figs. 6.30a and 6.30b), since steric factors dictate that alkylation occurs to the less hindered (*exo*-) convex face, leading to the most thermodynamically stable product. On the other hand, a closely related substrate gives predominantly (*endo*-) concave face alkylation, probably due to more favourable anti-stereoelectronic interactions in the intermediate enolate with the nitrogen lone pair (Fig. 6.30c). A similar system has been reported to efficiently alkylate with high stereoselectivity, but interestingly the expected products are not always obtained; thus, the *cis*-lactam can be successfully alkylated with MeI once to give the product in which the electrophile approaches from the less hindered *exo*-convex face and then again to give the dimethylated compound (Fig. 6.30d). The closely related *trans*-lactam can be alkylated once, but not subsequently (Fig. 6.30e). For the first alkylation, deprotonation of only H_α occurs, because only in the corresponding enolate is orbital overlap possible between the p orbital and the π-system

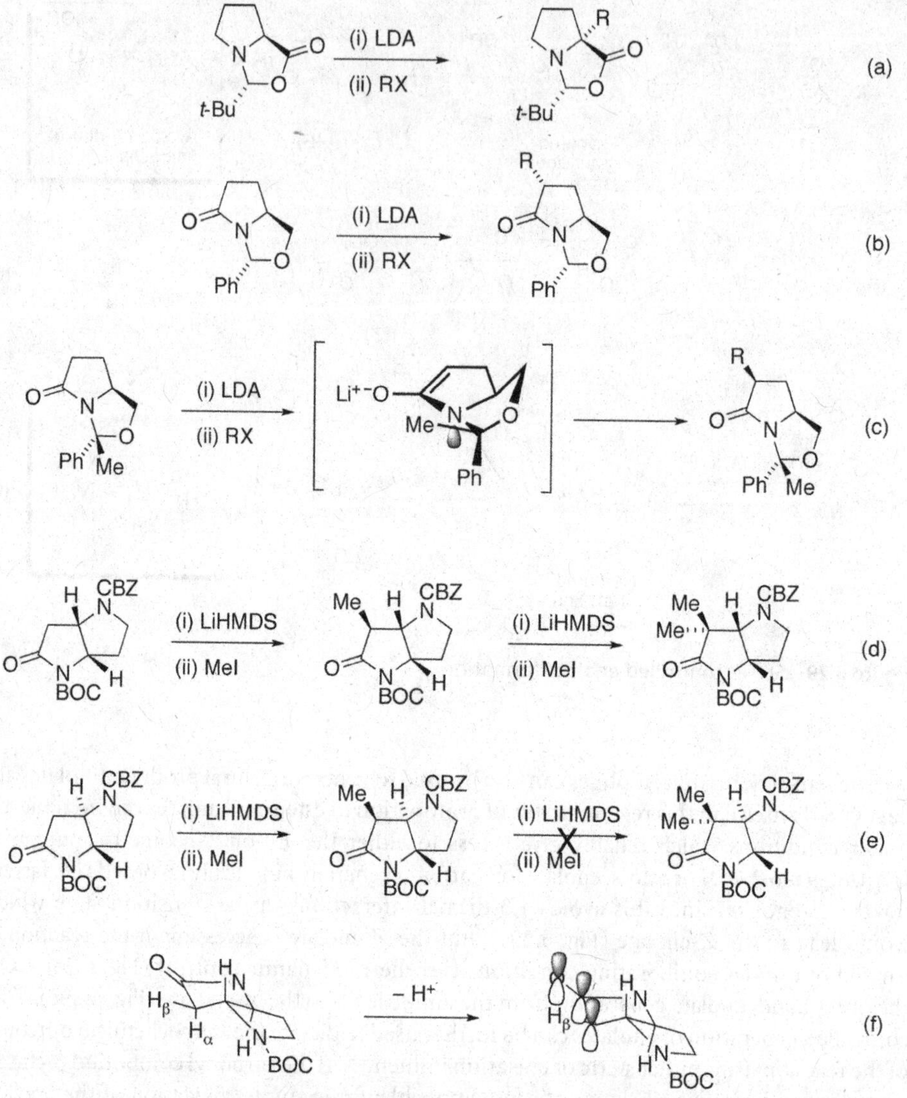

Figure 6.30 Stereocontrolled alkylation of carbonyl compounds.

of the carbonyl group. Deprotonation of H_β does not occur, since the p orbital and the π-system of the carbonyl group are now orthogonal, preventing orbital overlap (Fig. 6.30f).

In cases where the substrate does not have its own intrinsic chirality, it is possible to introduce stereochemistry by the use of chiral auxiliaries. This is particularly easy for carbonyl compounds, and some of the most widely used systems include the Enders hydrazone, Meyer's oxazolidine, Schollkopf system and Myer's amide (Figs. 6.31a–6.31d), and perhaps most important of all, the Evans oxazolidinone (Fig. 6.32). The Meyer's system behaves in a somewhat unusual manner, in which it is thought that chelation of the incoming alkyl halide with the lithium enolate directs attack to the more hindered side (Fig. 6.31b). A common

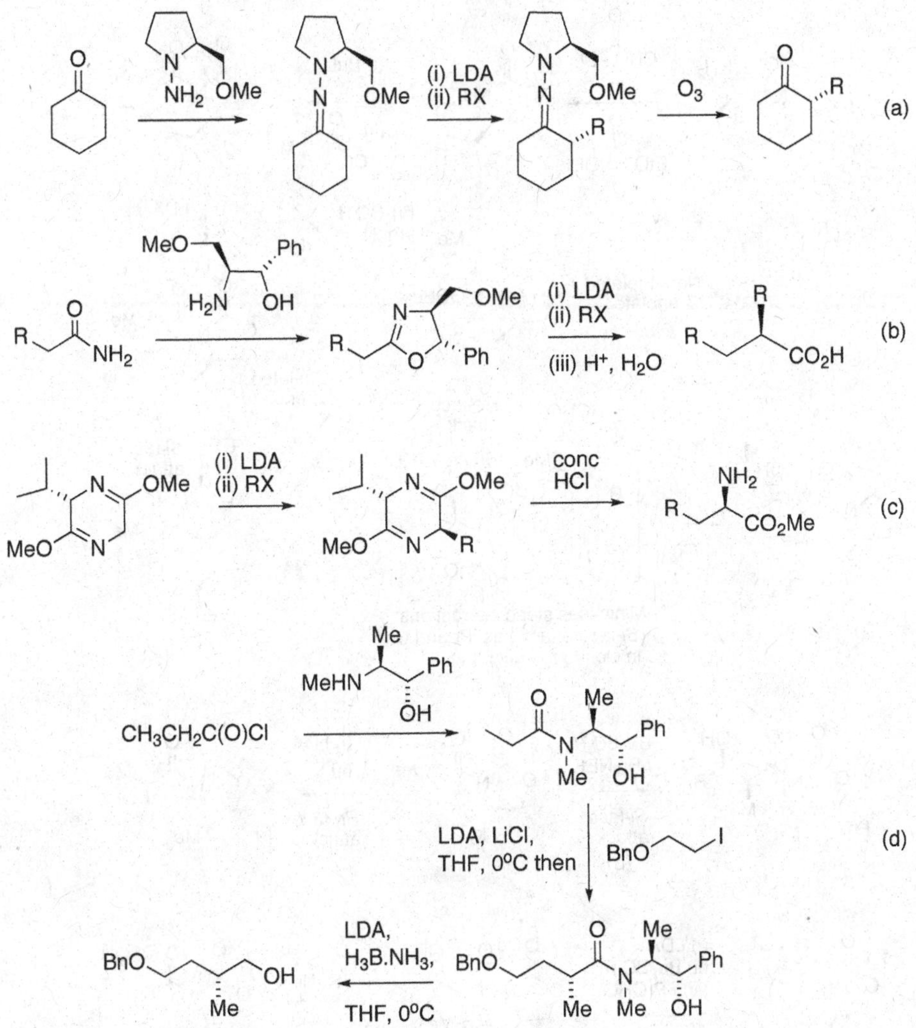

Figure 6.31 Chiral auxiliary-controlled nucleophilic substitution reactions.

theme of these approaches is the use of cyclic chiral auxiliaries, since for such systems steric demands and chelation effects are more easily predicted and therefore controlled. However, recently an approach based on (acyclic) pseudoephedrines has been shown to be effective; the Myer's system makes use of the amide derivatives of (1*S*,2*S*)-pseudoephedrine for a double deprotonation followed by enolate alkylation (Fig. 6.31d). Direct reduction of this amide can be used to access the corresponding alcohol. In the case of Evans system, the enolate generated using lithium diisopropylamide (LDA) is stabilised as the internally chelated *Z*-enolate, and approach of the electrophile from the less hindered side (i.e. away from the isopropyl group) leads to the stereoselective alkylation shown in Fig. 6.32a. On the other hand, the boron enolate, generated using dibutylboron triflate, is too hindered to form a chelated enolate (Fig. 6.32b) and instead alkylates in the opposite stereochemical sense.

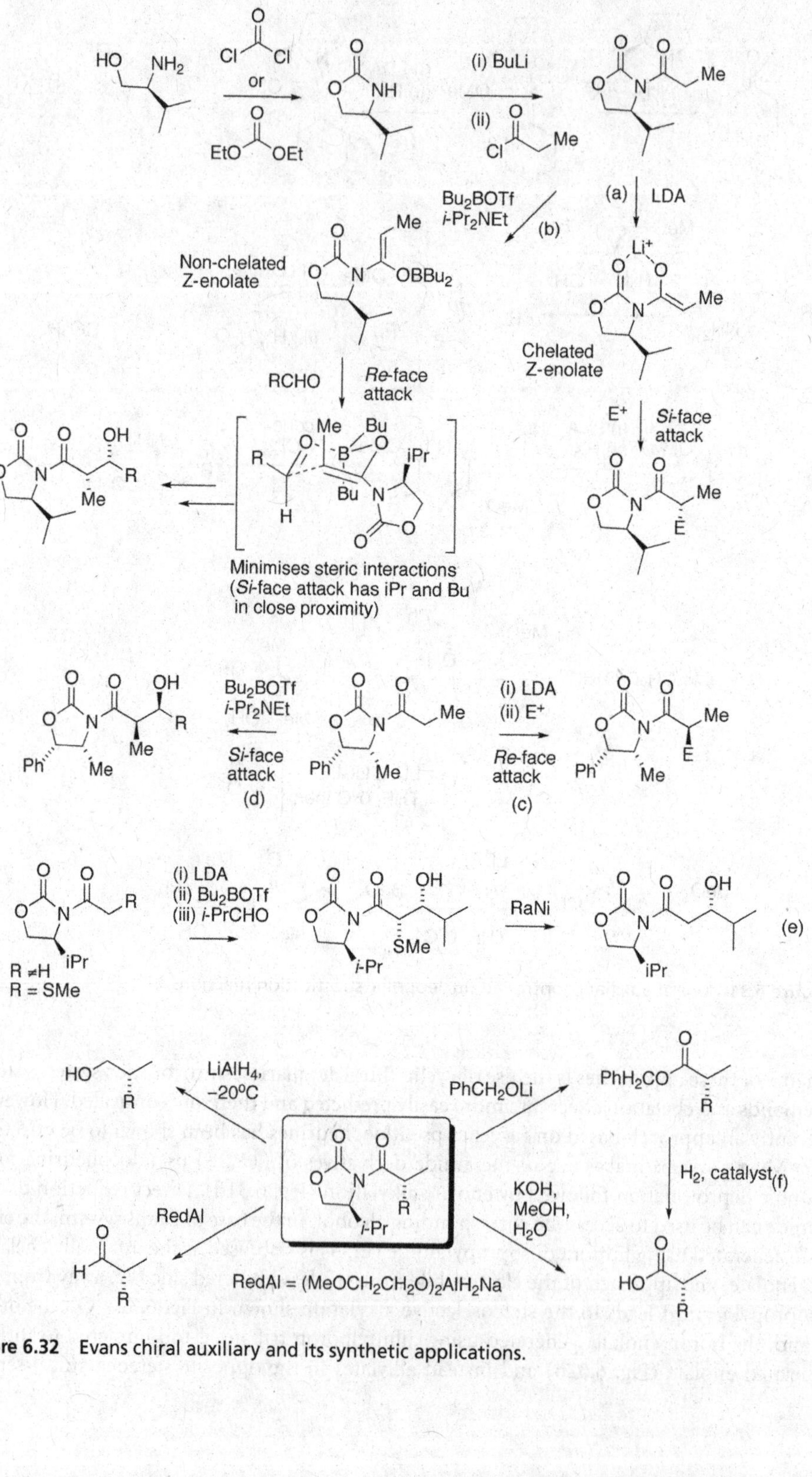

Figure 6.32 Evans chiral auxiliary and its synthetic applications.

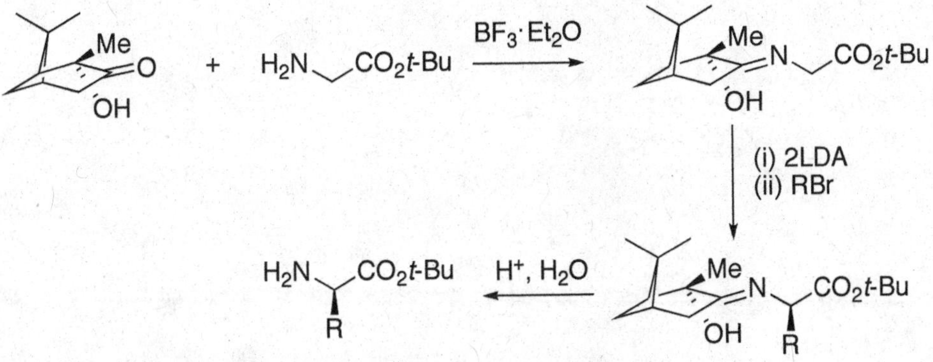

Figure 6.33 Chiral auxiliaries for amino acids.

Similar control of relative stereochemistry but in an opposite absolute sense can be achieved by appropriate choice of the stereochemistry of the auxiliary (Figs. 6.32a and 6.32b versus Figs. 6.32c and 6.32d). However, one drawback of this approach is that it is necessary to have a substituent on the acyl portion; if it is not required in the final product, it is possible to use SMe as a temporary group and then remove it later (Fig. 6.32e). The power of the Evans approach is that it provides access to enantiopure intermediates, which can be readily converted to a variety of final products (Fig. 6.32f). This type of approach, involving the conversion of an achiral starting material into a chiral one by attachment of a suitable auxiliary, can also be used for the synthesis of substituted amino acids (Fig. 6.33).

Chapter 7
Addition Reactions

In the previous chapter, the formation of new bonds by the displacement of a suitable leaving group by nucleophiles was described; in this chapter, alternative modes for bond formation by cleavage of multiple bonds, with simultaneous formation of new σ-bonds are described. Such processes are described as additions, since the new single bond is formed at the expense of the double bond.

7.1 Electrophilic addition reactions

The carbon–carbon double (and triple) bond, which is electron rich as a result of the diffuse electron density of its π-electron cloud formed by the overlap of two p orbitals on each of the respective carbon atoms which comprise the bond (Fig. 7.1), is also relatively weak (280 kJ mol^{-1}) and susceptible to reaction with electron-deficient reagents (electrophiles) of various types; in this process, the double bond is broken with concomitant formation of two new single bonds. Alkene reactivity is therefore typically as a nucleophile, and any substituents which enhance this nucleophilicity will also enhance its reactivity. The addition of a variety of electrophiles to carbon–carbon multiple bonds can occur, although the exact details of the mechanism can vary. An important complication in these reactions is that each end of the double bond may not be equivalent, and this means that isomeric products may be possible.

7.1.1 Addition of halogens

The reaction of chlorine, bromine and iodine with carbon–carbon double and triple bonds is commonly second order (first order in each of alkene and halogen). The reaction proceeds by initial polarisation of the halogen–halogen bond by the electron-rich double bond, followed (usually) by the formation of a three-membered halonium (chloronium, bromonium and iodonium) intermediate (Fig. 7.2). Halonium ions are not equally stable for all halogens: this intermediate is particularly important for bromine and iodine, but much less so for chlorine. This intermediate may be intercepted by the remaining halide anion or by the solvent if the reaction is conducted in a hydroxylic solvent (such as water, methanol or acetic acid). The nature of the substituents is known to profoundly influence the reaction rate; increasing numbers of electron-releasing substituents on the double bond lead to significant increases in reaction rate (Fig. 7.3). Conversely, electron-withdrawing groups (e.g. Br, CO$_2$H and CO$_2$Et) lower the rate of reaction.

Because of the structure of the halonium intermediate, the stereochemical outcome of the reaction is that the incoming nucleophile attacks at the opposite side, leading to *anti*-addition; that is, each of the halogens effectively reacts on opposite sides of the double bond (Fig. 7.2). This mode of reaction has some important consequences if the substrate is

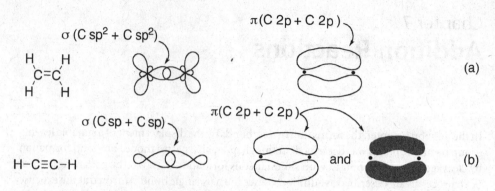

Figure 7.1 Bonding orbitals in alkenes and alkynes.

symmetrical; thus, the reaction of *cis*-2-butene gives a racemic mixture of the 2,3-dibromobutane product (Fig. 7.4a), but *trans*-2-butene gives the corresponding *meso* product (Fig. 7.4b). In the case of the cyclic substrate 4-*t*-butylcyclohexene, only two *trans* products are obtained, with none of the *cis* stereoisomer (Fig. 7.4c). However, this stereochemical control can be compromised in the case of aryl-substituted alkenes, since the stability of the intermediate halonium ion is lowered due to the increased stability of the corresponding carbocation in the open form of the intermediate, which is benzylic in character (Fig. 7.5); here, the nucleophile can approach from either side of the carbocation, leading to isomeric *syn*- and *anti*-products. The rate of reaction is greater relative to an unsubstituted alkene, as a result of the resonance stabilisation of the intermediate carbocation. Similarly, for the chloronium ion, the bridged species is much less important, and the stereoselectivity

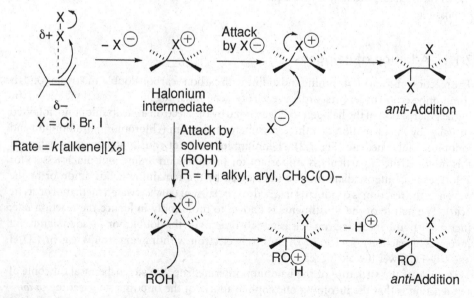

Figure 7.2 Electrophilic addition of halogens to alkenes.

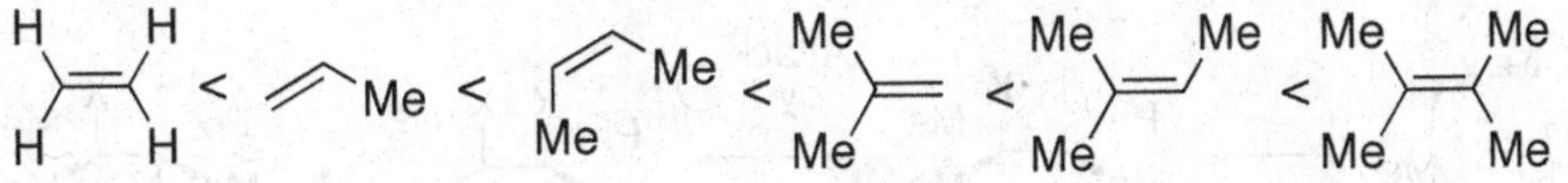

Figure 7.3 Rates of electrophilic addition of halogens to substituted alkenes (relative rate of reaction with Br$_2$).

of the reaction is usually lower than that in the case of the bromonium and iodonium species.

7.1.2 Addition of hydrogen halides

The reaction of hydrogen halides, and of water under strongly acidic conditions (e.g. H$_2$SO$_4$), with alkenes proceeds by an addition reaction, which is first order with respect to each reagent. Unlike the addition of halogens (see Section 7.1.1), initial protonation of the double bond leads to the formation of a carbocation intermediate, which is subsequently intercepted by the nucleophile in the reaction (Fig. 7.6a). Although both ends of the double bond may react with the proton, only that which leads to the formation of the more stable carbocation intermediate will in fact do so; the reaction is called *regioselective*, and the outcome of this type of addition can be predicted by the Markovnikov rule (which says that in the addition of HX to an alkene, the hydrogen adds to the less substituted end of the double bond, so as to generate the more stable carbocationic intermediate). Good examples of highly regioselective additions are those for which R = Me, Ph (Fig. 7.6a), since in these cases the carbocation is tertiary and benzylic respectively.

The rate of the addition reaction is known to depend on the nature of the substituents on the double bond, as might be expected; electron-releasing substituents (R = alkyl, Ph, MeO (Fig. 7.6a)) *increase* the rate of reaction, and electron-withdrawing substituents (R = Cl, CF$_3$) *decrease* the rate of reaction, consistent with the change in electron density

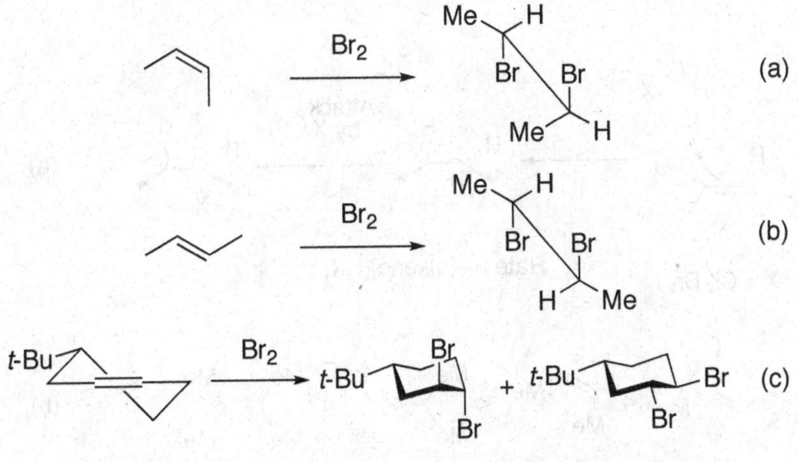

Figure 7.4 Stereochemical outcomes in electrophilic addition of halogens to alkenes.

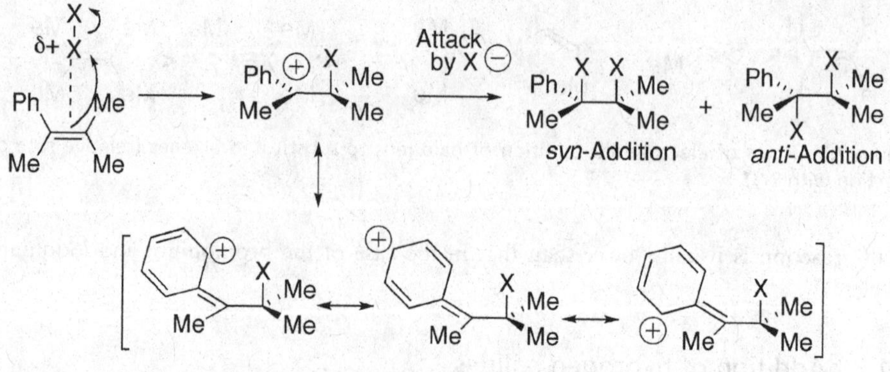

Figure 7.5 Electrophilic addition of halogens in arylalkenes.

of the carbon–carbon double bond that these substituents cause. Substantial increases in the rate of reaction of substituted double bonds is possible with increasingly substituted double-bond systems (Fig. 7.6b).

Because these addition reactions proceed with the formation of carbocationic interme- diates, rearrangements involving 1,2-carbon shifts (see Section 4.1.4) are possible; this can lead to the formation of multiple products (Fig. 7.7).

One important example of a reaction involving the addition of halogen across a double bond is the haloform reaction; in this case, the double bond results from the enolisation of the ketone and addition of halogen then occurs. Regeneration of the ketone by loss of HX, further enolisation and halogen addition, and then repetition of the entire sequence generates an α-trihaloketone. Because the reaction is run under basic conditions, addition– elimination of hydroxide then generates a carboxylic acid (Fig. 7.8).

The hydration of alkynes proceeds in an analogous manner; rapid tautomerisation of the first-formed enolic product generates a ketone, which is the observed product (Fig. 7.9).

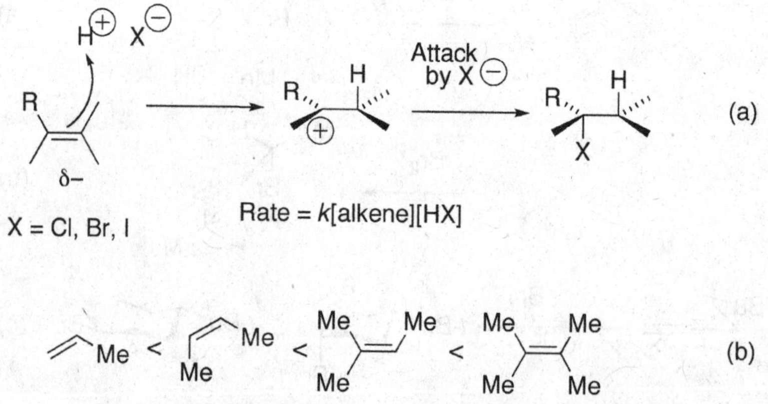

Figure 7.6 (a) Electrophilic addition of hydrogen halides to alkenes, and (b) effect of substitution on the relative rates of electrophilic addition of hydrogen halides to alkenes.

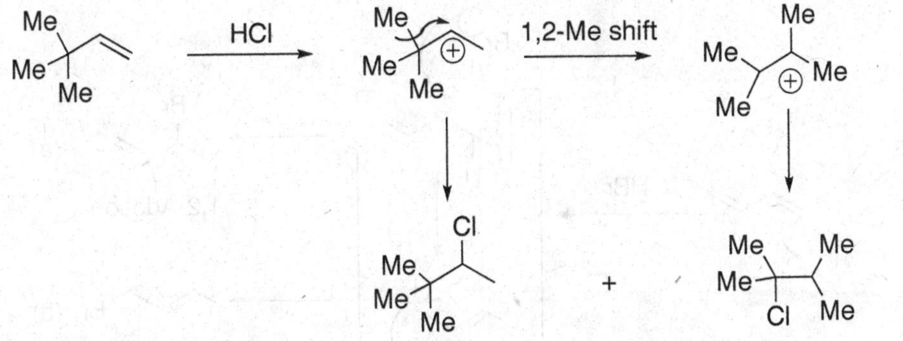

Figure 7.7 Rearrangements in the course of the electrophilic addition of hydrogen halides to alkenes.

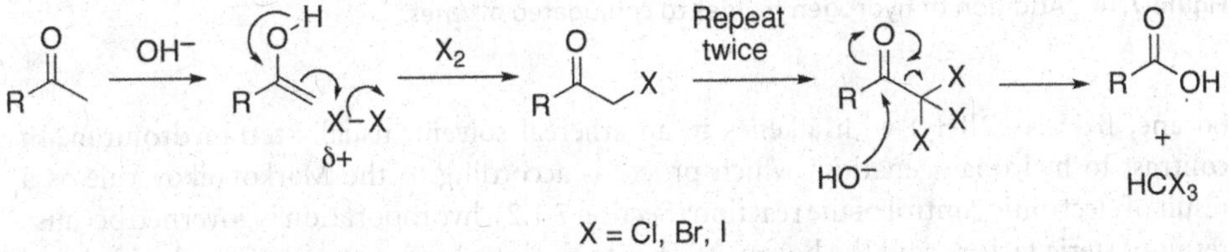

X = Cl, Br, I

Figure 7.8 Haloform reaction.

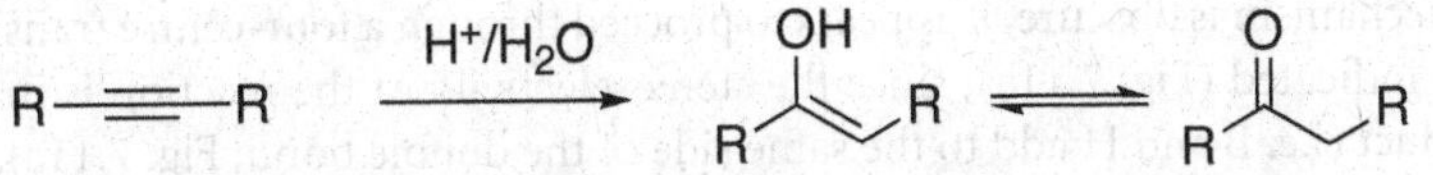

Figure 7.9 Electrophilic addition of water to alkynes.

7.1.3 Addition of hydrogen halides to conjugated dienes

In the case of the addition of hydrogen halides to conjugated dienes, the situation is more complicated since the first-formed carbocationic intermediate is resonance stabilised, and addition of the nucleophile in the second stage can lead to 1,2- or 1,4-addition (Fig. 7.10). Although the 1,2-mode of addition, attacking as it does the end of the diene system, is faster (i.e. kinetically preferred), it does not lead to the more stable outcome. The more stable product is in fact the one which results due to 1,4-addition, since the alkene product is the more substituted one. Provided that the reaction is conducted either for a short reaction time or at lower temperature, it is possible to favour the formation of the kinetic product, but if the reaction time is prolonged or the temperature increased, the thermodynamic outcome becomes preferred.

7.1.4 Addition of diborane (hydroboration)

The addition of the elements of boron and hydrogen to a carbon–carbon double bond leads to the corresponding alkylborane product (Fig. 7.11a); this reaction is very easy to conduct and is achieved by simply treating diborane (which exists in equilibrium with its monomer,

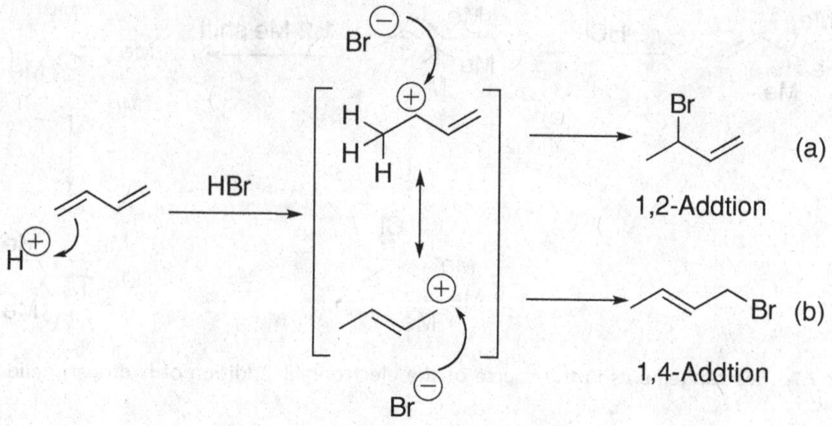

Figure 7.10 Addition of hydrogen halides to conjugated alkenes.

borane, $B_2H_6 \leftrightarrow 2BH_3$) with alkenes in an ethereal solvent, usually tetrahydrofuran. In contrast to hydrohalogenation, which proceeds according to the Markovnikov rule as a result of electronic control of the reaction (Section 7.1.2), hydroboration is governed because result of steric factors, and the boron attaches to the least hindered end of the double bond (which is sometimes called anti-Markovnikov addition); this regioselectivity becomes more marked when the R substituents on the boron are increasingly búlky (Fig. 7.11b). Although the exact mechanism is obscure, it appears to proceed through a four-centre transition state of the type indicated (Fig. 7.11a), since the stereoselectivity of the reaction is in favour of the *syn*-adduct (i.e. B and H add to the same side of the double bond; Fig. 7.11c).

The synthetic utility of these compounds arises not only from their ease of formation, but also from their participation in many useful reactions (Figs. 7.12a–7.12g). Since the

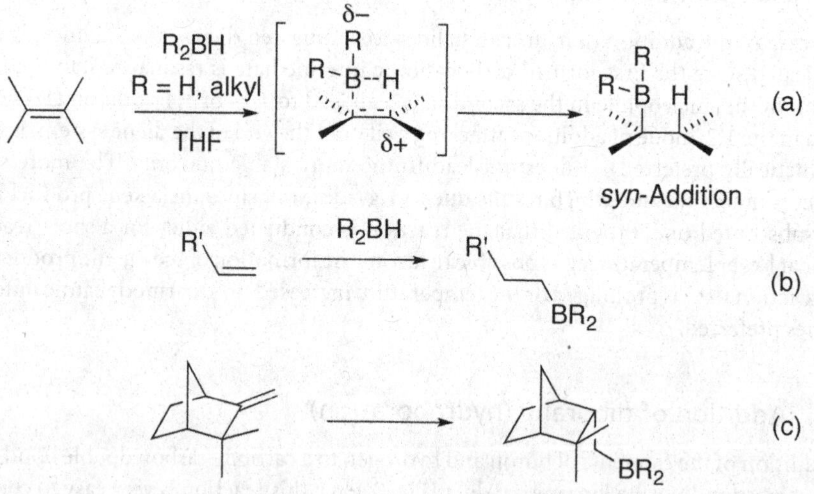

Figure 7.11 Hydroboration of alkenes: (a) mechanism, (b) regioselectivity and (c) stereoselectivity.

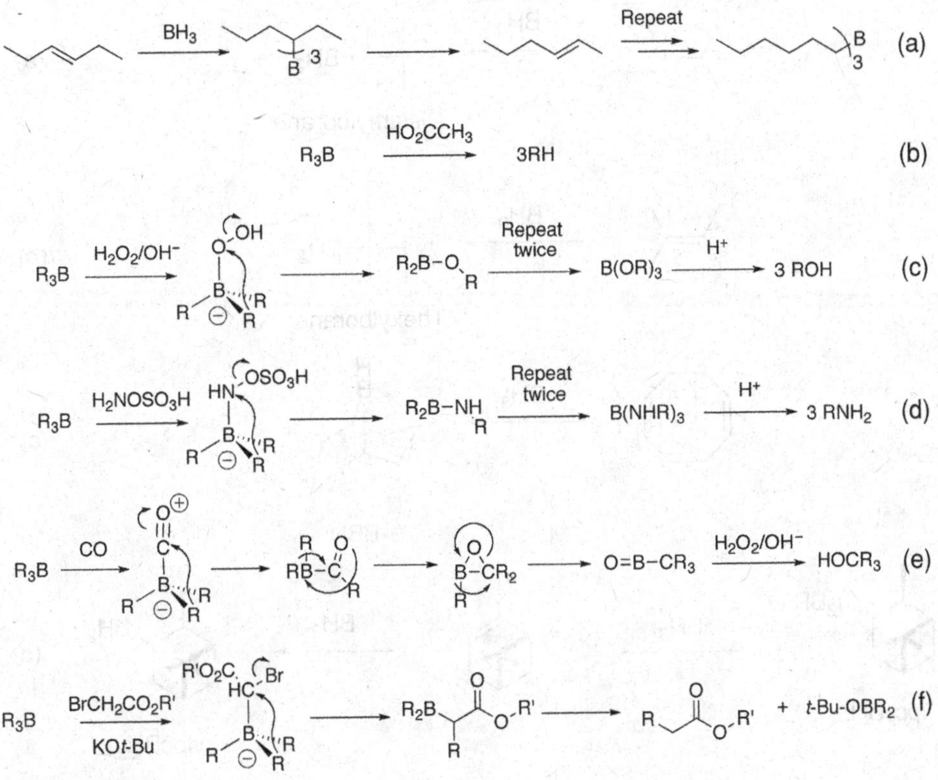

Figure 7.12 Reactions of alkylboranes.

hydroboration process is highly reversible, it is possible to achieve alkylborane isomerisation under these conditions (Fig. 7.12a); the initial reaction reverses to give the alternative alkene, and this then hydroborates again. The process continues until the boron resides at the end of the chain, in the least hindered position. In addition, cleavage of the C–B bond by protonolysis with mild acid (Fig. 7.12b), halogenolysis with Cl_2, Br_2 or I_2, oxidation with alkaline peroxide (Fig. 7.12c), amination with a suitable amine nucleophile (such as hydroxylamine O-sulfonic acid; Fig. 7.12d), carbonylation with carbon monoxide (Fig. 7.12e) and alkylation with α-bromoester enolates are possible (Fig. 7.12f). All of these reactions are characterised by the fact that the boron adduct is still electron deficient and that reaction therefore commences by nucleophilic addition to the boron to generate an eight-electron ate complex.

The intrinsic regio- and stereoselectivity of the hydroboration process has been extensively developed and proved to have substantial synthetic value. By using increasingly substituted alkenes in the hydroboration process, increasingly hindered alkylboranes can be prepared (Fig. 7.13), including disiamylborane (Fig. 7.13a), t-hexylborane (Fig. 7.13b), 9-borabicyclononane (9-BBN; Fig. 7.13c), as well as hindered enantiopure alkylboranes, such as mono-isopinocampheylborane (IpcBH$_2$; Fig. 7.13d) and diisopinocampheylborane (Ipc$_2$BH; Fig. 7.13d). Since all of these still contain a B–H bond, they are capable of reacting with another alkene, but now the existing substitution can be used to further control

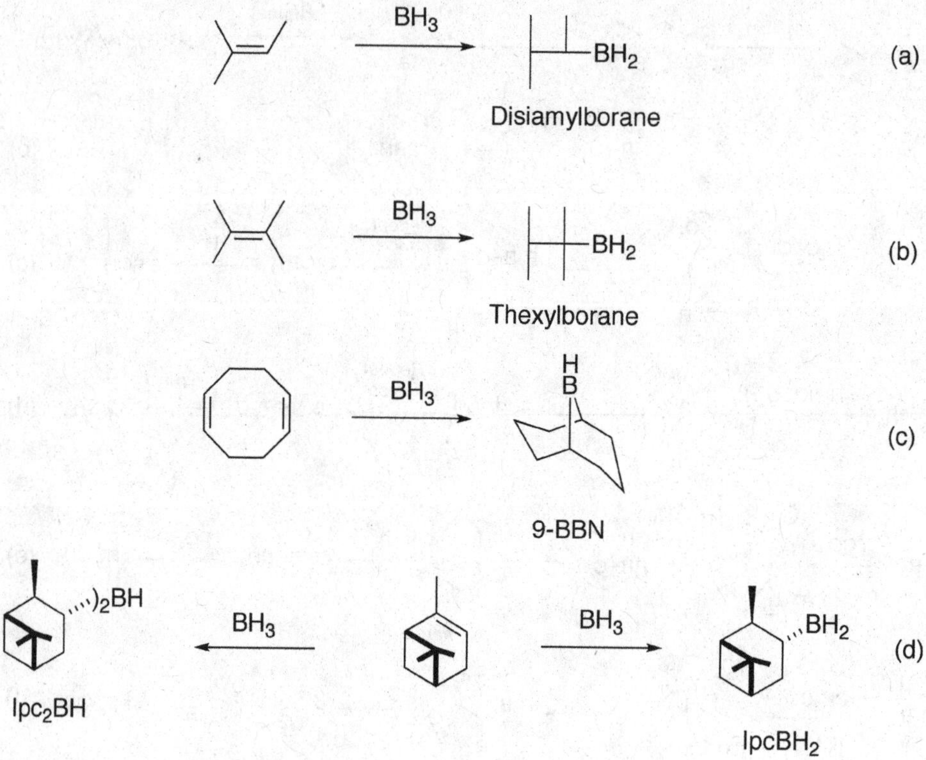

Figure 7.13 Hindered alkylboranes.

the hydroboration process. Furthermore, since the C–B cleavage reactions almost always occur with retention of stereochemistry, enantiocontrol of the overall outcome of the reaction procedure can be achieved. Thus, for example, hydroboration of *cis*-butene with Ipc$_2$BH and alkaline peroxide treatment directly leads to the alcohol product in high enantiopurity (Fig. 7.14).

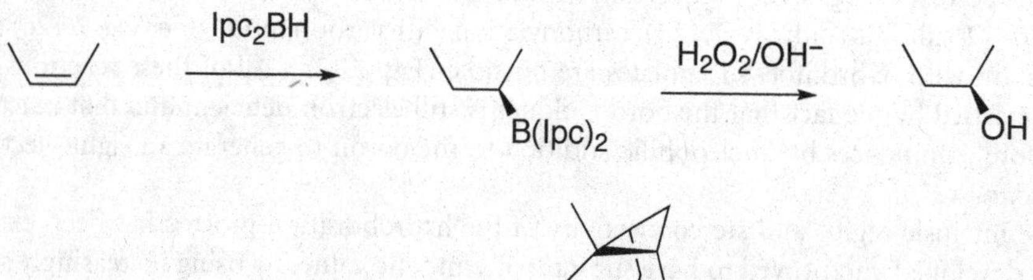

Figure 7.14 Reactions of hindered enantiopure alkylboranes.

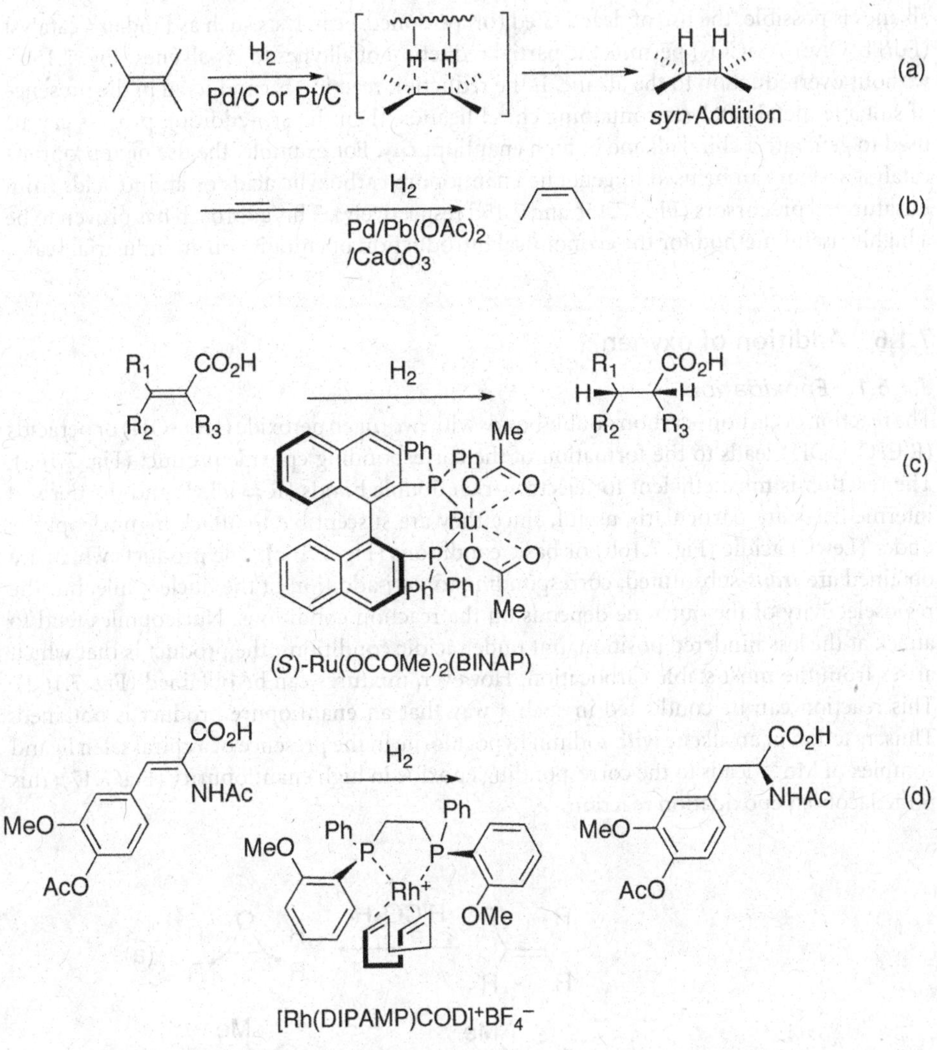

Figure 7.15 Reductions of alkenes and alkynes.

7.1.5 Addition of hydrogen

The addition of hydrogen to carbon–carbon multiple bonds leading to the formation of the reduced alkane products, and catalysed by a transition metal, such as Pt or Pd supported on charcoal, is a very important reaction (Fig. 7.15a). The metal catalyst is usually supported on charcoal or on an alternative solid support for convenience of handling, since only small quantities are required for reaction and the process is particularly attractive in industry, where the primary feedstock (hydrogen gas) is inexpensive. The reaction almost always leads to the *syn*-addition of hydrogen, and on this basis a mechanism involving a four-centre transition state has been proposed in which the hydrogen is initially adsorbed onto the catalyst surface. Although reduction of alkenes and alkynes to the corresponding alkanes is possible, the use of deactivated (or 'poisoned') catalysts such as Lindlar's catalyst ($Pd/Pb(OAc)_2/CaCO_3$) permits the partial reduction of alkynes to *cis*-alkenes (Fig. 7.15b), without overreduction to the alkane. If the reduction reaction is conducted in the presence of suitable metal catalysts containing chiral ligands, then the *syn*-addition process can be used to generate a chiral alkane in high enantiopurity. For example, the use of appropriate catalyst system can be used to generate enantiopure carboxylic acids or amino acids from unsaturated precursors (Figs. 7.15c and 7.15d respectively). This approach has proven to be a highly useful method for the economical introduction of chirality on an industrial scale.

7.1.6   Addition of oxygen

7.1.6.1   Epoxidation

The reaction of carbon–carbon double bonds with hydrogen peroxide (HO–OH) or peracids ($R'C(O)OOH$) leads to the formation of the corresponding epoxide product (Fig. 7.16a). The reaction is most efficient for electron-rich double bonds (R = alkyl) and the derived intermediates are particularly useful, since they are susceptible to attack by nucleophiles under (Lewis) acidic (Fig. 7.16b) or basic conditions (Fig. 7.16c). The products which are obtained are *trans*-substituted, corresponding to *anti*-addition of the nucleophile, but the regioselectivity of the outcome depends on the reaction conditions. Nucleophiles tend to attack at the less hindered position, but under acidic conditions, the product is that which arises from the most stable carbocation. However, mixtures can be obtained (Fig. 7.16d). This reaction can be conducted in such a way that an enantiopure product is obtained. Thus, reaction of an alkene with sodium hypochlorite in the presence of a chiral salen ligand complex of Mn^{3+} leads to the corresponding epoxide in high enantiopurity (Fig. 7.17); this is the Jacobsen epoxidation reaction.

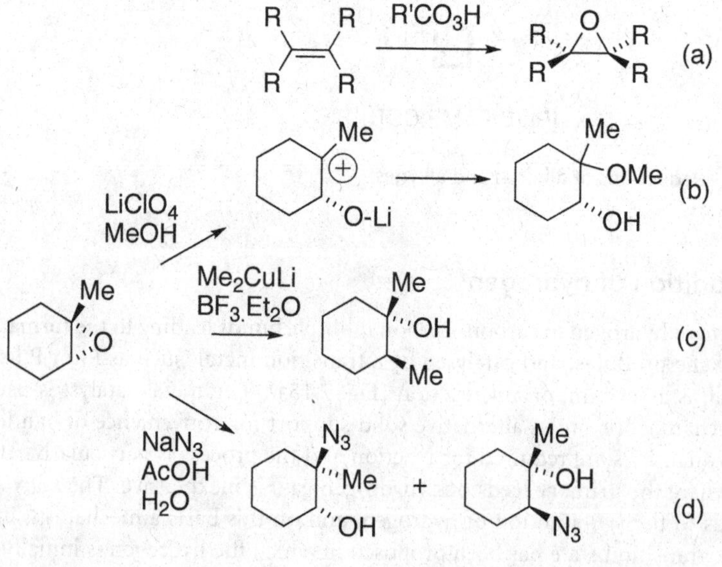

Figure 7.16   Epoxidations of alkenes.

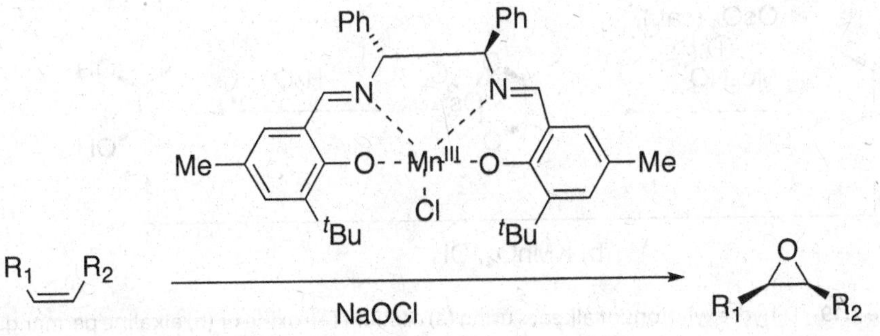

Figure 7.17 Epoxidation mediated by the Jacobsen salen ligand.

Another important variation of this reaction is the Sharpless asymmetric epoxidation of allylic alcohols (Fig. 7.18), which proceeds by treatment of an allylic alcohol with *t*-butyl hydroperoxide in the presence of titanium isopropoxide and either enantiomer of a tartarate ester; this provides convenient access to either enantiomer of the product epoxide as desired by simple choice of reagent. The reaction leads to high yields and enantiopurity of the epoxide and, just as importantly, is totally predictable using the mnemonic indicated in Fig. 7.18; there are no known exceptions to this rule.

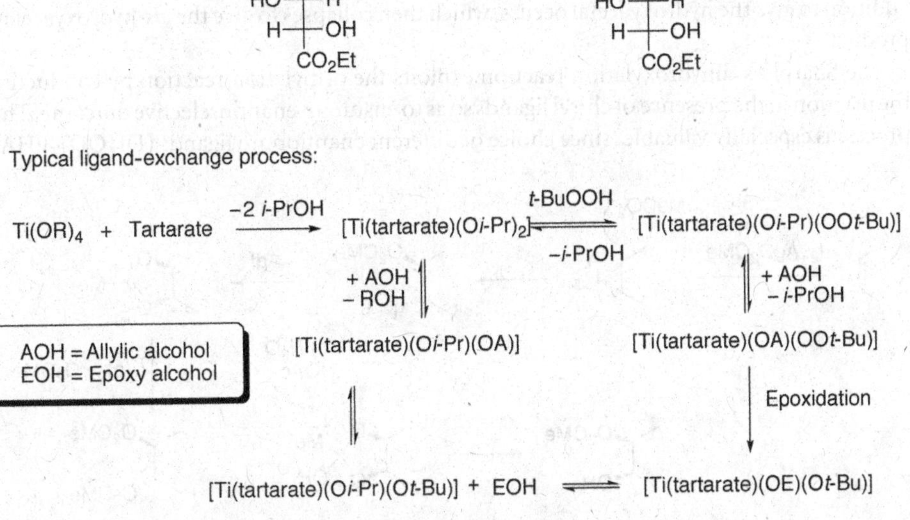

Figure 7.18 Sharpless asymmetric epoxidations of allylic alcohols.

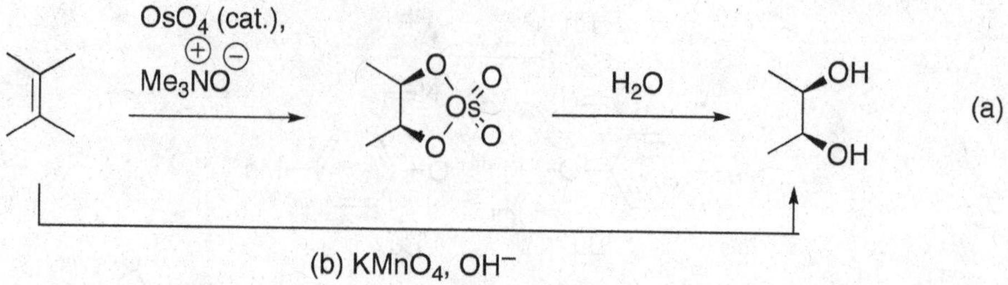

Figure 7.19 Dihydroxylations of alkenes using (a) osmium tetroxide or (b) alkaline permanganate.

7.1.6.2 Dihydroxylation

Alkenes when treated with osmium tetroxide give *syn*-1,2-diols; the use of *N*-amine oxides allows catalytic quantities of osmium tetroxide to be used, which is important for reasons of expense and safety (Fig. 7.19a), since under these conditions the metal is reoxidised during the course of the reaction. The mechanistic course of the reaction, which proceeds via a cyclic osmate ester, ensures that the product which is formed has a *cis*-diol relationship. Alkaline permanganate also gives *syn*-1,2-diols in a mechanism related to osmylation, but generally requires more careful control of conditions and as a result this reaction has not been developed in the same way as that of osmium tetroxide; however, it does have the advantage that the use of toxic osmium is avoided. Prévost and Woodward hydroxylations provide alternative means for accessing *trans*- or *cis*-diols from alkene starting materials; the stereochemical outcome is a consequence of the choice of reagents and therefore the mechanistic course of the reaction (Fig. 7.20). Initial addition of iodine and attack by acetate gives the *trans* adduct. In the presence of the silver cation, this intermediate is not stable and collapses to the indicated oxonium cation; in the case of Prévost hydroxylation (Fig. 7.20a), which is conducted under anhydrous conditions, this is opened by acetate in an *anti*-sense to give the *trans*-diacetate product. However, under the Woodward conditions, in which the reaction is conducted in the presence of aqueous acetic acid (Fig. 7.20b), addition to give the hydroxyacetal occurs, which then collapses to give the *cis*-hydroxyacetate product.

The Sharpless dihydroxylation reaction exploits the osmylation reaction by conducting the reaction in the presence of chiral ligands so as to ensure an enantioselective outcome. This process is especially valuable, since choice of different enantiopure ligands (DHQD)$_2$PHAL

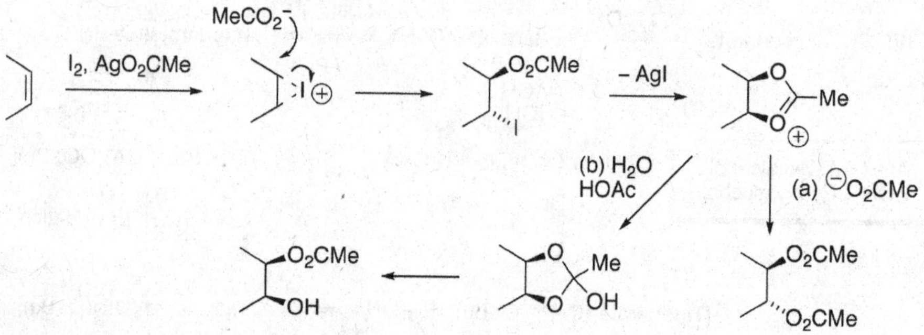

Figure 7.20 (a) Prévost and (b) Woodward hydroxylations.

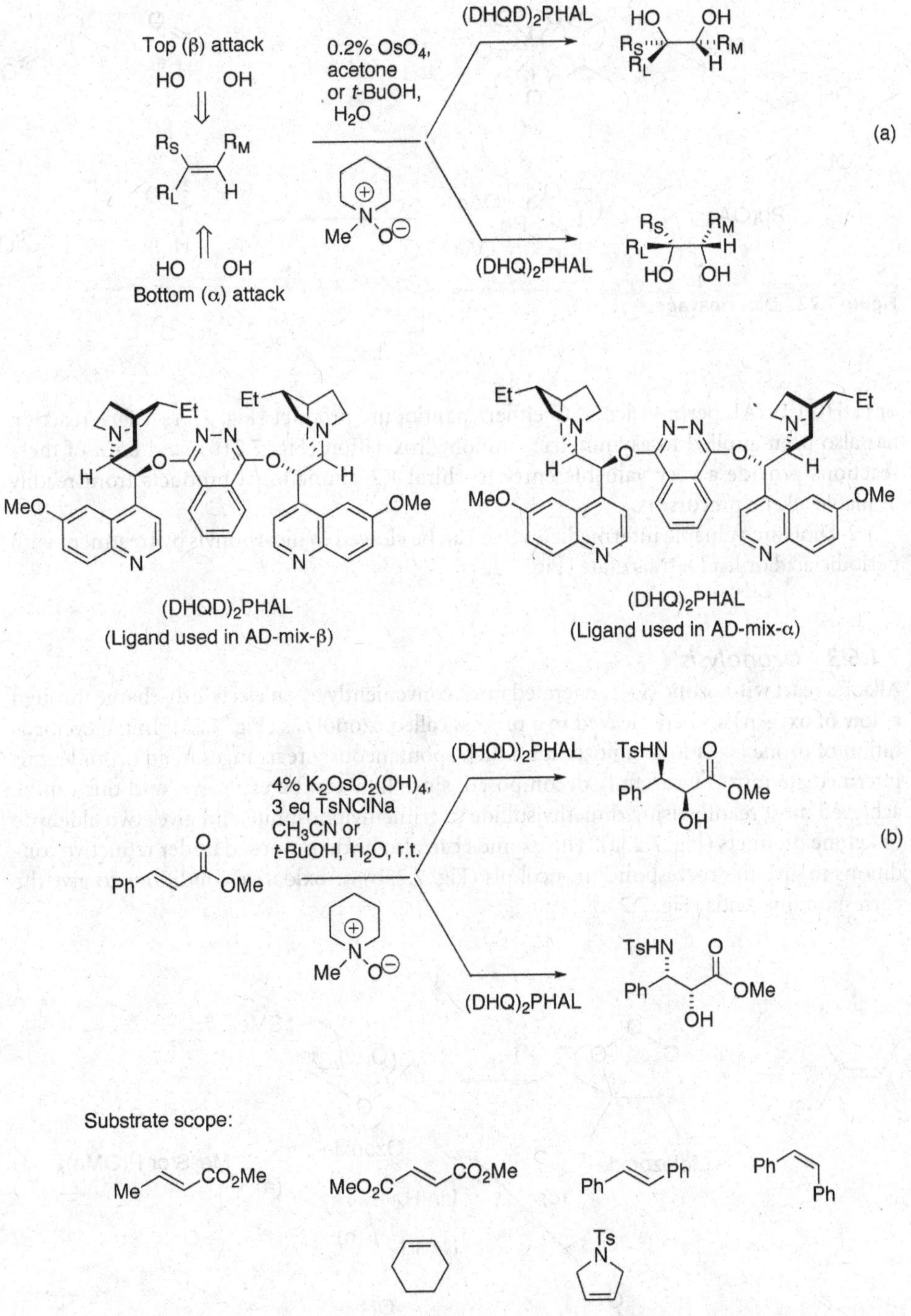

Figure 7.21 The Sharpless dihydroxylation (a) and (b) aminohydroxylation reaction.

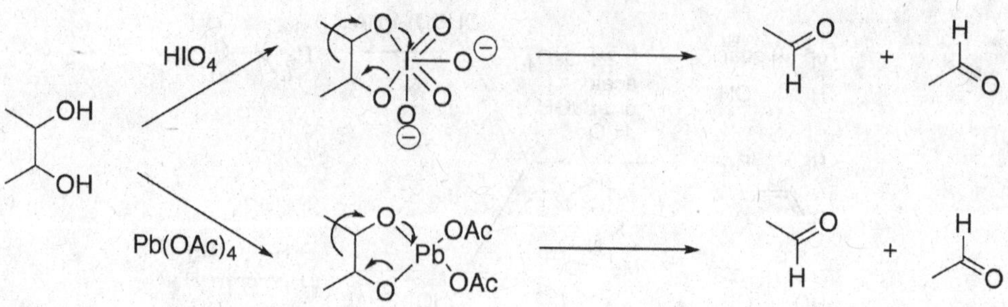

Figure 7.22 Diol cleavage.

or (DHQ)$_2$PHAL permits access to either enantiopure product (Fig. 7.21a). This reaction has also been applied to asymmetric aminohydroxylation (Fig. 7.21b), and both of these reactions provide a very valuable entry to chiral 1,2-difunctional products from readily available alkene precursors.

1,2-Diols are valuable intermediates that can be cleaved to dicarbonyls by treatment with periodic acid or lead tetraacetate (Fig. 7.22).

7.1.6.3 Ozonolysis

Alkenes react with ozone (O$_3$, generated most conveniently by an electric discharge through a flow of oxygen) and are cleaved in a process called ozonolysis (Fig. 7.23). Initial cycloaddition of ozone leads to a molozonide which spontaneously rearranges to an ozonide; this intermediate must be carefully decomposed, since they can be explosive, and this can be achieved most readily using dimethylsulfide or trimethylphosphite and gives two aldehyde or ketone products (Fig. 7.23a). The ozonide can also be decomposed under reductive conditions to give the corresponding alcohols (Fig. 7.23b) or oxidative conditions to give the corresponding acids (Fig. 7.23c).

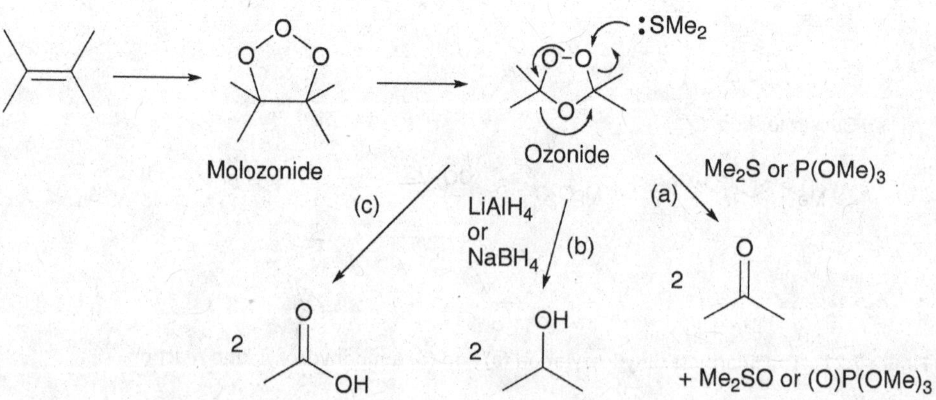

Figure 7.23 Ozonolysis of alkenes, and three types of work-up.

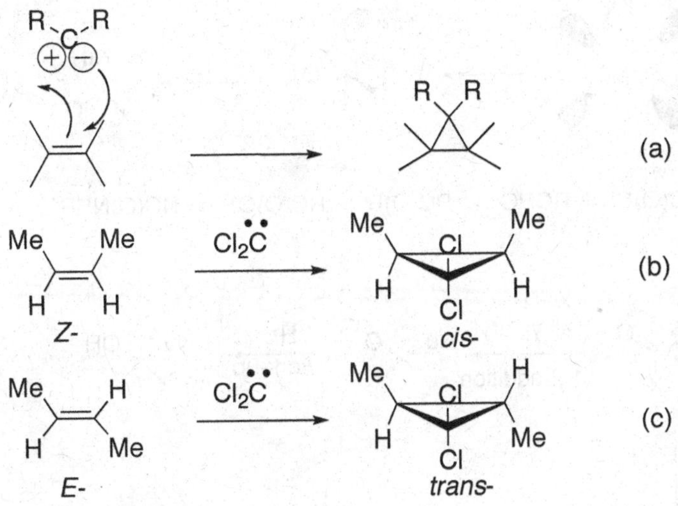

Figure 7.24 Additions of carbenes to alkenes.

7.1.7 Addition of carbon

The addition of carbon across a double bond can be achieved by the addition of a carbene to give a cyclopropane product (Fig. 7.24a); the reaction is a stereospecific *syn*-addition, and the stereochemistry of the product directly reflects that of the starting material. Thus, addition to a *Z*-alkene yields the *cis*-cyclopropane (Fig. 7.24b) and to an *E*-alkene, the *trans*-cyclopropane (Fig. 7.24c).

7.2 Nucleophilic addition reactions

The reverse of the process considered so far in this chapter – the addition of nucleophilic reagents to electrophilic substrates – is particularly common, since electrophilicity is readily conferred on compounds as a result of the presence of electron-withdrawing groups, in particular oxygen-containing functional groups. Thus, nucleophilic addition reactions to all types of carbonyl-containing compounds readily occur. Such additions result from interaction of the highest occupied molecular orbital (HOMO) of the nucleophile with the lowest unoccupied molecular orbital (LUMO) (the π^*) of the carbonyl group (Fig. 7.25a), which as a result of the geometry of the antibonding orbitals forces nucleophiles to approach along the Burgi–Dunitz angle (about 107°). When X is a hydrogen (i.e. the carbonyl compound is an aldehyde), the carbonyl is most reactive, since hydrogen does not sterically shield the carbon; if X is an alkyl or aryl group, the reactivity of the carbonyl group is reduced, since inductive and mesomeric effects, respectively, reduce the magnitude of the electrophilic carbon. The approximate order of reactivity is therefore typically as shown in Fig. 7.25b. The addition of a nucleophile to a carbonyl compound is given by the general reaction of Fig. 7.25c. This carbonyl compound may be an aldehyde, ketone, ester, acid chloride or amide, and the nucleophile may be hydride (H^-), cyanide (CN^-), a carbanion (Grignard or organolithium reagent) or a heteroatom-containing functional group

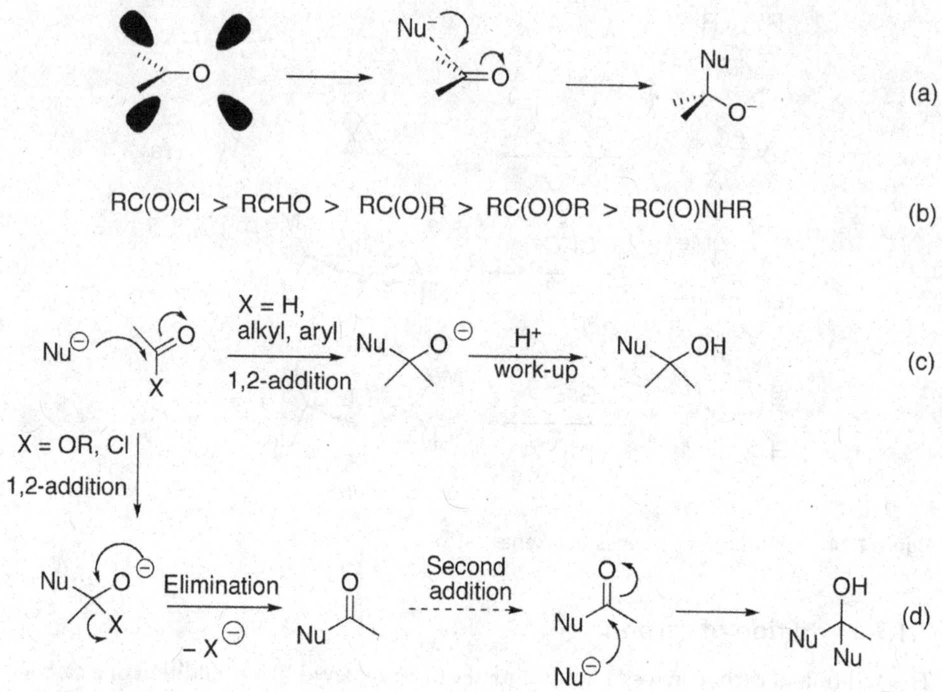

Figure 7.25 Nucleophilic additions to carbonyl groups.

(O, N or S). Aldehydes and ketones give products from direct (or 1,2-) addition (Fig. 7.25c), while esters and acid chlorides give products which result from initial direct addition immediately followed by elimination, since in these cases X is a good leaving group (X = OR, Cl respectively; Fig. 7.23d); this process is called addition–elimination, and this will be covered in more detail in Chapter 10. Depending on the nucleophile, further direct addition might occur to give products containing two added nucleophilic residues, as indicated; this can be important for aldehydes and ketones, and results in overall addition–substitution. Some examples of the processes which occur are shown in Fig. 7.26: simple addition of nucleophiles is shown in Figs. 7.26a and 7.26b, addition–elimination in Figs. 7.26c and 7.26d and addition–substitution in Fig. 7.26e.

7.2.1 Irreversible nucleophilic addition

Irreversible nucleophilic addition to aldehydes, ketones, esters and amides of a variety of nucleophiles, including hydride, organometallics and enolates, is possible. Hydride is typically delivered by metal hydrides (e.g. $LiAlH_4$, $NaBH_4$, $LiBH_4$ and variously substituted alkoxymetal hydrides) to give an alcohol product, in a reaction which is accelerated by coordination of the lithium or sodium cation to the oxygen of the carbonyl group (Fig. 7.27a). The order of reactivity of carbonyl groups follows the usual pattern so that chemoselective reduction of different functional groups in the same molecule can be achieved; thus, aldehydes are more readily reduced than ketones, than esters and than amides. This order of reactivity can be further changed by altering the nature of the ligands on the metal; bulky or

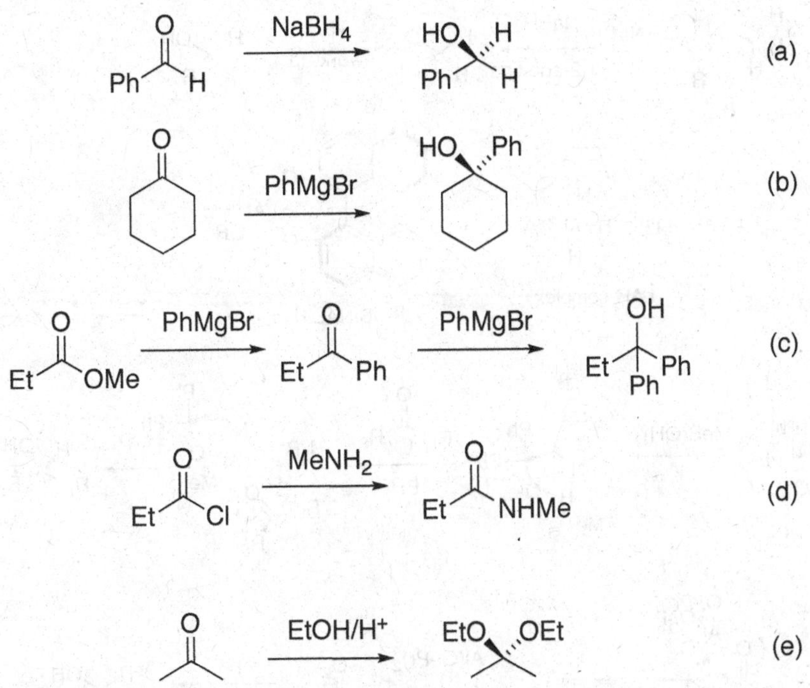

Figure 7.26 Some products arising from nucleophilic addition reactions to a carbonyl group.

electron-withdrawing ligands slow down hydride transfer. For example, LiAlH(OR)$_3$-type reagents are much less reactive than lithium aluminium hydride, on account of the electron-withdrawing effect of the oxygen, but lithium triethylborohydride is more reactive than sodium borohydride, on account of the inductively releasing effect of the alkyl groups, and sodium cyanoborohydride is known for its selective reduction of imines. Because these reagents are a source of hydride, they must be conducted in rigorously dried solvents, such as ether or tetrahydrofuran. Appropriate modification of the ligands of these reagents can be used to generate reducing agents which are sources of chiral hydride and which can be used for enantioselective reductions, for example, lithium aluminium hydride modified with a chiral diamine or 1,1'-bi-2-naphthol (BINOL) ligand sets (Fig. 7.27b). The reduction of ketones using BH$_3$ is also possible, and furthermore by using a chiral boron catalyst, the stereochemical outcome can be reliably achieved and is rationalised by the transition state, which is shown in Fig. 7.27c. The coordination of oxophilic Lewis acids, in particular cerium trichloride, enables the selective reduction of ketones over aldehydes and of 1,2-reduction over 1,4-reduction in conjugates systems; this is called Luche reduction. Reductions of carbonyl compounds can be achieved with reagents other than metal hydrides; sources of organic hydride will do just as well, and the Meerwein–Pondorf–Verley reaction is an important example (Fig. 7.27d) for which the reaction can be easily forced in the forward or reverse directions by the use of the Le Chatelier principle; removal of the acetone by-product can be used to drive the reaction in the forward direction. Aluminium triisopropoxide in triisopropanol provides the source of hydride-reducing agent. This reaction is, however, not

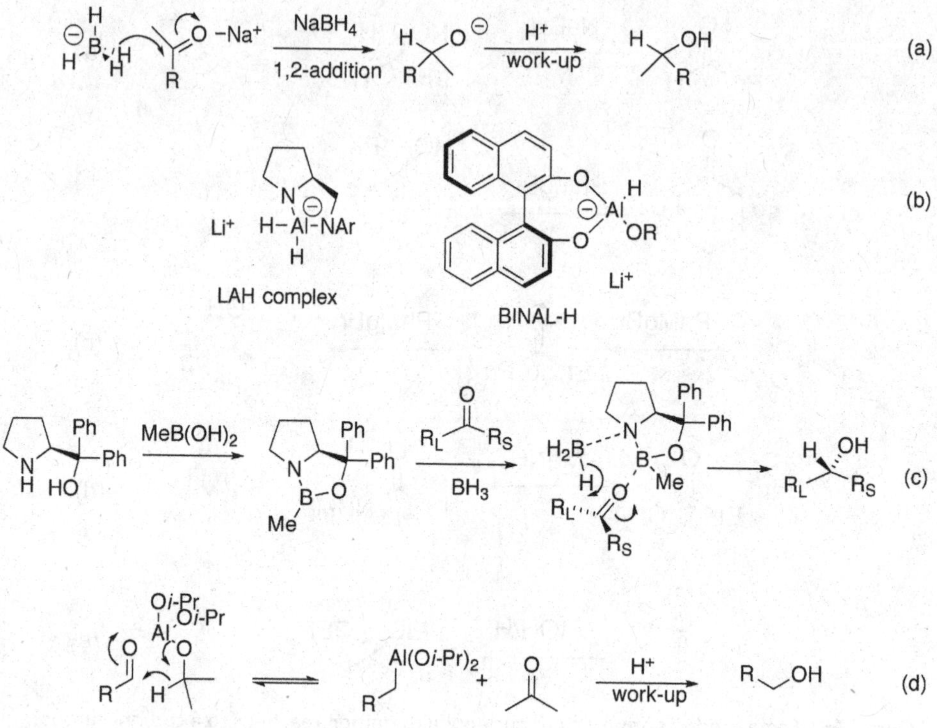

Figure 7.27 Reductions in aldehydes and ketones.

used so much these days, since more effective and simpler processes are available, such as the metal hydrides.

Organolithium and organomagnesium (Grignard) reagents are so reactive as nucleophiles with carbonyl compounds that the reaction is fast and irreversible, to give the corresponding alcohol products (Fig. 7.28a). However, side reactions are possible: Because of the basicity of these reagents, any carbonyl groups with α-hydrogens can be readily enolised, thereby inhibiting further addition, and reduction of the carbonyl group is possible if the Grignard reagent has a β-hydrogen (Fig. 7.28b). Organometallic additions to carbonyl groups can also be achieved in the presence of chiral ligands, and this has been most successful with Grignard or dialkylzinc reagents (Figs. 7.28c and 7.28d). The reaction of Grignard reagents is significantly enhanced in the presence of binaphthyl-type ligands, giving alcohols in high enantiopurity. Mechanistically, the reaction with zinc (Fig. 7.28d) is of interest, since in addition to controlling the stereochemical outcome of the reaction, the ligands are responsible for enforcing on the normally linear diorganozinc structure, which of course has no net dipole and is therefore unreactive – a bent arrangement, in which there is an overall molecular dipole and therefore nucleophilic reactivity is induced.

These reactions when conducted on a molecule which possesses a chiral centre may be diastereoselective; that is, they lead to a product in which one diastereomer is preferred over another. In the case of organometallic additions to a carbonyl group, the outcome can be rationalised using Cram's rule, or the Felkin–Anh model (Fig. 7.29). Cram's rule, established on the basis of empirical evidence, defines the direction of facial attack of a

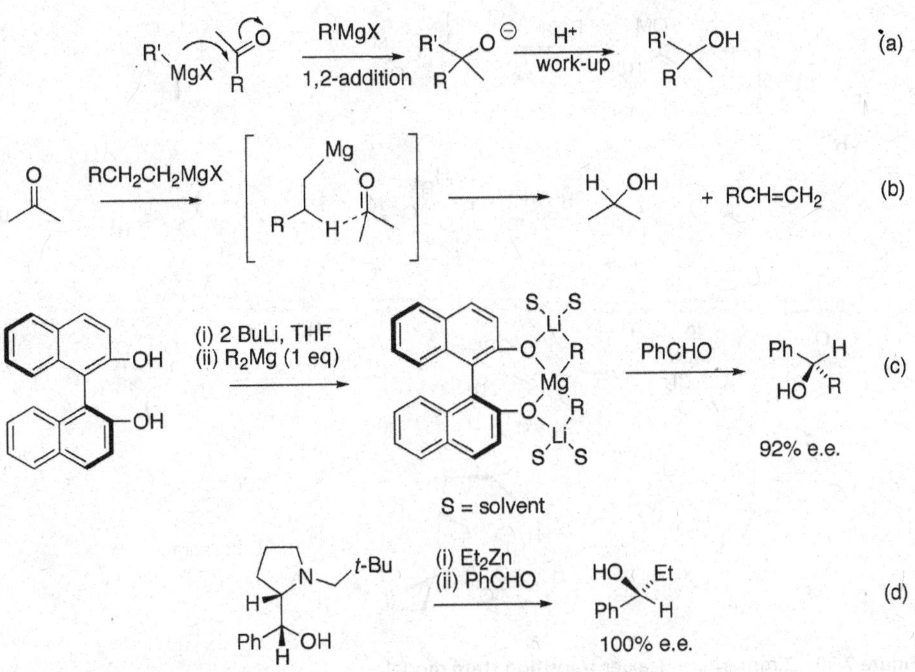

Figure 7.28 Organometallic additions in aldehydes and ketones.

nucleophile to a carbonyl from the less hindered side of a conformation which places the oxygen and largest substituent in an *anti*-relationship. This rule applies for carbonyl groups with an α-chiral centre and small (S), medium (M) and large (L) substituents, and states that (i) the substrate can be considered to exhibit a preferred reacting conformation in which the carbonyl oxygen is located between the small and medium groups for reasons of

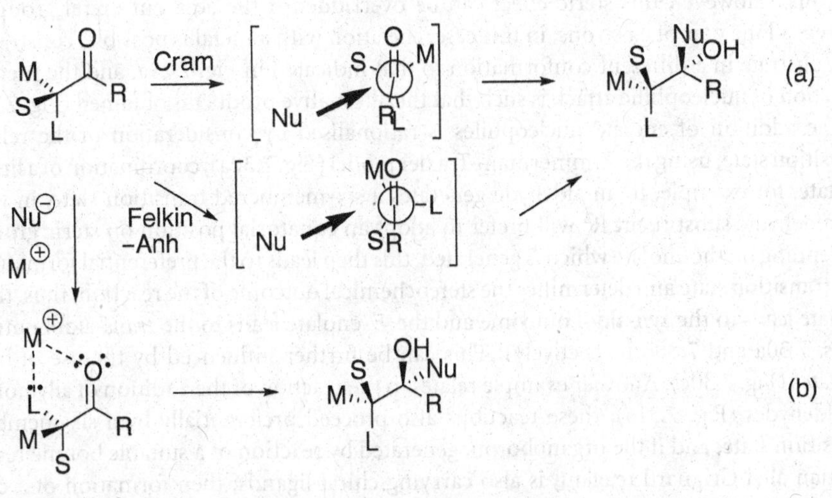

Figure 7.29 Additions of nucleophiles to chiral carbonyl groups.

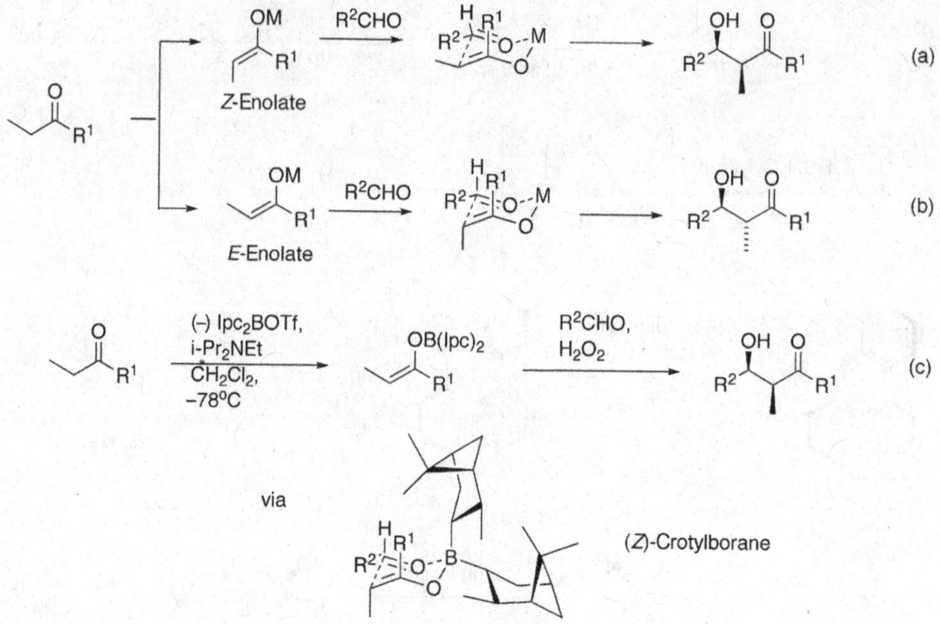

Figure 7.30 Zimmerman–Traxler transition state model.

minimising steric congestion; (ii) in this conformation, a nucleophile will attack from the side of the smallest substituent, again for reasons of steric interactions; (iii) the preferred stereochemical outcome is as indicated. The Felkin–Anh model assumes that the preferred conformation is the one in which the σ^*-orbital of the C–L bond is orthogonal to the C=O π-bond, due to favourable orbital overlap, and in this circumstance, the favoured trajectory of nucleophilic attack is between the small and medium groups to give the same observed outcome. However, this steric effect can be overridden if the adjacent chiral group also possess a lone pair of electrons; in this case, chelation with a metal is possible, constraining the substrate in a different conformation to that indicated in Fig. 7.29a, and the preferred direction of nucleophilic attack is such that the alternative product is obtained (Fig. 7.29b).

The addition of enolate nucleophiles is rationalised by consideration of the relevant transition state, using the Zimmerman–Traxler model (Fig. 7.30a); coordination of a lithium enolate, for example, to an aldehyde generates a six-membered transition state, in which the aldehydic substituent R^2 will prefer to adopt an equatorial position on steric grounds. Depending on the enolate which is generated, this then leads to the preferential formation of one transition state and determines the stereochemical outcome of the reaction; thus, the Z-enolate leads to the *syn*-aldol outcome and the E-enolate leads to the *trans*-aldol outcome (Figs. 7.30a and 7.30b respectively). This can be further influenced by the use of boron enolates (Fig. 7.30c). Another example relates to the reaction of the addition of allylboranes to aldehydes (Fig. 7.31a); these reactions also proceed preferentially by a six-membered transition state, and if the organoboron, generated by reaction of a suitable borane reagent with an allyl Grignard reagent, is also carrying chiral ligands, then formation of a chiral homoallylic alcohol with high enantiopurity occurs. Note that the stereochemistry of the

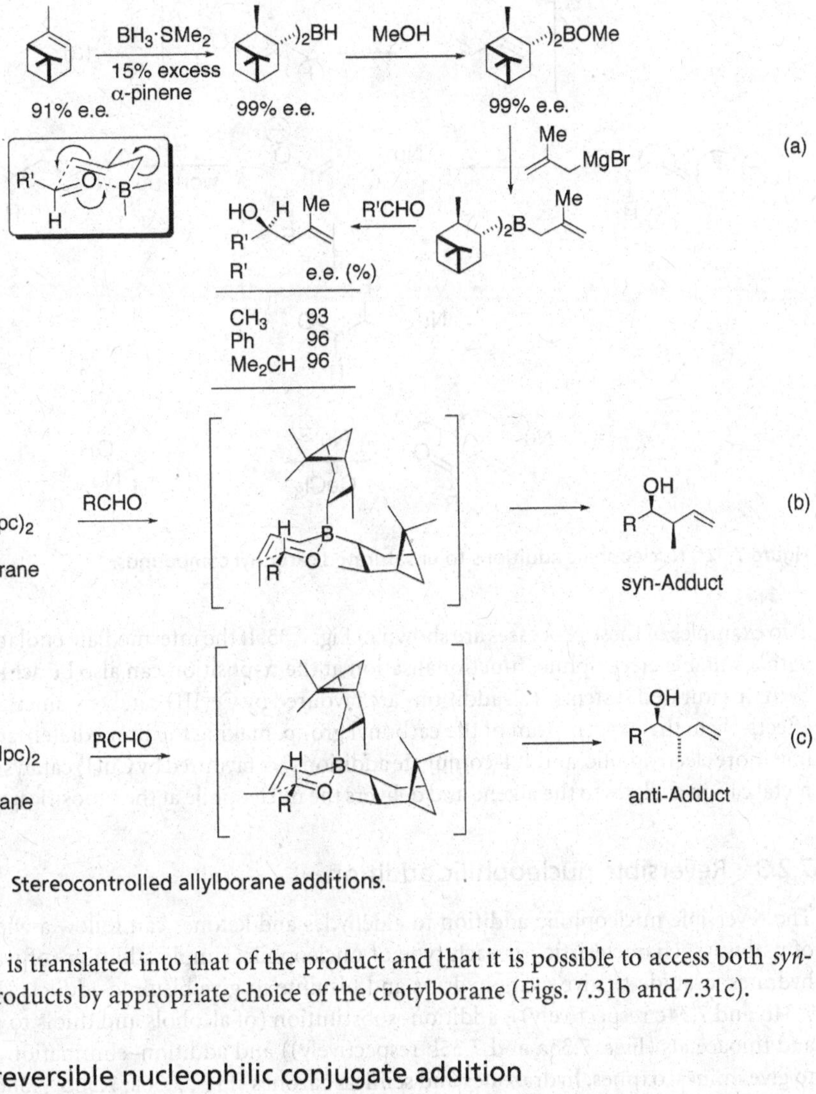

Figure 7.31 on part (a):

$$\text{91\% e.e.} \xrightarrow[\substack{15\% \text{ excess} \\ \alpha\text{-pinene}}]{\text{BH}_3 \cdot \text{SMe}_2} \text{(...)}_2\text{BH, 99\% e.e.} \xrightarrow{\text{MeOH}} \text{(...)}_2\text{BOMe, 99\% e.e.}$$

Me /\ MgBr

$$\text{HO} \ \text{H} \ \text{Me} \xleftarrow{\text{R'CHO}} \text{(...)}_2\text{B}$$

R'

R'	e.e. (%)
CH$_3$	93
Ph	96
Me$_2$CH	96

(a)

(Z)-Crotylborane, B(Ipc)$_2$ $\xrightarrow{\text{RCHO}}$ → syn-Adduct (b)

(E)-Crotylborane, B(Ipc)$_2$ $\xrightarrow{\text{RCHO}}$ → anti-Adduct (c)

Figure 7.31 Stereocontrolled allylborane additions.

allylborane is translated into that of the product and that it is possible to access both *syn*- and *anti*-products by appropriate choice of the crotylborane (Figs. 7.31b and 7.31c).

7.2.2 Irreversible nucleophilic conjugate addition

Nucleophilic additions can also occur if a carbonyl group is conjugated with one or more double bonds; in this case, the polarisation of the carbonyl group, which results from the presence of the highly electronegative oxygen atom, is transmitted along the carbon skeleton, so that the terminal position of the double bond is also electrophilic, as can been seen from a consideration of the relevant resonance structures (Fig. 7.32a). Since this is an exact reversal of the normal polarity of an unsubstituted alkene, such α, β-unsaturated alkenes will react with nucleophiles (Nu$^-$; Fig. 7.32b) to generate an intermediate enolate, which after work-up with acid generates the product in a reaction which formally adds the components of Nu and H across the alkene double bond and leads to a β-substituted carbonyl compound. This reaction, called the Michael reaction, or more generally, conjugate addition, is applicable to variety of nucleophiles, including alcohols, amines, thiols and cyanide, as well as carbanions, including enolates and organometallic derivatives such as Grignard reagents (Fig. 7.32b).

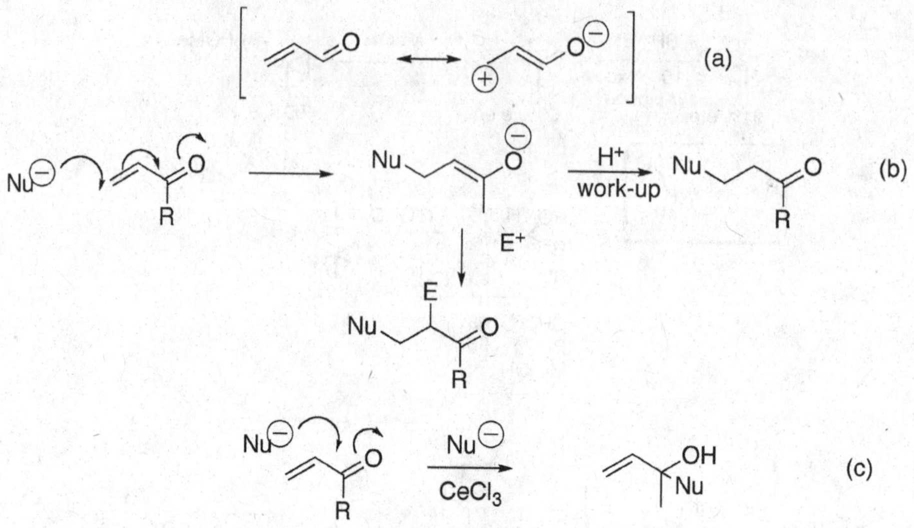

Figure 7.32 Nucleophilic additions to unsaturated carbonyl compounds.

Two examples of these processes are shown in Fig. 7.33. If the intermediate enolate is trapped with a suitable electrophile, functionalisation at the α-position can also be achieved.

In unsaturated systems, 1,2-additions are favoured by Ce(III) catalysis, since this chelates effectively to the oxygen atom of the carbonyl group, making the immediately adjacent carbon more electrophilic, and 1,4-conjugate additions are favoured by Cu(I) catalysis, since this metal cation chelates to the alkene and delivers the nucleophile at the 4-position (Fig. 7.32c).

7.2.3 Reversible nucleophilic addition

The reversible nucleophilic addition to aldehydes and ketones can follow a wide diversity of pathways, characteristic for each type of nucleophile: simple addition (of water to give hydrates, cyanide to give cyanohydrins and bisulfite to give bisulfite adducts (Figs. 7.34a, 7.34b and 7.34c respectively)), addition–substitution (of alcohols and thiols to give acetals and thioacetals (Figs. 7.35a and 7.35b respectively)) and addition–elimination (of amines to give imines, oximes, hydrazones and semicarbazones (Figs. 7.36a, 7.36b, 7.36c and 7.36d

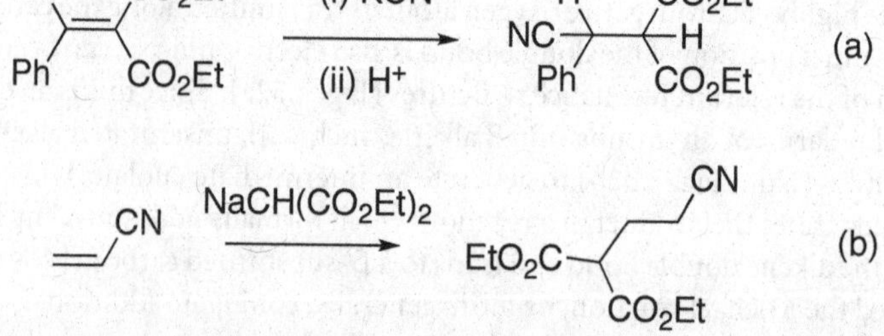

Figure 7.33 Examples of nucleophilic additions to unsaturated carbonyl compounds.

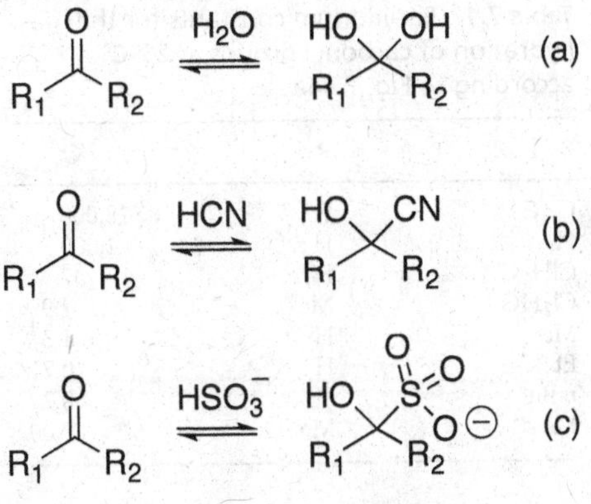

Figure 7.34 Addition of nucleophiles to carbonyl compounds.

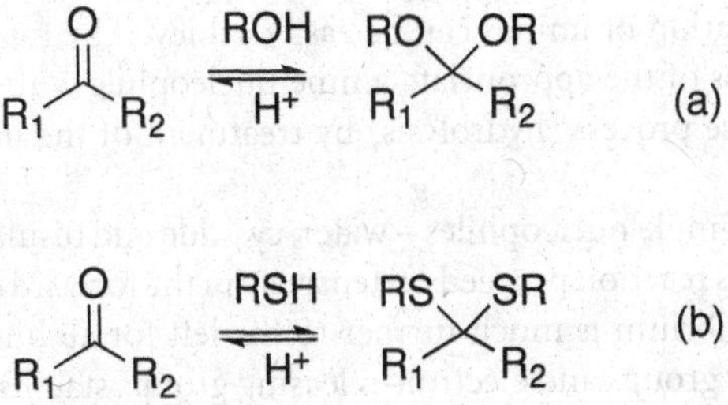

Figure 7.35 Addition–substitution of carbonyl compounds.

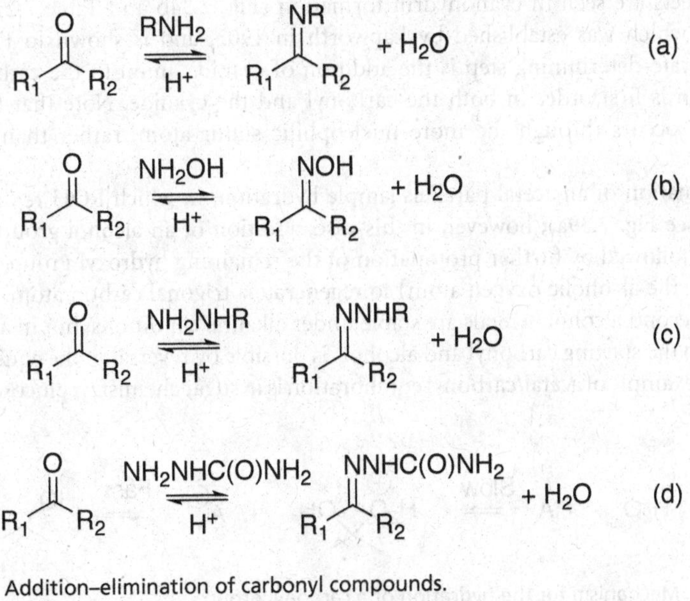

Figure 7.36 Addition–elimination of carbonyl compounds.

Table 7.1 Equilibrium constants for the hydration of carbonyl groups at 25°C according to Fig. 7.34a

R_1	R_2	K_{eq}
Cl_3C	H	28,000
H	H	2,000
ClH_2C	H	37
Cl_2HC	Me	2.9
Me	H	1.3
Et	H	0.71
t-Bu	H	0.24
Me	Me	0.002

respectively)). In these cases, the free reversibility of these processes means that the forward or reverse processes can be accessed by appropriate application of the Le Chatelier principle; for example, formation of imines can be easily achieved by treatment of the carbonyl substrate with an excess of the appropriate amine nucleophile with simultaneous removal of water and the reverse process (hydrolysis) by treatment of the imine with excess water (Fig. 7.36).

The addition of the simple nucleophiles – water, cyanide and bisulfite – leads to direct addition to a carbonyl; this reaction proceeds extensively in the forward direction for aldehydes ($K > 1$), but the equilibrium is much further to the left for all but the most reactive for ketones ($K < 1$). Bulky groups and electron-releasing groups stabilise the starting carbonyl group either sterically or electronically, respectively, but the carbonyl group is activated by the presence of electron-withdrawing groups or location of the carbonyl group in a strained ring (Fig. 7.34a and Table 7.1); a mechanism for this process is given in Fig. 7.37. Similar effects are seen in cyanohydrin formation (Fig. 7.34b and Table 7.2), the mechanism for which was established by Lapworth in 1903, and is shown in Fig. 7.38, for which the rate-determining step is the addition of cyanide anion to the carbonyl group; the reaction is first order in both the carbonyl and the cyanide. Note that the addition of bisulfite occurs through the more nucleophilic sulfur atom, rather than the oxygen (Fig. 7.34c).

The formation of an acetal parallels simple hydration, in which ROH replaces H_2O in Fig. 7.37 (see Fig. 7.39a); however, in this case, addition of an alcohol group under acid catalysis is followed by further protonation of the remaining hydroxyl group, elimination (assisted by the alcoholic oxygen atom) to regenerate a trigonal carbon atom, followed by entry of a second alcohol. Acetals are stable under alkaline conditions, but in acid solution reversion to the starting carbonyl and alcohols is possible by reversal of the equilibrium. An important example of acetal/carbonyl equilibration is in sugar chemistry; glucose in aqueous

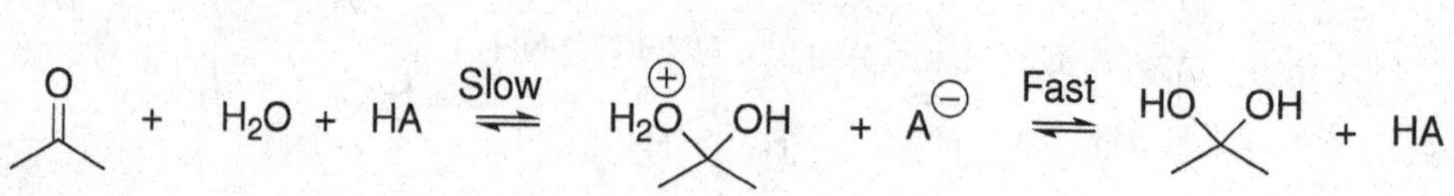

Figure 7.37 Mechanism for the hydration of a carbonyl group.

Table 7.2 Equilibrium constants for the hydrocyanation of carbonyl groups at 25°C according to Fig. 7.34b

R_1	R_2	K_{eq}
$NO_2C_6H_4$	H	1820
Ph	H	220
$MeOC_6H_4$	H	33
Me	i-Pr	65
Me	Et	38
Me	Me	33
C_6H_{12}		7.8
C_5H_{10}		1000
C_4H_8		48

solution exists largely in the pyranose form, but as a mixture of α- and β-anomers, with only a tiny proportion of open-chain form present (0.003%) (Fig. 7.39b). In a suitable glucose substrate with a good leaving group at the anomeric position, it is possible to generate a glycosyl cation and this may be intercepted by another donor species, in such a way that a disaccharide is formed (Fig. 7.39c).

Thiols very effectively add to carbonyl groups, and the equilibrium constant for this process is usually substantially higher than that for the equivalent alcohol addition (Fig. 7.40); the process differs from acetal formation in that the initial addition of thiol can be base catalysed, because thiols are more acidic than alcohols, but this condensation is also often catalysed by Lewis acids like zinc chloride. The initial equilibrium which generates thiolate then leads to a slow addition to the carbonyl component.

The addition of amines once again follows a very similar process (Fig. 7.36); initial addition generates a carbinolamine, which although isolable in rare cases, normally rapidly eliminates water to give the corresponding imine, which are also called Schiff bases. Note that elimination in this case, unlike acetals, is possible because of the trivalency of nitrogen. Depending on the nature of the nitrogen nucleophile, various adducts are possible: imines are derived from amines, oximes from hydroxylamine, hydrazones from hydrazines, semicarbazones from semicarbazide and 2,4-dinitro-phenylhydrazones from 2,4-dinitro-phenylhydrazine. The rate of addition of the amine is very sensitive to pH, being

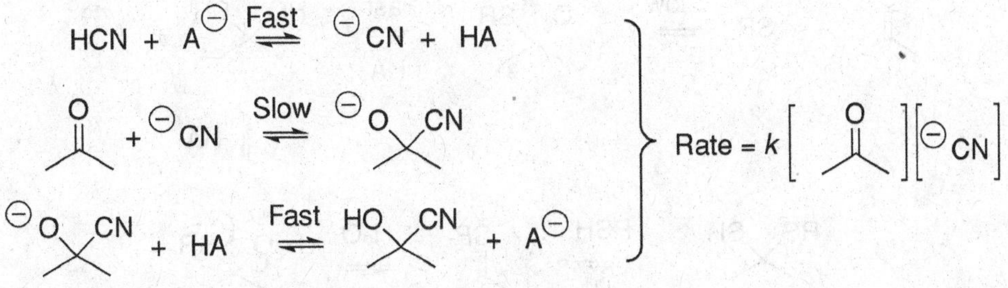

Figure 7.38 Cyanohydrin formation from carbonyl compounds.

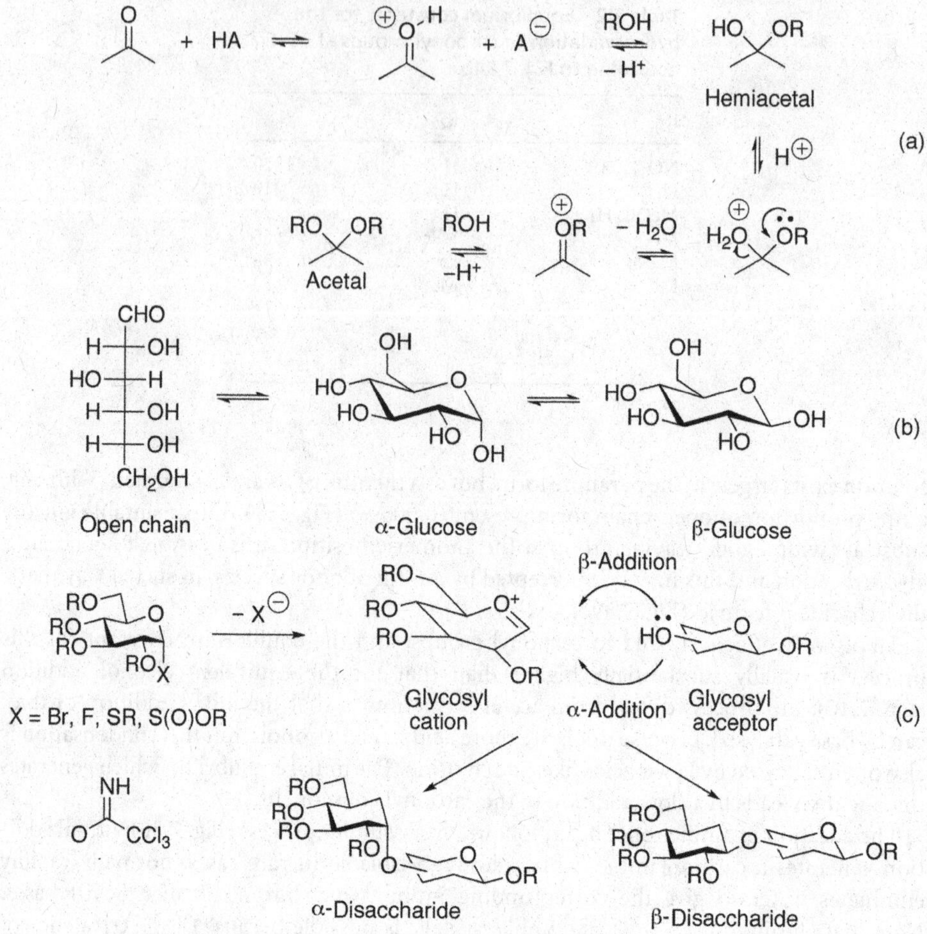

Figure 7.39 Reaction mechanism for hemiacetal and acetal formation.

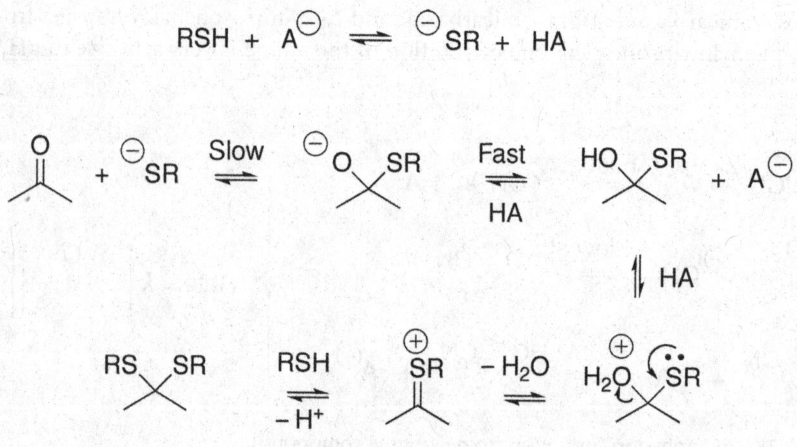

$$RSH + A^{\ominus} \rightleftharpoons {}^{\ominus}SR + HA$$

Figure 7.40 Mechanism of the formation of thioacetals.

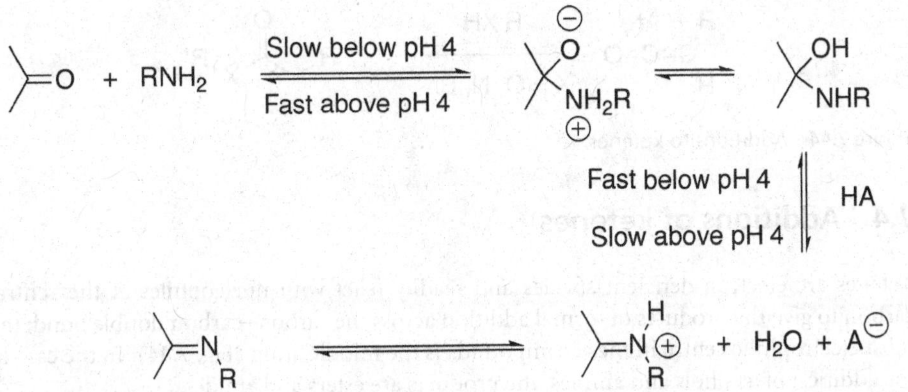

Figure 7.41 Reaction mechanism for imine formation.

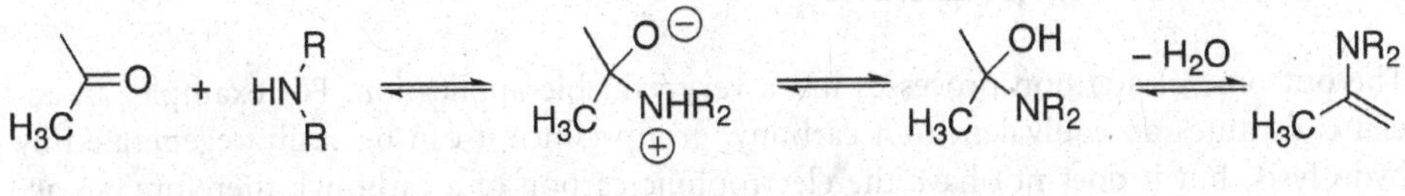

Figure 7.42 Reaction mechanism for enamine formation with secondary amines.

maximised at about pH 5, for which the medium is acidic enough to facilitate the necessary proton transfers in the reaction, but not so acidic that the reacting amine itself becomes protonated and therefore non-nucleophilic. The reaction proceeds according to the mechanism shown in Fig. 7.41. However, if this reaction is done with a secondary amine, the final deprotonation is not possible, and in this case elimination of water can occur only in such a way as to generate an enamine product (Fig. 7.42); enamines are neutral equivalents to enolates and will react with reactive electrophiles typically as nucleophiles in substitution reactions.

7.3 Additions to electron-deficient alkenes

Although the characteristic reactions of alkenes are as nucleophiles (i.e. they react with electrophiles), under some circumstances they can react in an opposite sense, that is, as electrophiles. This most commonly occurs when a carbon–carbon double bond is substituted with inductively electron-withdrawing groups (e.g. fluorine) (Fig. 7.43) or conjugated with an electron-withdrawing group, such as carbonyl, cyano or nitro groups (see Figs. 7.33a and 7.33b).

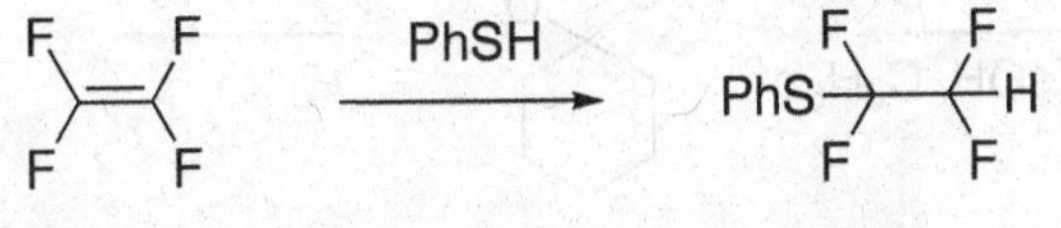

Figure 7.43 Examples of nucleophilic additions to electrophilic alkenes.

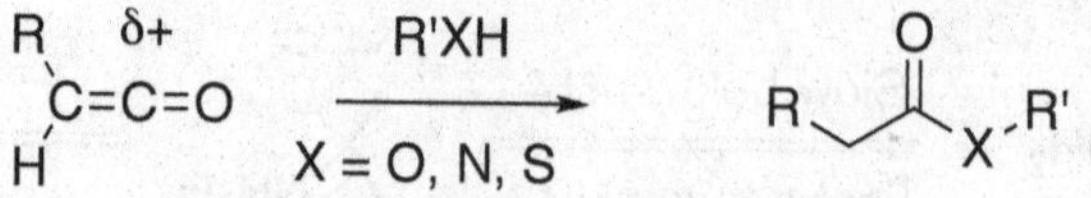

Figure 7.44 Additions to ketenes.

7.4 Additions of ketenes

Ketenes are electron-deficient species and readily react with nucleophiles at the central carbon to give the products of formal addition across the carbon–carbon double bond; the most electrophilic centre in these compounds is the middle atom (Fig. 7.44). In the case of the addition of alcohols and amines, the products are esters and amides.

7.5 Synthetic applications

The outcome of addition processes has a very valuable application. For example, an acetal constitutes an equivalent to a carbonyl group (since it can be easily regenerated by hydrolysis) but it does not have the electrophilic carbon of a carbonyl; therefore, an acetal can be considered to be a protecting group for a carbonyl, since it can be added and then later removed in a synthetic sequence if that carbonyl group needs to be 'hidden' as a result of its potential reactivity (Fig. 7.45a). A wide variety of acetal protecting groups have been devised, which differ in the conditions that can be applied for their removal; the most usual conditions are the use of dilute aqueous acid. However, alternative conditions for deprotection can be devised and, for example, it is possible to prepare an acetal from a system that can be cleaved by the use of a reductive elimination (Fig. 7.45b). This strategy is also applicable to the protection of alcohols for which unsymmetrical acetals can be used (Fig. 7.46). Thus, tetrahydropyran (THP; Fig. 7.46a), methoxymethyl (MOM; Fig. 7.46b), methoxyethoxymethyl (MEM; Fig. 7.46c) and benzyloxymethyl (BOM; Fig. 7.46d) are all widely used and can be readily introduced and selectively removed under

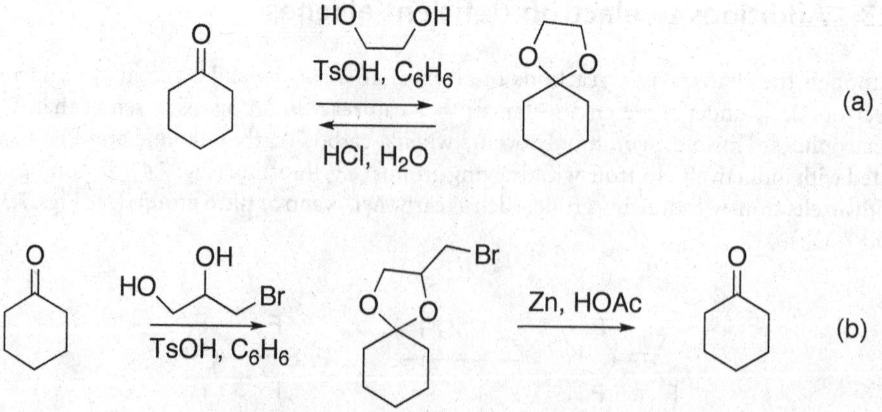

Figure 7.45 Protection of carbonyls as acetals.

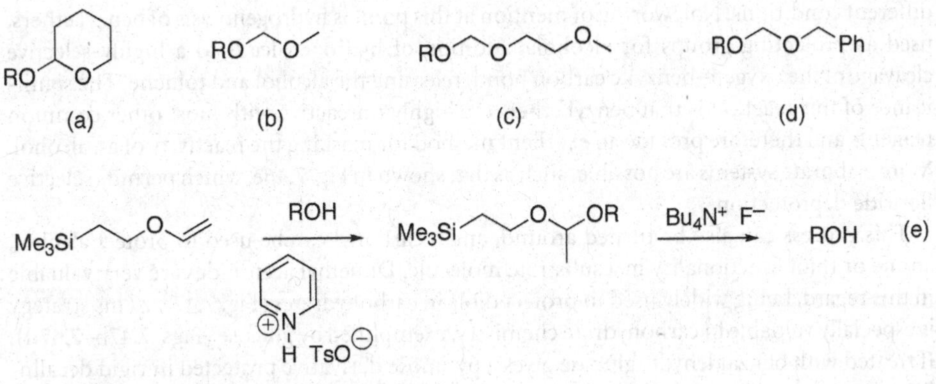

Figure 7.46 Protection of alcohols as acetals.

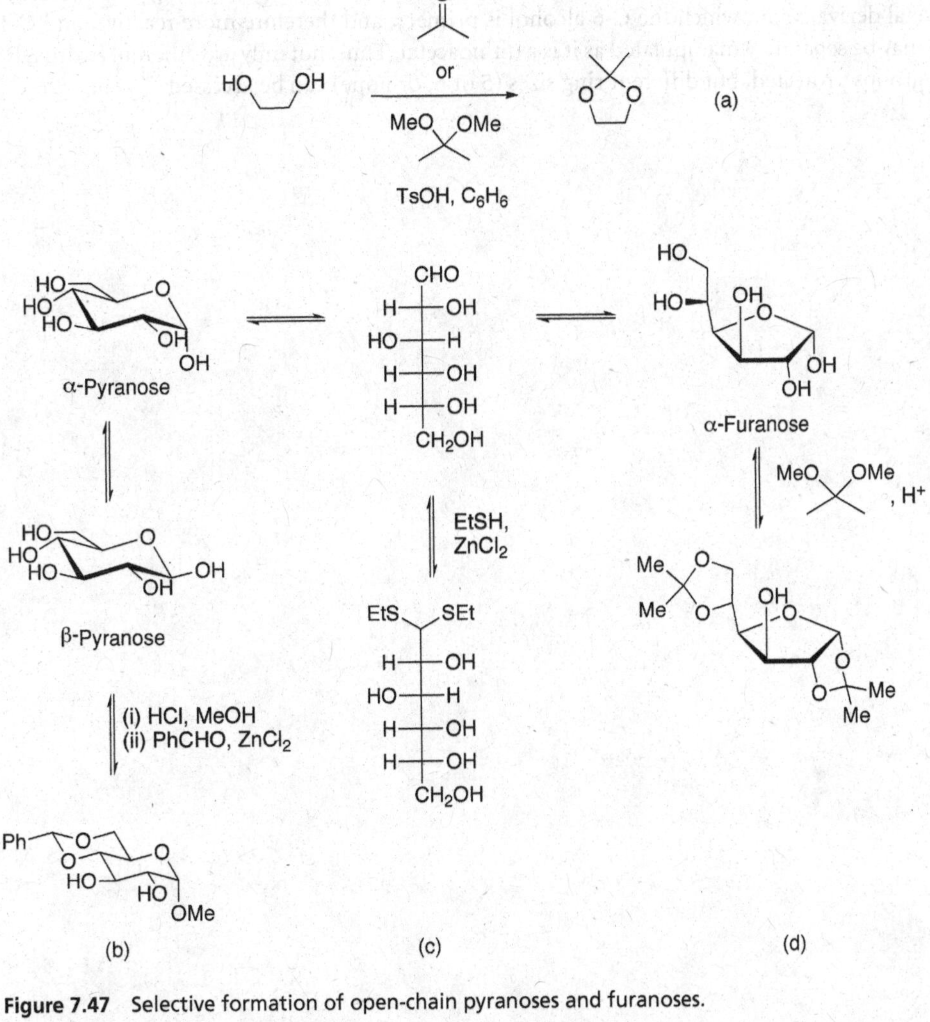

Figure 7.47 Selective formation of open-chain pyranoses and furanoses.

different conditions. Noteworthy of mention at this point is hydrogenolysis of benzyl ethers, used as protecting groups for alcohols; addition of hydrogen leads to a highly selective cleavage of the oxygen–benzylic carbon bond, releasing the alcohol and toluene. The significance of this reaction is that benzyl ethers are highly unreactive with most other common reagents and therefore provide an excellent method for masking the reactivity of an alcohol. More elaborate systems are possible, such as that shown in Fig. 7.46e, which permits selective fluoride deprotection.

This process can also be turned around, and a carbonyl can be used to protect alcohol, amine or thiol functionality in a substrate molecule. Dimethyl acetonides are very valuable in this regard, being widely used to protect diols in carbohydrates (Fig. 7.47a). This strategy is especially valuable in carbohydrate chemistry, exemplified by glucose (Figs. 7.47b–7.47d). If treated with benzaldehyde, glucose gives a pyranose derivative protected in rigid decalin-like structure, leaving the C-2 and C-3 hydroxyls free for further reaction. If treated with dimethoxypropane, it however gives a tricyclic furanose derivative, in which only the C-3 hydroxyl is free for further reaction. Finally, if treated with ethanedithiol, it gives a dithioacetal derivative, in which the C-6 alcohol is primary, and therefore more reactive, and C-1 may be separately manipulated as it is a dithioacetal. Thus, not only are different hydroxylic groups protected, but different ring sizes (5 or 6, or none) can be accessed.

Chapter 8
Elimination Reactions

In the previous chapters, we have seen how the formation of new σ-bonds can be achieved in various ways. In this chapter, we will examine in details means by which new π-bonds are formed.

8.1 Elimination

In their simplest form, eliminations are very common and involve the removal of the elements of HX (X = leaving group) from a substrate to generate a double or triple bond (Figs. 8.1a and 8.1b). There are three principal classes, two proceeding through intermediates (E_1 and E_1CB) and one through a transition state (E_2), although the products are the same in each case; what differs is the mechanistic pathway leading to those products. These are discussed in detail below.

8.1.1 E_1 reactions

Eliminations involving the stepwise elimination of HX from an alkyl halide, with initial slow ionisation by loss of Cl^- (the rate-determining step), lead to formation of a carbocation intermediate (Fig. 8.2); up to this point, the reaction is in fact identical to an S_N1 reaction (compared with Fig. 8.3a). This step is, like the S_N1 reaction, rate determining, with the reaction rate directly proportional to the concentration of the alkyl halide, and so the elimination reaction is classified as E_1 (elimination, unimolecular). However, in the absence of a suitable nucleophile and in the presence of a good base, removal of hydrogen in the β-position from the carbocation leads to the formation of an alkene. In fact, in solvents of low solvating power, dissociation of the carbocation and leaving group is incomplete, and the cation/anion pair is held together in a solvent cage as an ion pair. Under these circumstances, the leaving group can act as a base, thereby generating the alkene product, rather than the one of substitution. An example of this process is the elimination of cyclopentyl bromobenzene sulfonate in hexafluoroisopropyl alcohol/water, which mostly gives cyclopentene along with some cyclopentanol (Fig. 8.3).

8.1.2 E_1CB reactions

An alternative stepwise elimination of HX from an alkyl halide, called E_1CB (elimination, unimolecular, conjugate base), is possible, with initial slow, and therefore rate-determining, ionisation by loss of H^+ as the rate-determining step. The reaction rate is directly proportional to the concentration of the starting alkyl halide. Not surprisingly, this process is favoured if the resulting carbanion intermediate is relatively stabilised, for example, by inductive or more importantly resonance effects from adjacent substituents, such as ester

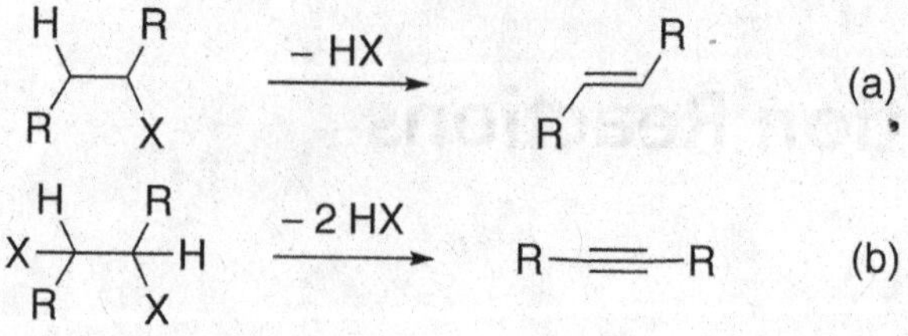

Figure 8.1 Elimination reactions.

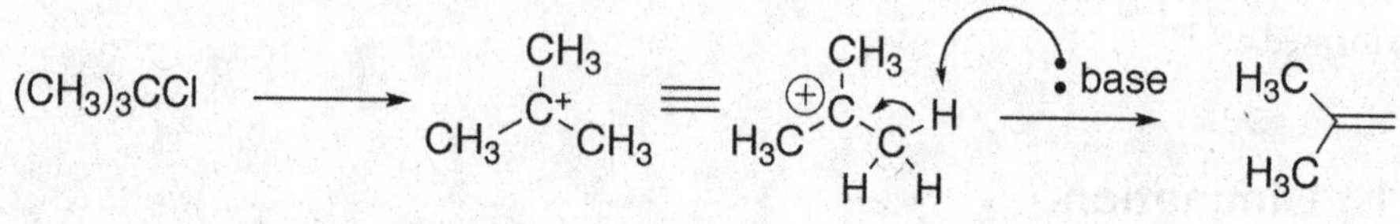

Figure 8.2 An E_1 elimination reaction.

Bs = p-BrC$_6$H$_4$SO$_2$

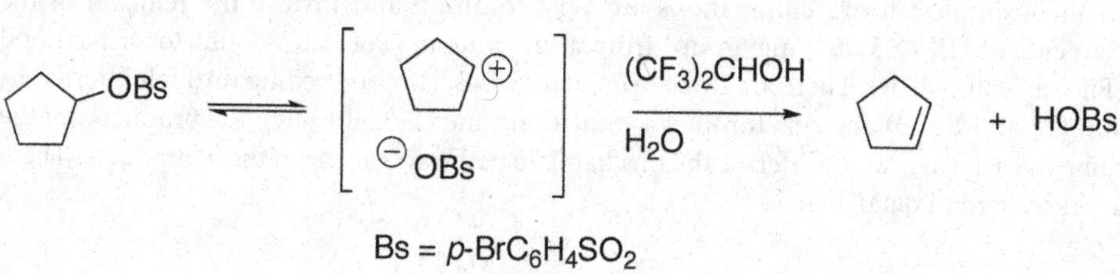

Figure 8.3 Elimination involving tight ion pairs.

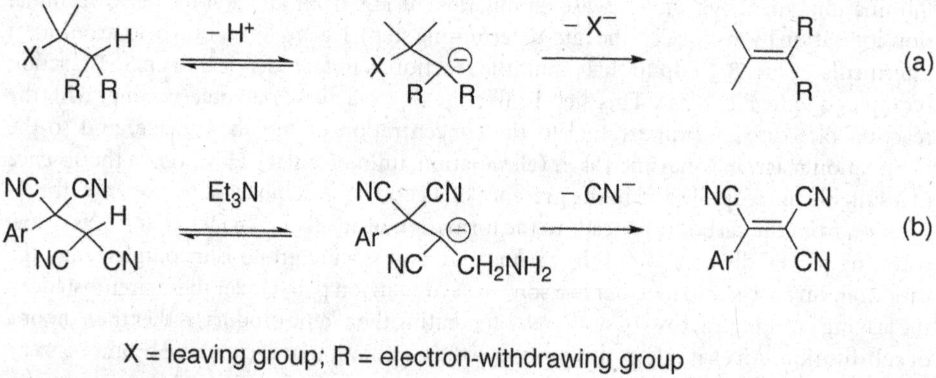

X = leaving group; R = electron-withdrawing group

Figure 8.4 E_1CB eliminations.

or cyano groups (Fig. 8.4a). This is then followed by departure of the leaving group; an E_1CB process is favoured by a relatively poor leaving group. A specific example is given in Fig. 8.4b.

8.1.3 E_2 reactions

An alternative to both of the above processes is the E_2 reaction, which proceeds by the simultaneous elimination of HX; in this case, the bond breaking of the C–H and C–X

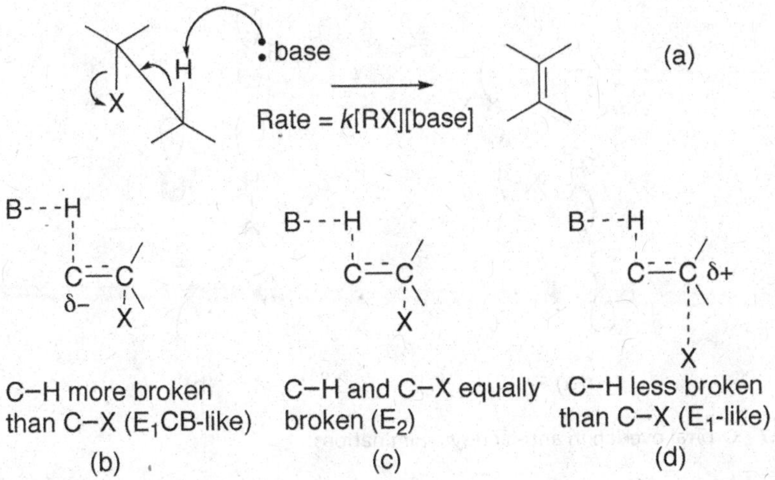

Figure 8.5 Transition states in the E_2 reaction.

bonds occurs in *concert*, and simultaneously with the formation of the double bond, via the transition state indicated in Fig. 8.5a. The reaction is called E_2 (elimination, biomolecular), with a rate equation shown in Fig. 8.5a; the rate of reaction is first order each in base and substrate, that is, second order overall, and is dependent on the leaving-group ability of the group X (see Section 6.1). In addition, if the hydrogen which is removed in the reaction is replaced by a deuterium, then the rate of elimination reaction is decreased as a consequence of the kinetic isotope effect (C–D bond is stronger than a C–H bond as a result of its lower zero-point energy, and if this is the bond that is broken in the rate-determining step, then the overall reaction rate is decreased). However, in the transition state, the relative extent of C–H and C–X bond cleavage and of C=C bond formation may not exactly correspond, and so some charge separation may occur in the transition state, as shown in Figs. 8.5b–8.5d; this may impart some ionic character to the reaction.

In order to minimise the energy (steric and electronic) of this transition state, *anti*-elimination, in which the antiperiplanar β-hydrogen and X group are removed, is preferred as indicated in Fig. 8.6a. Note that in this process, the proton being removed, the leaving group X and the electrons involved in the bond-making and -breaking processes are all coplanar. However, if this is not possible, *syn*-elimination, in which the *syn*-periplanar β-hydrogen and X group are removed, can occur, although it is less favourable (Fig. 8.6b) on steric grounds.

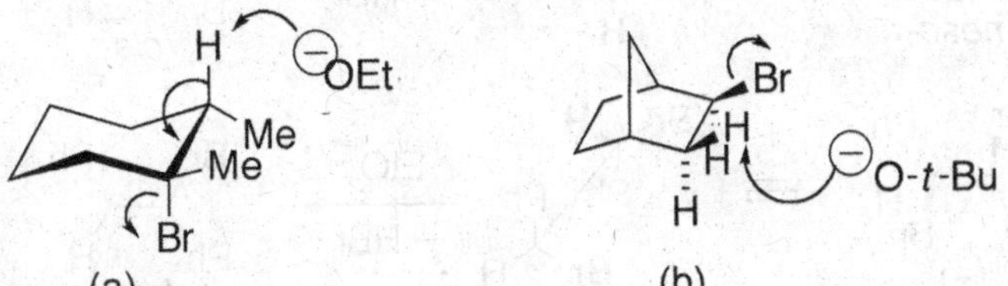

Figure 8.6 *Anti-* and *syn*-eliminations.

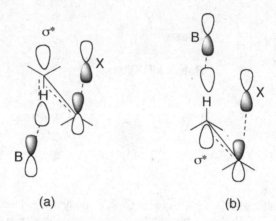

Figure 8.7   Orbital overlap in *anti-* and *syn*-eliminations.

The reason for the preference for *anti*-elimination is that orbital overlap is maximised throughout the course of the reaction (Fig. 8.7). Thus, in *anti*-elimination, donation of electron density from the base B into the σ* (lowest unoccupied molecular orbital, LUMO) orbital of the C–H bond weakens it and permits overlap of the electrons of the C–H σ-bond to be donated in turn into the σ* (LUMO) orbital of the C–X bond with formation of a πg(–C) bond; the antiorbital overlap also permits a fully staggered arrangement and keeps the electron-rich base and leaving groups as far apart as possible. The next best mode for elimination is a *syn*-arrangement, which although maintaining a planar arrangement of orbitals allows only for a weaker σ (C–H) highest occupied molecular orbital (HOMO) and σ∗ (C–X) lowest unoccupied molecular orbital (LUMO) orbital interaction. This preferred *anti*-orientation has important consequences for the stereochemistry of elimination reactions; for example, the *meso*-dibromide (Fig. 8.8a) reacts with ethoxide base to give the *cis*-bromostilbene, but racemic stilbene dibromide under the same conditions gives the *trans*-bromostilbene (Fig. 8.8b).

However, this preference for *anti*-eliminations does not mean that no other processes are possible; *syn*-eliminations will occur if geometric constraints mean that only such an elimination is possible (see Fig. 8.6b). Alternatively, they will also occur if a *syn*-related hydrogen is more activated to elimination (Fig. 8.9); in this example, although *anti*-elimination

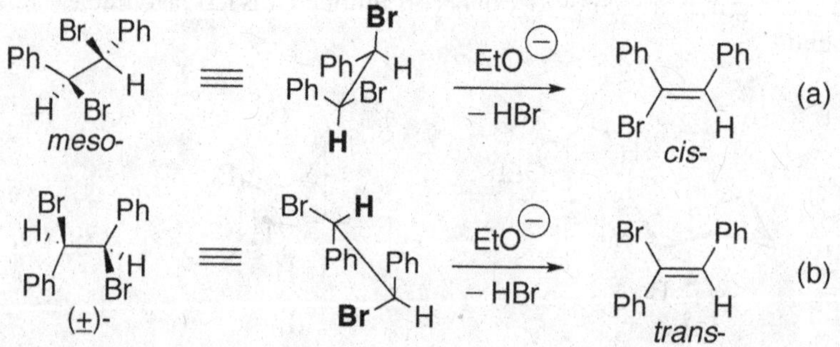

Figure 8.8   Eliminations giving stereoisomeric products.

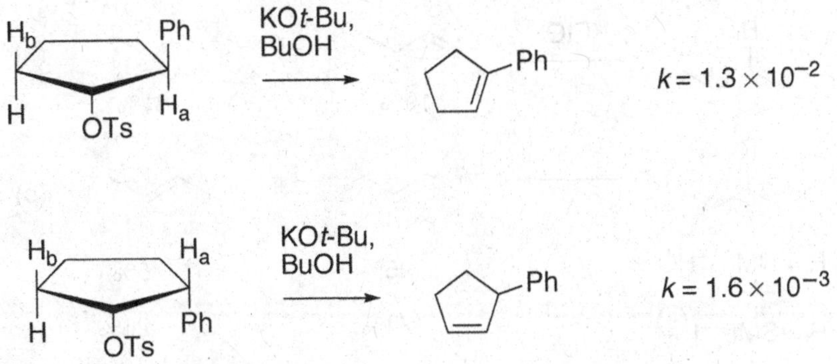

Figure 8.9   Competition between *anti-* and *syn-*elimination.

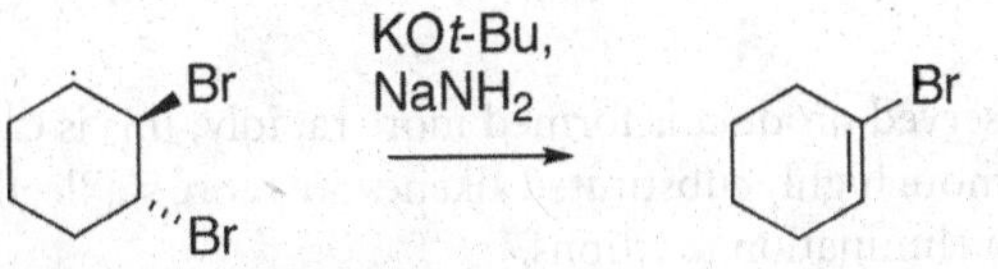

Figure 8.10   Base dependence of eliminations.

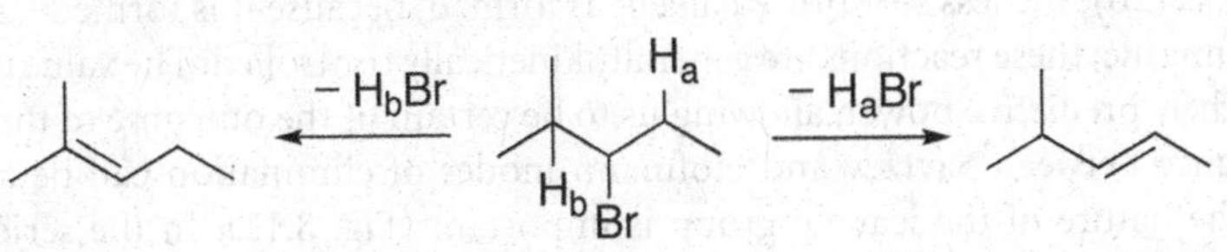

Figure 8.11   Eliminations giving isomeric products.

of proton H_b and the tosylate is intrinsically preferred, *syn-*elimination of proton H_a predominates, since it is activated to removal as it is benzylic and stabilises very effectively by resonance any charge which develops at that position. In an alkane substrate, the favoured elimination process may also be altered by the choice of base; for example, in the elimination of *trans-*1,2-dibromocyclohexane with the base sodamide-potassium *t*-butoxide, the product of *syn-*elimination (1-bromocyclohexene) is formed preferentially (Fig. 8.10).

8.1.4   Eliminations leading to isomeric products

An additional complication related to elimination is that there is often more than one β-hydrogen in an alkyl halide, and different products arise depending on which one is removed; for example, in Fig. 8.11, elimination of either H_a or H_b along with the bromide leaving group will lead to two different alkene products. In fact, it is possible to control the outcome of these reactions, at least to some extent, by choice of the leaving group of the reaction substrate.

Thus, in the elimination of an alkyl halide or tosylate (Fig. 8.12a), the more substituted alkene is formed. This is the Saytzev rule, which is particularly common for the E_1 mechanism. This outcome arises because the transition state leading to product formation has some double-bond character, and therefore any factor which stabilises the alkene will also stabilise the transition state. This therefore lowers the activation energy of the elimination

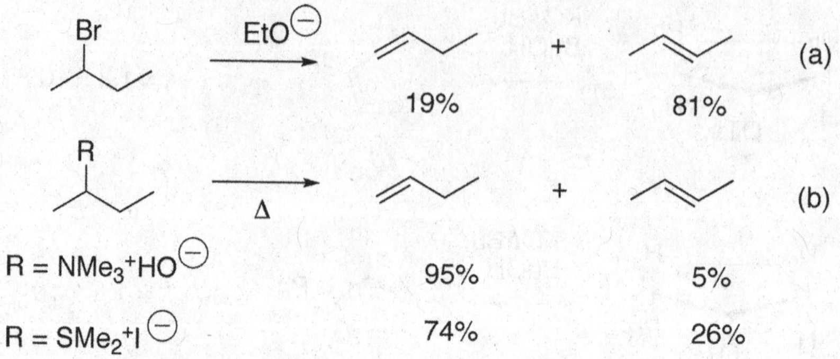

| | | Br | EtO⊖ | | + | | (a) |
| | | | 19% | | | 81% | |

R = NMe₃⁺HO⊖   95%   5%

R = SMe₂⁺I⊖   74%   26%

Figure 8.12 Saytzev and Hofmann eliminations.

reaction, and the observed product is formed more rapidly; this is called product develop-ment control. Since more highly substituted alkenes are more stable, they tend to be formed preferentially in such elimination reactions.

On the other hand, in the elimination of a substrate with a positively charged leaving group (Fig. 8.12b), the less substituted alkene is formed, because it is formed faster. This is the Hofmann rule; these reactions are generally kinetically controlled. The value of these two rules is in their predictive power, allowing us to be certain of the outcome of the reaction.

The balance between Saytzev and Hofmann modes of elimination can be altered. For example, the nature of the leaving group is important (Fig. 8.13). In the series fluoride, chloride, bromide and iodide, electron-withdrawing character decreases and leaving-group ability increases, and this simultaneously lowers the acidity of the β-hydrogens and weakens the C–X bond, and so Saytzev elimination becomes dominant for the iodide. Alternatively, increased basicity of the leaving group can favour Hofmann-type elimination; under these conditions, the most acidic protons, which are typically the least hindered ones, are removed fastest internally by the leaving group, and the reaction is not controlled by the stability of the product which is formed (Fig. 8.14). On the whole, stronger bases (such as KOt-Bu), elevated temperature (typically in refluxing solvent) and longer reaction times tend to favour Saytzev elimination, giving the more stable product.

The thermolysis of esters or xanthates is a process which proceeds by a concerted *syn*-elimination in a six-membered transition state to give the products of Hofmann elimination;

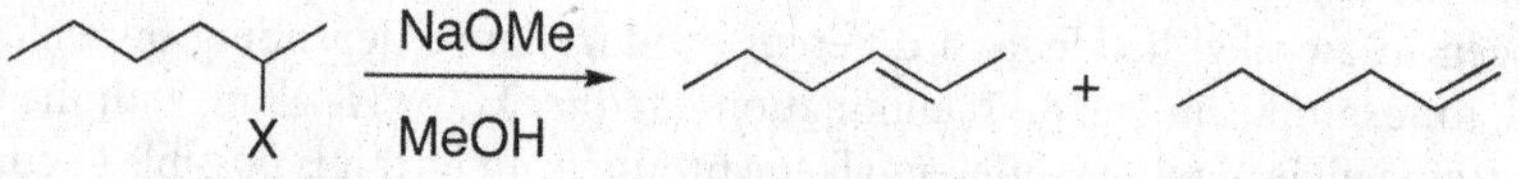

X	Saytzev	Hofmann
F	0.4	1
Cl	2.0	1
Br	2.6	1
I	4.2	1

Figure 8.13 Product ratio for Saytzev versus Hofmann elimination.

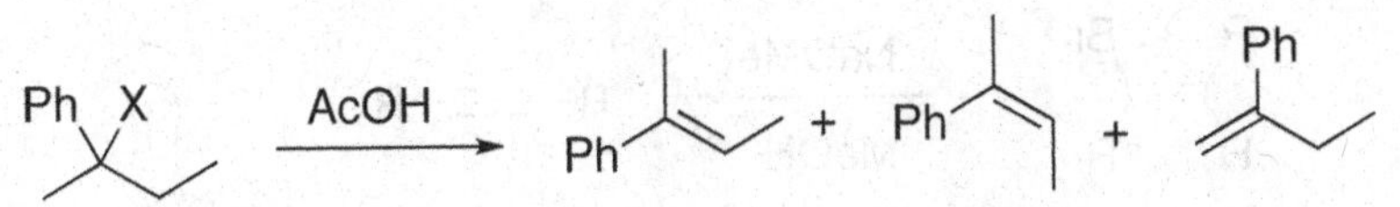

X	cis-	trans-	1-Alkene
Cl	68	9	23
OAc	53	2	45
NHNH$_2$	40	0	60

Figure 8.14 Percent product formed for different leaving groups.

xanthates pyrolyse at a lower temperature and as a result tend to be higher yielding and more synthetically useful (Fig. 8.15).

Vinyl halides, and dihaloalkanes, can also eliminate in a synthetically useful manner to alkyne products, as shown in Fig. 8.16, but of course in these cases no geometrical isomerism is possible in the products.

The control of *E*- and *Z*-selectivity in elimination reactions is not always guaranteed; in principle, provided an elimination proceeds cleanly by either a *syn*- or *anti*-process, a well-defined stereochemical outcome in terms of alkene geometry should result (that is, the reaction should be stereoselective). However, in many cases such selectivity is not obtained in practice, and this arises from product development control of the reaction; factors which destabilise the formation of the *cis*-product (typically steric hindrance between the substituents on the double bond) also tend to destabilise the transition state leading to its formation, and this leads to an intrinsic preference for the formation of *trans*-alkenes, since these are more stable.

It is possible for *syn*- and *anti*-diols to be eliminated stereospecifically to the *trans*- or *cis*-alkenes using the Corey–Winter reaction (Fig. 8.17). In this process, the diol is first converted to a cyclic thiocarbonate and then heated in the presence of trimethylphosphite. Extrusion of carbon dioxide and formation of a P=S bond provide a potent driving force for the reaction.

8.1.5 Competition between substitution and elimination

Since both nucleophilic substitution reactions and elimination reactions involve the reaction of alkyl halides with nucleophiles or bases, it is not surprising that they should compete.

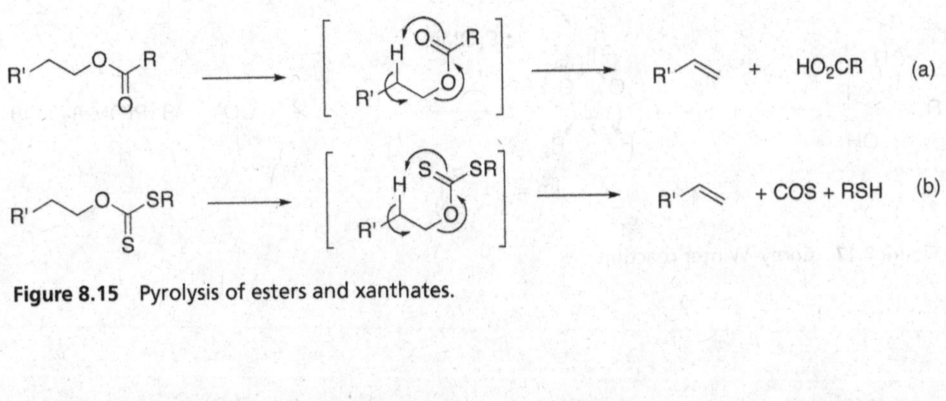

Figure 8.15 Pyrolysis of esters and xanthates.

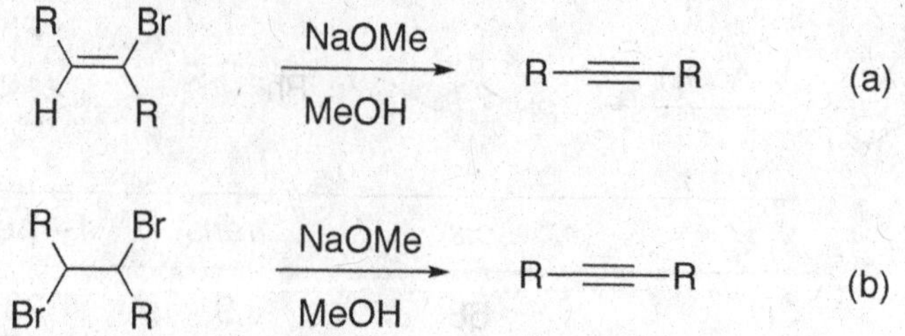

Figure 8.16 Elimination of vinyl halides and dihaloalkanes to give alkynes.

Tertiary alkyl halides are most likely to give elimination under all conditions, since they are hindered to attack by a nucleophile and possess multiple β-hydrogen substituents capable of removal by the base, while primary alkyl halides give predominantly substitution, since they are not so hindered. Secondary alkyl halides will react by either substitution or elimination, depending on the reaction conditions (Fig. 8.18). The balance between the two possibilities of substitution and elimination can be shifted in favour of elimination by ensuring that the reaction is done in the presence of a strong, hard base (such as ethoxide (EtO$^-$) or *t*-butoxide (*t*-BuO$^-$)) or at elevated temperature (refluxing ethanol) or both. Amide bases such as lithium diisopropylamide (LDA), lithium 2,2,6,6-tetramethylpiperidide (LiTMP) and lithium hexamethyldisilazide (LiHMDS) (Fig. 8.19) also favour elimination.

8.1.6 The leaving group

The nature of the leaving group is clearly critical in these types of reactions, just as we saw earlier for nucleophilic substitution reactions, and the reaction is best in those cases where the leaving-group ability is high. However, it is possible to convert a group of poor leaving-group ability to one of substantially higher ability, and a good example of this is the alcohol group. Treatment of an alcohol with toluenesulfonyl chloride (tosyl chloride, TsCl), methanesulfonyl chloride (mesyl chloride, MsCl) or trifluoromethanesulfonyl

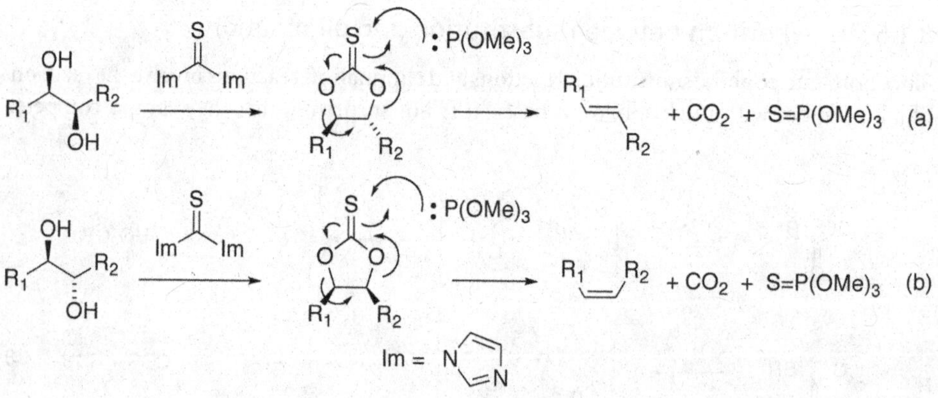

Figure 8.17 Corey–Winter reaction.

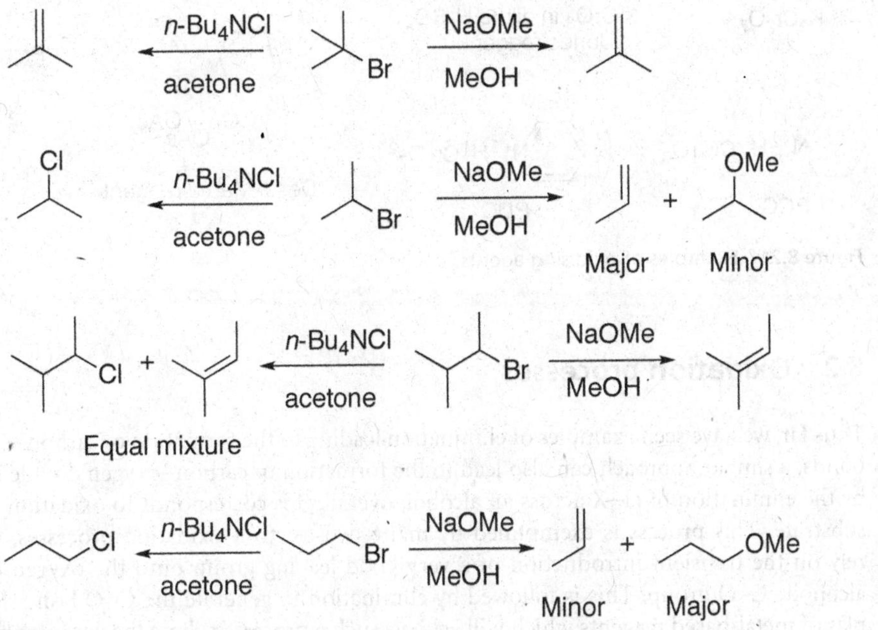

Figure 8.18 Substitution versus elimination of alkyl halides.

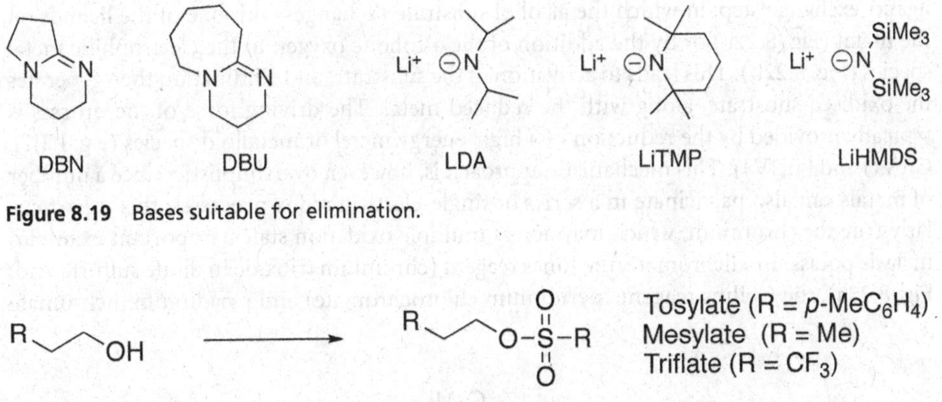

Figure 8.19 Bases suitable for elimination.

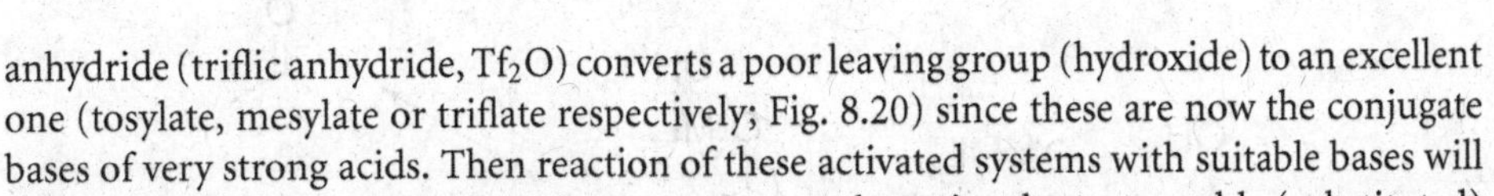

Tosylate (R = p-MeC₆H₄)
Mesylate (R = Me)
Triflate (R = CF₃)

Figure 8.20 Activation of alcohols by conversion to sulfonates.

anhydride (triflic anhydride, Tf₂O) converts a poor leaving group (hydroxide) to an excellent one (tosylate, mesylate or triflate respectively; Fig. 8.20) since these are now the conjugate bases of very strong acids. Then reaction of these activated systems with suitable bases will lead to elimination, normally under Saytzev control, to give the more stable (substituted) alkenes. The advantage of such good leaving groups is that the elimination reaction may be conducted using milder bases, including 1,5-diazabicyclo(4.3.0)non-5-ene (DBN) and 1,8-diazabicycloundec-7-ene (DBU).

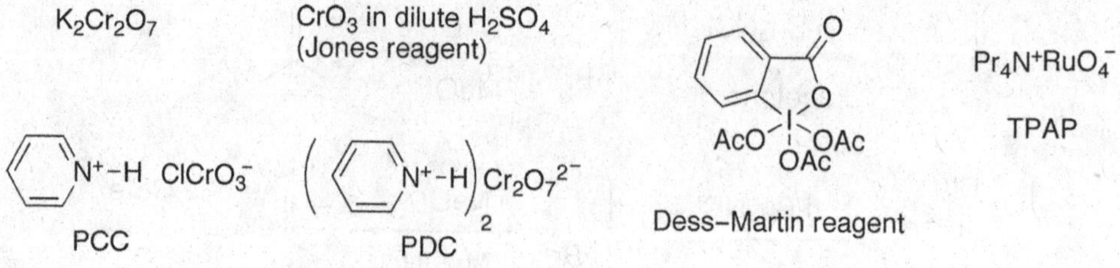

K_2Cr_2O_7   CrO_3 in dilute H_2SO_4
(Jones reagent)

$Pr_4N^+RuO_4^-$

TPAP

PCC   PDC

Dess–Martin reagent

Figure 8.21 Examples of oxidising agents.

8.2 Oxidation processes

Thus far, we have seen examples of elimination leading to the formation of carbon–carbon bonds; a similar approach can also lead to the formation of carbon–oxygen double bonds by the elimination of H–X across an alcohol; overall, this corresponds to oxidation of the substrate. This process is exemplified by many two-electron oxidation processes, which rely on the transient introduction of a very good leaving group onto the oxygen of the alcoholic C–O group. This is followed by elimination to generate the C=O bond. Examples of metal-based reagents which will achieve such a process include the commonly used potassium dichromate, chromium trioxide and periodate, and less commonly manganese dioxide and lead tetraacetate (Fig. 8.21). It is possible to generalise the mechanism of action of each of these oxidation processes as shown in Fig. 8.22; the reaction commences with a ligand-exchange step, in which the alcohol substrate exchanges with one of the ligands on the metal (Fig. 8.22a) or by the addition of the alcoholic oxygen to the electrophilic metal species (Fig. 8.22b). This leads to activation of the substrate, and elimination then generates the oxidised substrate, along with the reduced metal. The driving force of the process is typically provided by the reduction of a high-energy metal or metalloid species (e.g. I(III), Cr(VI) and Pb(IV)). This mechanistic approach is, however, oversimplistic, since a number of metals can also participate in a series of single-electron redox processes; this is particularly true for chromium, which may access multiple oxidation states. Important examples include potassium dichromate, the Jones reagent (chromium trioxide in dilute sulfuric acid; Fig. 8.23a), the Collins reagent (pyridinium chlorochromate) and pyridinium dichromate

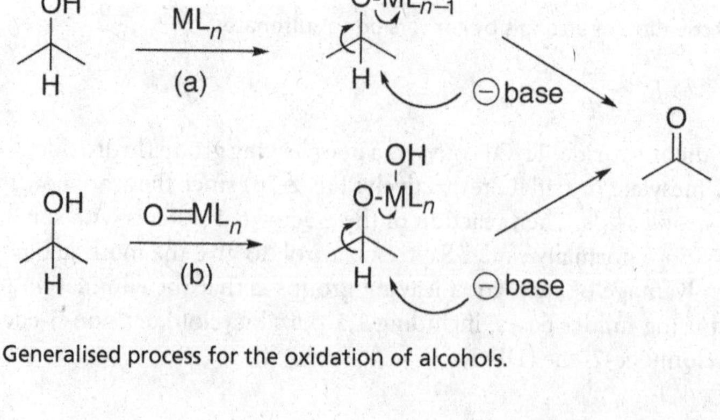

Figure 8.22 Generalised process for the oxidation of alcohols.

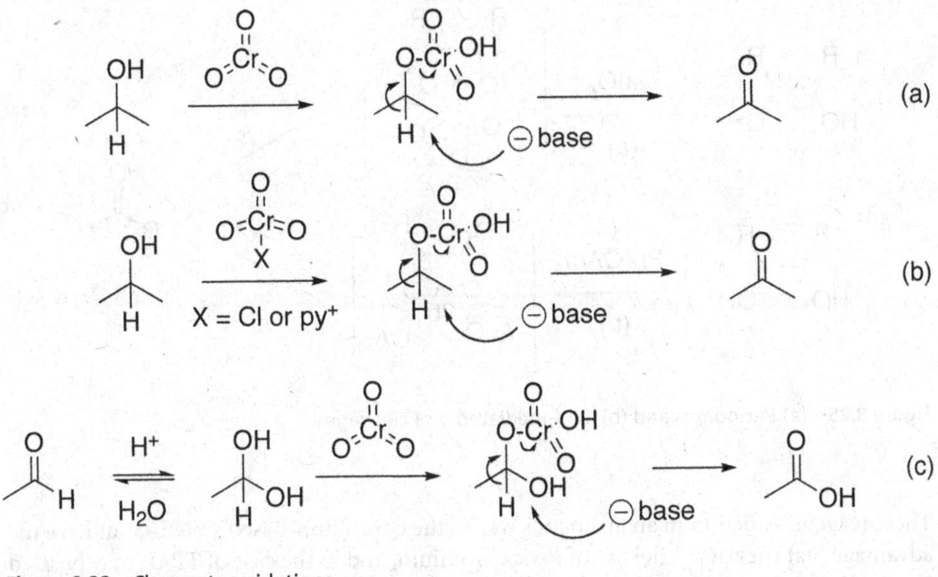

Figure 8.23 Chromate oxidations.

(PDC; Fig. 8.23b). Mechanistically, these processes proceed by formation of a chromate (Cr(VI)) ester, which collapses by elimination with loss of a lower oxidation state of chromium (Cr(IV)) and generation of the carbonyl product. Provided these reactions are conducted in the absence of water, aldehydes may be isolated; however, in the presence of water, the initially formed aldehyde is hydrated, and further oxidation to the carboxylic acid then becomes possible (Fig. 8.23c).

Recent developments in oxidation chemistry of these reagents have identified some highly selective reagents for the oxidation of alcohols, and good examples include the Dess–Martin periodinane and the tetra-*n*-propylammonium perruthenate (TPAP) reagent (Fig. 8.24).

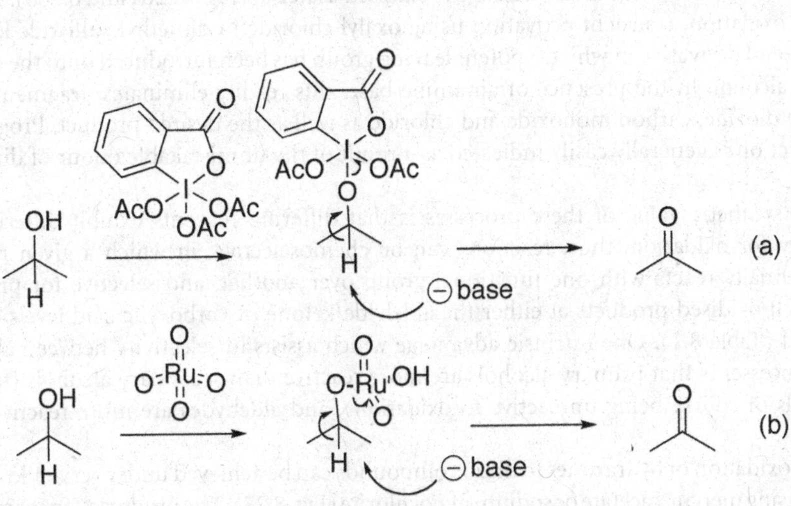

Figure 8.24 (a) Iodine- and (b) ruthenium-mediated oxidations.

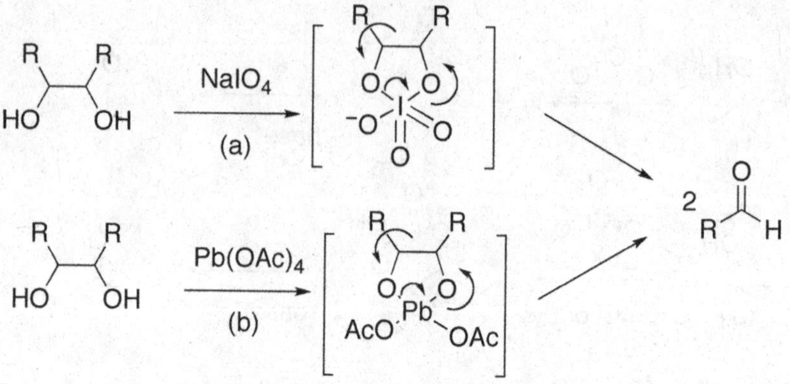

Figure 8.25 (a) Periodate- and (b) lead-mediated diol cleavages.

These reagents function in an analogous way to the chromium-based systems, but have the advantage that they avoid the use of toxic chromium, and in the case of TPAP, may be used in a catalytic amount, provided a stoichiometric amount of a suitable reoxidant (such as N-methylmorpholine oxide) is present.

Sodium periodate and lead tetraacetate are potent oxidants and will cleave 1,2-diols to the corresponding dicarbonyl products; the reason that cleavage occurs in this way is because both the iodine and lead nucleus can accept two ligands of the substrate to form a cyclic intermediate. The intermediates that are formed collapse as shown (Fig. 8.25) in a reaction that is driven by the two-electron reduction of the iodine or lead atoms.

It is possible to achieve fully analogous reaction outcomes, but without using metal oxidants. The driving force for the oxidation process can come from collapse of a suitably activated substrate; in the case of the Moffatt and Swern oxidations, the driving force comes from a combination of enthalpic and entropic effects, which result from the generation of stable and volatile products, as well as the oxidised material (Figs. 8.26a and 8.26b). For the Swern oxidation, transient activation using oxalyl chloride in dimethyl sulfoxide leads to the alcohol derivative in which a potent leaving group has been introduced onto the oxygen of the alcohol; in the presence of an amine base, this readily eliminates, fragmenting to carbon dioxide, carbon monoxide and chloride as well as the desired product. Progress of the reaction is generally easily indicated as a result of the unmistakable odour of dimethyl sulfide.

The synthetic value of these processes is that differing reagents exhibit differing selectivity for oxidation; thus, reactions can be chemoselective, in which a given reagent preferentially reacts with one functional group over another, and selective for product, in which oxidised products at either the aldehyde/ketone or carboxylic acid levels can be isolated (Table 8.1). One intrinsic advantage which assists in selectivity between alternative processes is that primary alcohols are more reactive than secondary alcohols (tertiary alcohols of course being unreactive to oxidation), and aldehydes are more reactive than ketones.

The oxidation of hydrazones to diazo compounds can be achieved under very mild conditions, using mercury acetate or sodium hypochlorite (Fig. 8.27). The products can themselves be further eliminated to give carbenes (Section 8.3).

Table 8.1 Selectivity in oxidations

Substrate	Product	Reagent(s) or reaction
Alcohol (primary)	Aldehyde	Swern, Moffatt, Collins, TPAP, Dess–Martin periodinane, PDC
Alcohol (secondary)	Ketone	PCC, PDC, Swern
Alcohol (primary)	Carboxylic acid	$K_2Cr_2O_7$, CrO_3
Aldehyde	Carboxylic acid	$NaClO_2$, H_2O_2
Ketone	Carboxylic acid	PCC, PDC
Benzylic or allylic alcohols	Aldehyde	MnO_2, TPAP

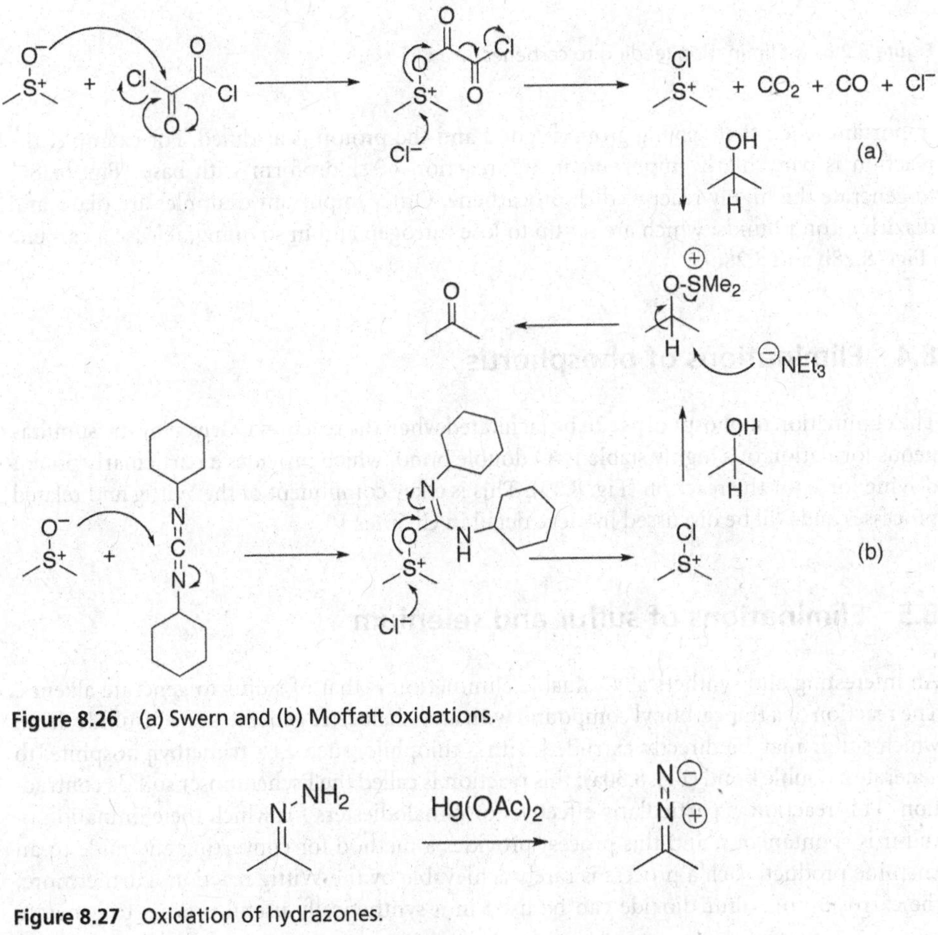

Figure 8.26 (a) Swern and (b) Moffatt oxidations.

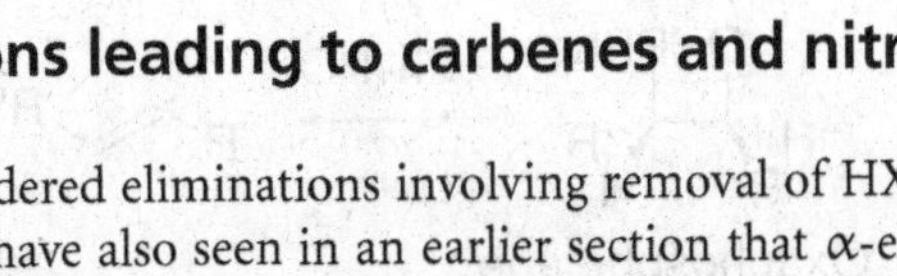

Figure 8.27 Oxidation of hydrazones.

8.3 α-Eliminations leading to carbenes and nitrenes

Thus far, we have considered eliminations involving removal of HX that occur in a β- or 1,2-sense; however, we have also seen in an earlier section that α-eliminations can occur, and in this case the products are carbenes or nitrenes (Section 4.5). This reaction requires removal of a leaving group and a proton from the same atom, and the reaction is particularly

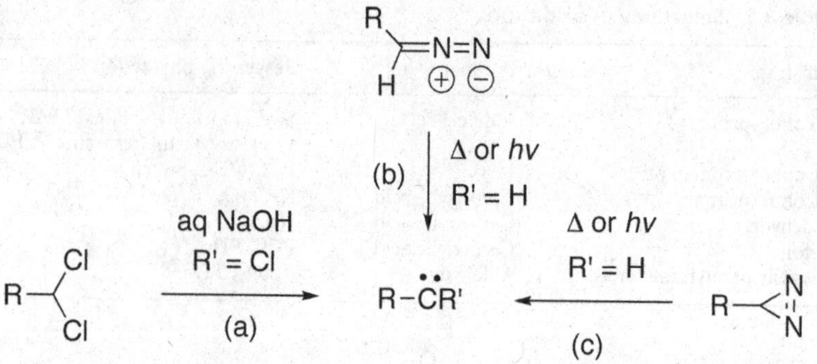

Figure 8.28 α-Elimination leading to carbenes.

important when that leaving group is good and the proton is acidified. For example, the reaction is particularly important in the reaction of chloroform with base (Fig. 8.28a) to generate the highly reactive dichlorocarbene. Other important examples are diazo and diazirine compounds, which are set up to lose nitrogen and in so doing, release a carbene (Figs. 8.28b and 8.28c).

8.4   Eliminations of phosphorus

The elimination of phosphorus can be facilitated when the reaction is driven by the simultaneous formation of a highly stable P=O double bond, which provides a particularly potent driving force for the reaction (Fig. 8.29). This is a key component of the Wittig and related processes and will be discussed in more detail in Chapter 10.

8.5   Eliminations of sulfur and selenium

An interesting and synthetically valuable elimination is that of sulfur to generate alkenes. The reaction of a thiocarbonyl compound with an α-haloester generates an episulfide, from which sulfur may be directly extruded with a thiophile, such as a trimethylphosphite, to generate a double bond (Fig. 8.30a); this reaction is called the Eschenmoser sulfide contraction. This reaction is particularly effective for α-halodiesters for which the elimination of sulfur is spontaneous, and this process provides a method for converting an amide to an enamine product; such a process is rarely achievable by the Wittig reaction. Furthermore, the extrusion of sulfur dioxide can be used in a synthetically useful process to generate

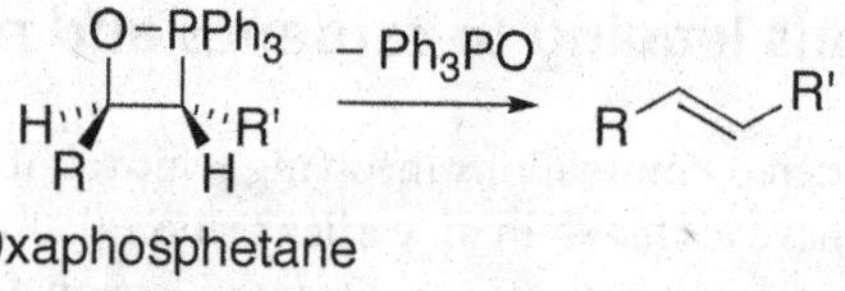

Oxaphosphetane

Figure 8.29 Elimination of triphenylphosphine oxide.

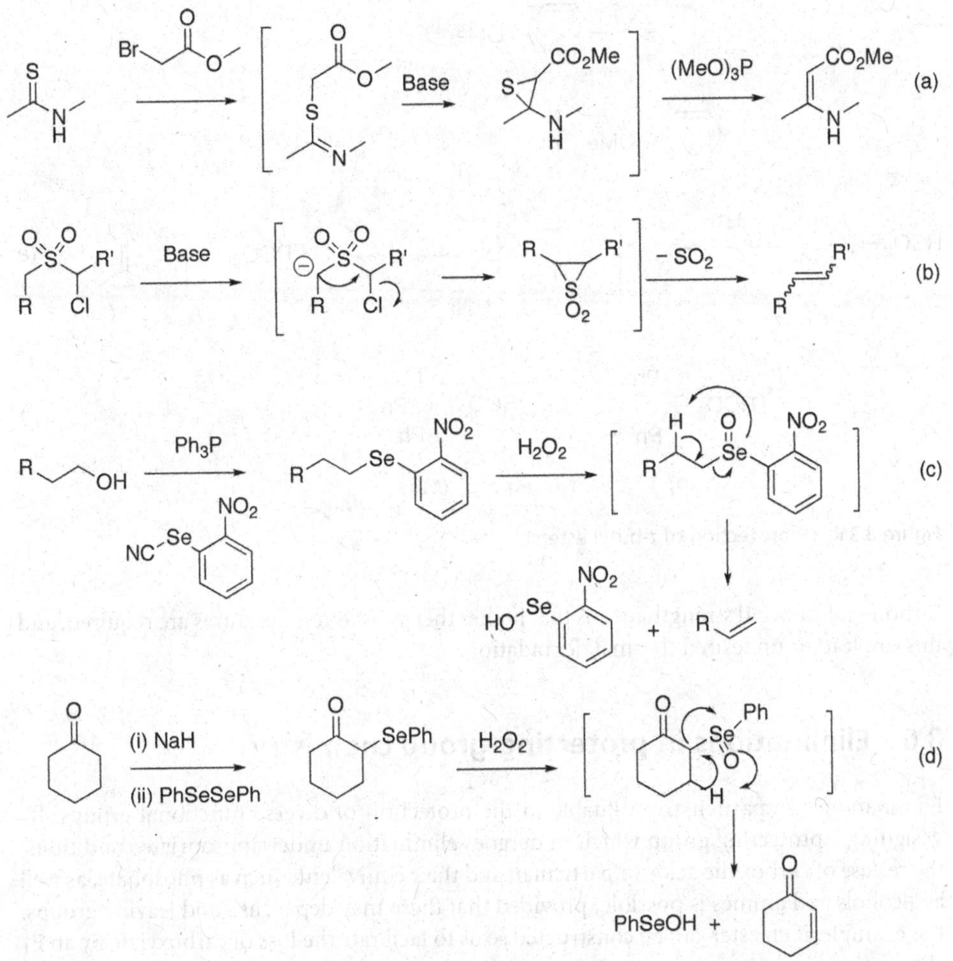

Figure 8.30 Elimination of sulfur by (a) Eschenmoser contraction, (b) Ramberg–Backlünd reaction and (c, d) elimination of selenoxides.

carbon–carbon double bonds. This reaction, called the Ramberg–Backlünd reaction, involves treating an α-halosulfone with a base; deprotonation and internal S_N2 reaction generate a cyclic intermediate which spontaneously eliminates sulfur dioxide to give an alkene (Fig. 8.30b). This process is not always stereocontrolled, and mixtures of E- and Z-alkenes can result.

Elimination of selenoxides under thermal conditions also provides a very effective method for the insertion of a double bond (Figs. 8.30c and 8.30d). The necessary substrates may be prepared from the alcohol by substitution or by reaction of an enolate. The reaction is sufficiently facile that it will often proceed at room temperature or below and is advantageous in that the selenoxide often does not need to be isolated. Rather, oxidation of a selenide generates the desired selenoxide, which spontaneously eliminates to give the product. Similar reactions are possible using the analogous sulfur-containing compounds, but the greater

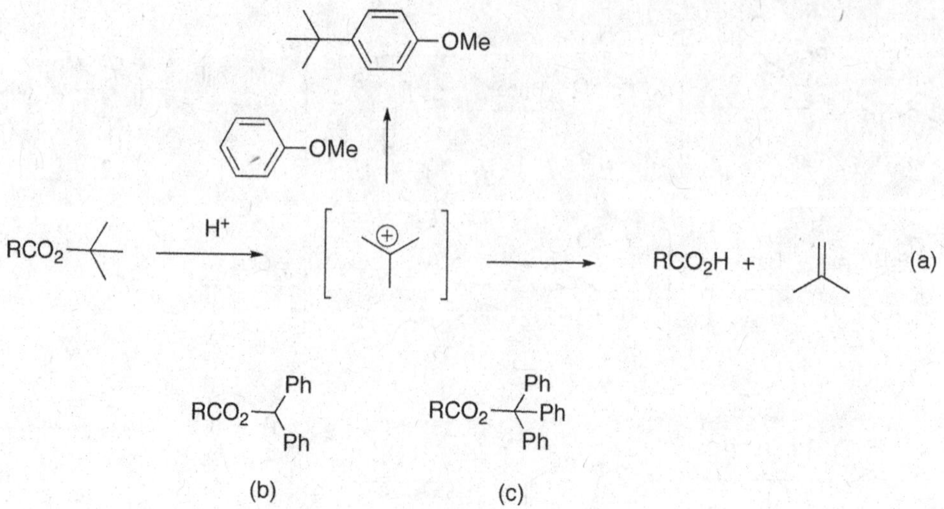

Figure 8.31 Deprotection of *t*-butyl esters.

carbon–sulfur bond strength means that higher thermolysis temperatures are required, and this can lead to undesired thermal degradation.

8.6 Eliminations in protecting-group chemistry

Eliminations are particularly valuable in the protection of diverse functional groups. By designing a protecting group which undergoes elimination under appropriate conditions, the release of carboxylic acids in particular, and their equivalents such as phosphate, as well as alcohols and amines is possible, provided that these may depart as good leaving groups. For example, if an ester can be constructed so as to facilitate the loss of carboxylate by an E_1 elimination process under appropriate conditions, the carboxylate can be released at will; the ester can therefore be considered to be a useful protecting group for the carboxylic acid. This is most simply illustrated for the *t*-butyl ester of a carboxylate, which can be hydrolysed very readily by treatment with strong acid (Fig. 8.31a), which protonates the carbonyl group, converting it into a good leaving group and promoting elimination of isobutylene. If the reaction is done in the presence of anisole, the intermediate carbocation is trapped in an S_E Ar reaction, and other undesirable reactions of the carbocation can be avoided. Collapse of benzhydryl and trityl esters is similarly possible (Figs. 8.31b and 8.31c), although in these cases the resulting highly stabilised carbocation cannot eliminate.

The application of eliminations in deprotecting reactions is very broadly applicable and can be used for the release of acids, alcohols and amines in systems suitably designed for E_1-, E_2- and E_{1CB}-type eliminations; some examples are shown in Fig. 8.32. An excellent example is the 2-cyanoethyl ester; this group is cleanly removed under mild alkaline conditions, releasing the carboxylic acid (Fig. 8.32a). A similar strategy can be applied for the protection of alcohols and amines, but in this case the reaction can often be improved by using a carbonate or carbamate as the linking group. On collapse, this regenerates the alcohol or amine, but the reaction is assisted since the formation of carbon dioxide provides an additional

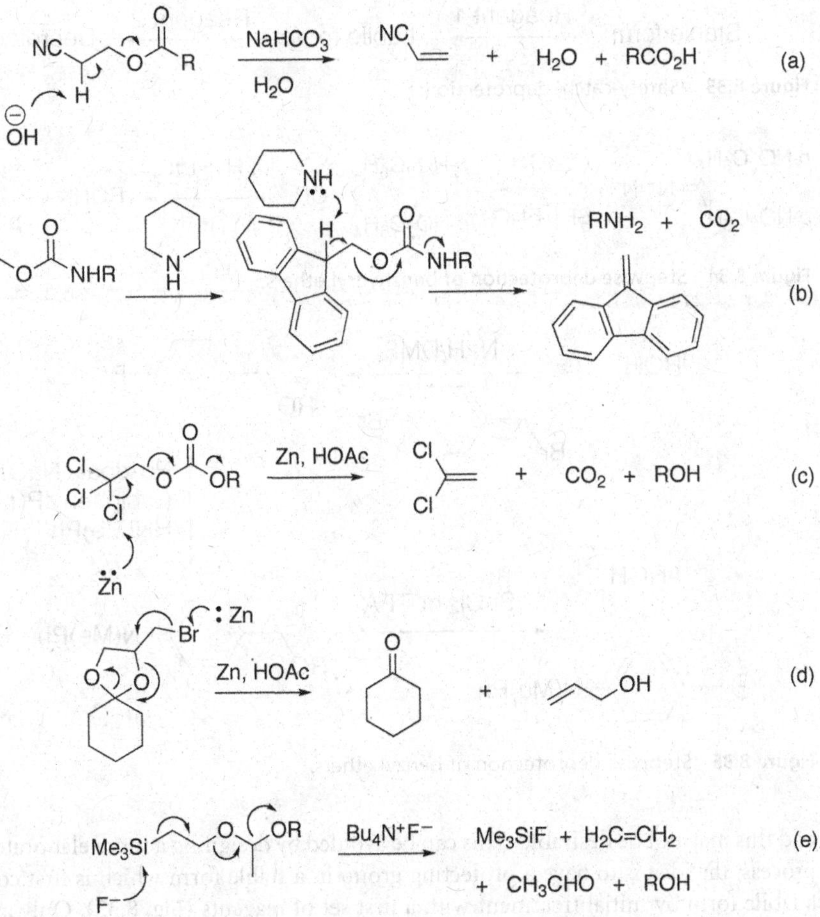

Figure 8.32 Deprotecting groups relying on eliminations for their removal.

enthalpic and entropic driving force, for example, the 9-fluorenylmethoxycarbonyl protecting group (Fig. 8.32b), which is selectively removed under mild base conditions using an E_1CB-type elimination. However, eliminations involving loss of a proton are not
the only ones which can be used to release carboxylic acids; the elimination process can
be driven by reactions involving heteroatoms too. For example, the trichloroethyl group
can be used to protect acids or alcohols as the carbonate, and release is achieved reductively using zinc in acetic acid or methanol (Fig. 8.32c). A similar group can even be
used to protect ketones as a cyclic ketal, and release is by analogous reduction, this time
giving the ketone and allyl alcohol (Fig. 8.32d). The trimethylsilylethyl group is similar
(Fig. 8.32e), but is normally released using fluoride in a reaction giving the deprotected alcohol and which is driven by the formation of a strong Si—F bond, as well as two other stable
molecules.

However, one limitation to this process relates to the fact that the capacity to conduct an
elimination critically depends on the presence of suitably disposed functions in the protecting group. However, this of course makes the molecule unstable under certain conditions,

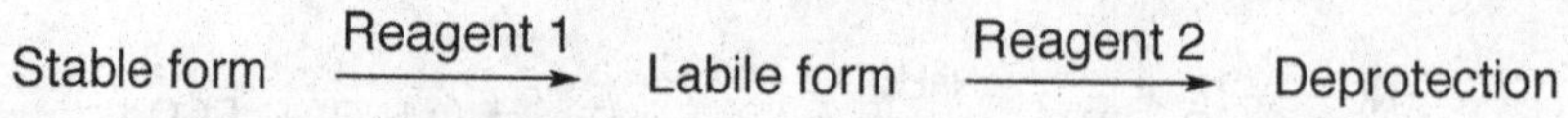

Stable form $\xrightarrow{\text{Reagent 1}}$ Labile form $\xrightarrow{\text{Reagent 2}}$ Deprotection

Figure 8.33 'Safety-catch' deprotection.

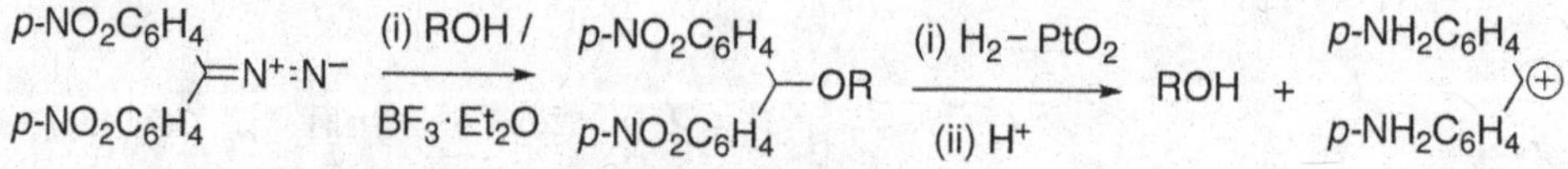

Figure 8.34 Stepwise deprotection of benzhydryl ethers.

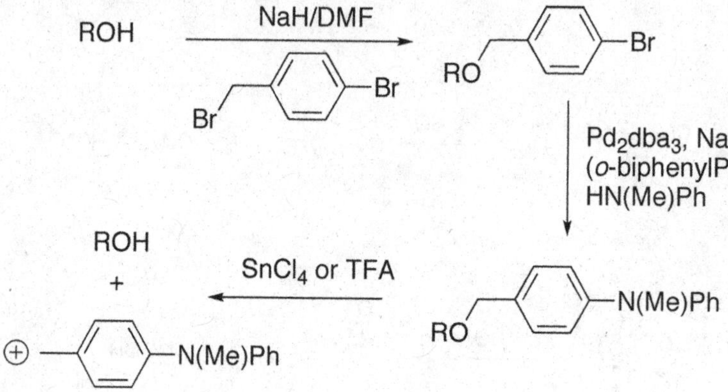

Figure 8.35 Stepwise deprotection of benzyl ethers.

and this may not be desirable. This can be avoided by designing a more elaborate two-stage process; the idea is to have a protecting group in a stable form which is first converted to a labile form by initial treatment with a first set of reagents (Fig. 8.33). Only at this stage does the group become capable of removal by treatment with a second reagent; this process has been devised since it offers an additional level of protection and avoids unwanted functionality in a protecting group until immediately before deprotection. For example, the rapid S_N1 cleavage of diaminobenzhydryl ethers becomes possible when the dinitro starting material is reduced to the diamine product using catalytic hydrogenation; the resulting compound fragments to release the desired alcohol as a result of the simultaneous formation of the highly stabilised diaminobenzhydryl cation (Fig. 8.34). Alternatively, the elimination of p-bromobenzyl ethers is possible by firstly converting the bromo function under palladium(0) coupling conditions to the strongly electron-releasing amino group and then, secondly, treating this intermediate directly with a Lewis or Brønsted acid; this results in an S_N1-type cleavage of the ether function which proceeds via the stabilised carbocation intermediate (Fig. 8.35).

This process can be extended to elimination processes. For example, a methylthioethyl carbamate group is highly stable, but on oxidation with hydrogen peroxide it is readily converted to the corresponding sulfone (Fig. 8.36a). In this intermediate, the adjacent position to the sulfone is strongly acidified, and so elimination by treatment with hydroxide liberates the carboxylic acid in a fragmentation process which liberates an α,β-unsaturated sulfone, carbon dioxide and the desired amine. The formation of these stable products provides an

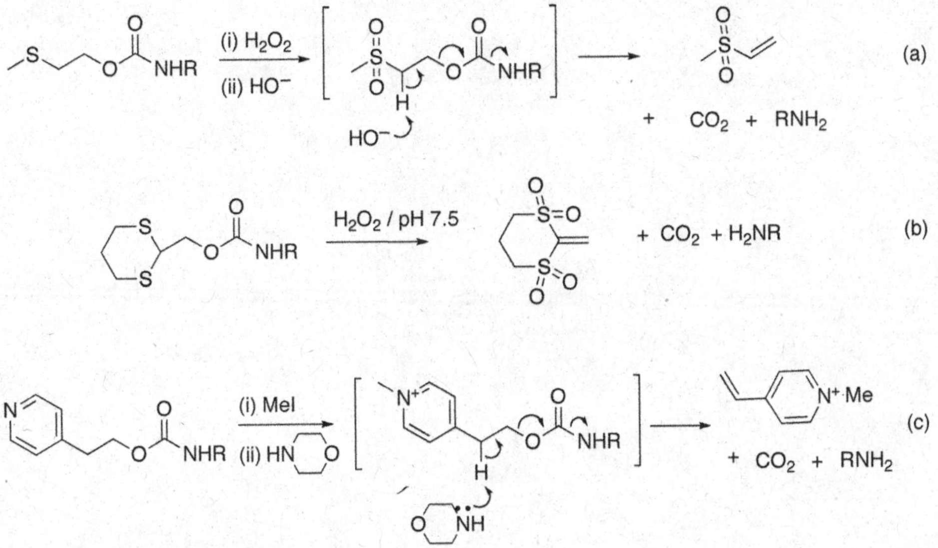

(a)

(b)

(c)

Figure 8.36 Examples of 'safety-catch' deprotection.

enthalpic driving force, and a number of these products, an entropic driving force, and so the overall free-energy change for the reaction is favourable. A similar strategy applies for the dithiane carbamate, which can be similarly activated (Fig. 8.36b). The 2-pyridinylethyl is slightly different, since activation is achieved not by oxidation, but by methylation of the pyridyl nitrogen. The overall result of this, however, is the same, since the acidity of the β-hydrogens is increased and elimination under mild conditions is then possible (Fig. 8.36c).

Chapter 9
Aromatic Substitution

In the previous chapters, we have considered in large part the types of reactions which single functional groups undergo; however, we saw in Chapter 7 that conjugation of a functional group through an adjacent π-system leads to a modification of the observed reactivity. This is particularly important for aromatic systems, in which conjugation of three carbon—carbon double bonds in a ring, exemplified by benzene, leads to a particularly stabilised system as a result of aromaticity.

9.1 Aromaticity

9.1.1 Benzene

Benzene and its derivatives were first isolated from coal, and this class of compounds was called aromatic because of its characteristic odour. The structure of these compounds was assigned to be three double bonds in a six-membered ring, represented as the Kekulé formula (Fig. 9.1a), but it was quickly appreciated that this characterisation left a problem in the rationalisation of their reactivity; their observed reactivity differed from what might be expected of a compound containing three double bonds, since the compounds were more stable and did not give addition products as readily as simple trienes, but in fact tended to give products of substitution. The unexpected stabilisation of benzene is called *aromaticity* and results from highly stabilising conjugation found in this arrangement of double bonds (Fig. 9.1b); the phenomenon is not specific to benzene, but to all aromatic systems containing $4n + 2$ (where $n = 1, 2, 3, \ldots$) π-electrons, as given by the Hückel rule. The phenomenon of aromaticity can be readily rationalised by a consideration of the molecular orbitals of benzene (Fig. 9.1c), in which all of the bonding orbitals are filled, but none of the antibonding orbitals; this provides about 150 kJ mol^{-1} of stabilisation energy relative to cyclohexatriene. This type of stabilisation may occur in diverse systems, so long as the $(4n + 2)$ rule is held; the cycloheptatrienyl (tropylium, $n = 1$) and cyclopropenyl cations $(n = 0)$ and the cyclopentadienyl anion $(n = 1)$ are examples (Fig. 9.1d).

9.1.2 Heteroaromatics

The Hückel $(4n + 2)$ rule is of course satisfied in carbon-derived cyclic systems with the appropriate number of electrons, but it can also be valid in heteroatom-derived systems too. Thus, the five-membered ring systems (furan, pyrrole, thiophene, indole and imidazole) and the six-membered systems (pyridine, quinoline and isoquinoline) are all aromatic (Fig. 9.2), being Hückel systems with $n = 1$, although in these cases the compounds are all higher in reactivity than benzene since the aromatic resonance energy is not as significant as that in benzene.

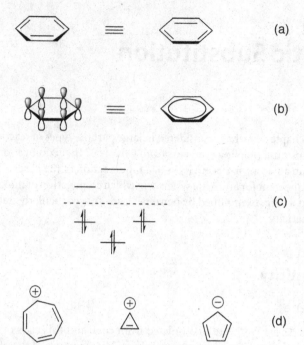

Figure 9.1   (a) Benzene, (b) π-delocalisation in benzene, (c) molecular orbitals of benzene and (d) other aromatic ring systems.

Furan, pyrrole and thiophene are five-membered analogues of benzene, and all exhibit aromatic behaviour (Fig. 9.2). In this case, the Hückel ($4n + 2$) rule requires the electron pair of the O, N or S heteroatom to achieve the stable sextet system. Thus, the aromatic sextet comprises four π- and two lone-pair electrons. In these cases, therefore, the lone pair is fully delocalised into the aromatic system, and it is much less available for reaction

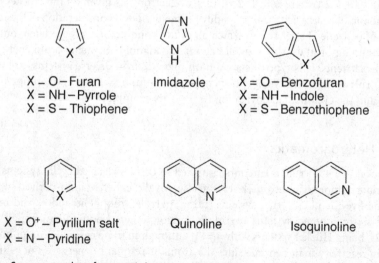

Figure 9.2   Some examples of aromatic heterocyclic systems.

with electrophiles or acids than in a simple amine. A similar analysis still applies for indole, although in this case the five-membered ring is annulated onto a benzene system. Imidazole is slightly different; in this case, one of the nitrogens is involved in the π-system, just as in pyrrole, but the lone pair of the other nitrogen is not and is free to be protonated.

The six-membered systems (pyridine, quinoline and isoquinoline) are all different to the cases considered above (Fig. 9.2), since in these cases, the nitrogen lone pair is not required for satisfaction of the Hückel $(4n + 2)$ rule, and therefore the nitrogen of these systems retains its basicity.

9.2 Reactions

The existence of aromaticity in these types of systems has a profound effect on their characteristic reactions so that they behave in a way which is quite different from that of a triene.

9.2.1 Acidity and basicity

One immediate chemical consequence of aromaticity is the effect on acidity and basicity of ring substituents. If deprotonation leads to an anion whose electrons are stabilised by the aromatic system, then the acidity of the parent system is enhanced, as might be expected, but if protonation disrupts an aromatic system, then basicity is reduced. For example, phenol is surprisingly acidic ($pK_a = 9.95$), because the resulting phenolate anion is stabilised by resonance with the adjacent aromatic ring; it is about 6 pK_a units more acidic than a simple alcohol (Fig. 9.3a). On the other hand, aniline is of reduced basicity compared to a simple amine, since in this case, protonation leads to an anilinium salt in which the resonance stabilisation of the amine with the aromatic system has been destroyed (Fig. 9.3b); the anilinium salt is therefore a stronger acid than the corresponding ammonium ion ($pK_a = 4.62$ vs 9.25). Pyrrole behaves similarly and is a weak base (Fig. 9.3c). Pyridine, quinoline and isoquinoline are all quite basic, since the nitrogen lone pair is not required for the π-aromatic system and is free to protonate (Fig. 9.3d). Dimethylaminopyridine, in particular, is a very important nucleophile and base, since the resonance resulting from the dimethylamino group in the protonated system is not disrupted (Fig. 9.3e).

9.2.2 Electrophilic aromatic substitution

The single most characteristic reaction exhibited by aromatic systems is electrophilic aromatic substitution (S_EAr; Fig. 9.4); in this reaction, a group X on the aromatic ring is substituted by an electrophile E; normally, X would be a hydrogen, but it is possible for other groups to be substituted, and if this occurs, the reaction is called ipso substitution. A group which is commonly substituted in this way is silicon group. The mechanism of the process has been studied in great detail and has been shown to proceed by initial formation of a π-complex of the aromatic substrate with the electrophile, followed by formation of a σ-complex; this complex is highly resonance stabilised (Fig. 9.5) and is considered to be a true intermediate in the reaction sequence, although the π-complex is a transition state, occurring at a relative energy maximum. The energetics of the transformation are shown in

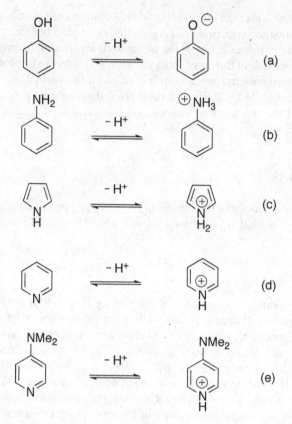

Figure 9.3 (a) Acidity of phenol and (b) basicity of aniline, (c) pyrrole, (d) pyridine and (e) dimethylaminopyridine.

Fig. 9.6. Formation of a new π-complex allowing the release of the proton enables regeneration of the aromatic system. Lewis acids play an important role in the formation of the electrophile prior to the attack on the aromatic system.

A wide range of electrophiles is suitable for this reaction, and these are considered in more detail below.

9.2.2.1   Protonation/deuteration

It is possible to substitute a proton for a deuteron using electrophilic aromatic substitution (Fig. 9.4; X = H, E = D) and this is valuable as it provides a reliable method for the incorporation of deuterium into organic substrates.

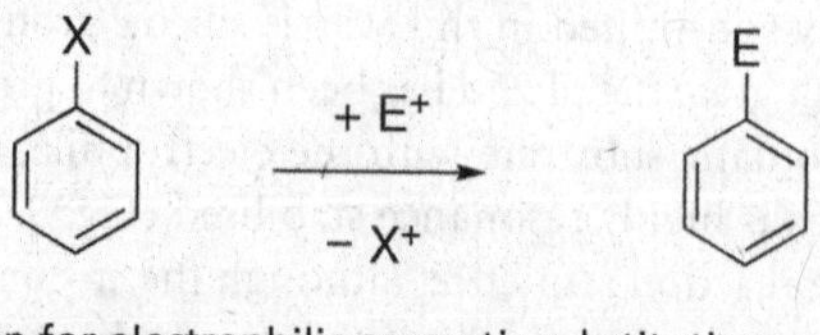

Figure 9.4   General reaction for electrophilic aromatic substitution.

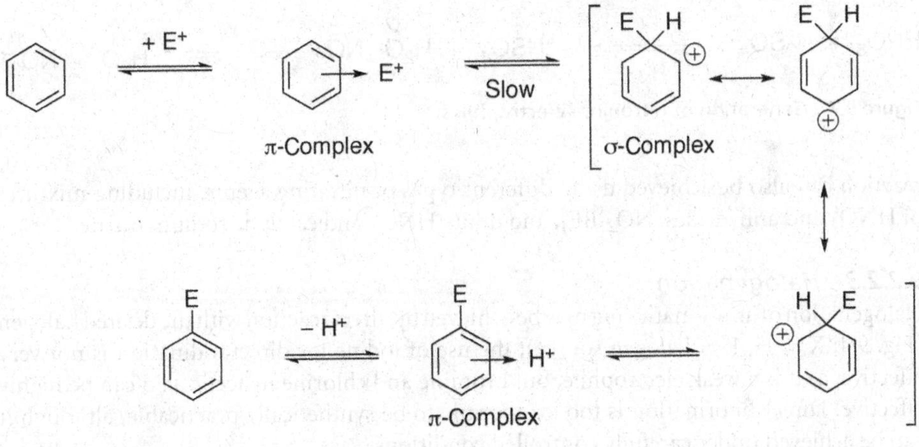

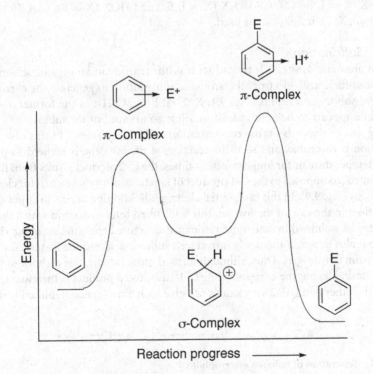

Figure 9.5 Mechanism for electrophilic aromatic substitution.

9.2.2.2 Nitration

When an aromatic substrate is treated with a mixture of concentrated nitric acid and sulfuric acid, that is, a source of NO_2^+ (the so-called nitronium cation), irreversible electrophilic aromatic substitution to generate the corresponding product of nitration results (Fig. 9.4; $X = H$, $E = NO_2$); the generation of the nitronium species arises as shown in Fig. 9.7. This

Figure 9.6 Energetics of electrophilic aromatic substitution.

$$HNO_3 + H_2SO_4 \rightleftharpoons HSO_4^- + H_2\overset{\oplus}{O}-NO_2 \rightleftharpoons H_2O + \overset{\oplus}{NO_2}$$

Figure 9.7 Generation of nitronium electrophiles.

reaction can also be achieved using different types of nitrating agents, including mixtures of HNO_3 and anhydrides, $NO_2 \cdot BF_4$, and dilute HNO_3 and catalytic sodium nitrite.

9.2.2.3 Halogenation

Halogenation of an aromatic ring may be achieved by direct reaction with the desired halogen (Fig. 9.4; X = H, E = halogen source); the use of iodine for direct iodination is not very effective, as it is a weak electrophile, but bromine and chlorine in acetic acid can be highly effective. Direct fluorination is too exothermic to be synthetically practicable, although it can be achieved under carefully controlled conditions.

Halogenation by reaction with a mixture of a halogen (bromine or iodine) and a Lewis acid catalyst is particularly common, being applicable for a wide variety of substrates; the Lewis acid (such as ZnX_2, $AlCl_3$ or $FeCl_3$) assists with the polarisation of the X–X bond to make this species more electrophilic. For X = I and Br, the Lewis acid breaks this bond and generates X^+ according to Fig. 9.8. In the case of chlorine, Cl_2 is sufficiently electrophilic that it will react directly without the formation of Cl^+.

Other sources of electrophilic halogen, other than X–X, include iodine monochloride (ICl), as well as species in which the halogen is the cationic component of a salt (e.g. CH_3CO_2X (X = I, Br, Cl), CF_3CO_2X (X = I, Br) and HOX (X = Br, Cl)). In the case of fluorination, XeF_2 or XeF_4 can be used.

9.2.2.4 Sulfonation

When an aromatic substrate is treated with sulfur trioxide in an organic solvent or concentrated sulfuric acid, electrophilic aromatic substitution to generate the corresponding product of sulfonation results (Fig. 9.4; X = H, E = SO_3H); in the former, the reactive electrophile appears to be SO_3, at least in dilute solutions, but for sulfuric acid the exact attacking species depends on the concentration and water content. In this case, however, the reaction is reversible, and so if the reaction of naphthalene is allowed to proceed at elevated temperature or for longer reaction times, the C-2 product results from thermodynamic control, as opposed to the C-1 product of kinetic control more usual for electrophilic substitutions (Fig. 9.9). In this case, initial electrophilic addition occurs to either the C-1 or C-2 position; in the case of the former, this leads to an intermediate in which the positive charge may be stabilised by one more resonance structure before the aromatic ring is disrupted, but also in which the newly introduced sulfonic acid residue is sterically hindered by the proximal hydrogen. Thus, although formed more rapidly, this carbocation is not so sterically desirable, but the corresponding C-1-susbtituted product is therefore the kinetic one. On the other hand, C-2 attack does not give such a resonance-stabilised carbocation,

$$AlCl_3 + X_2 \longrightarrow AlCl_3X^- + X^+$$

Figure 9.8 Generation of halogen electrophiles.

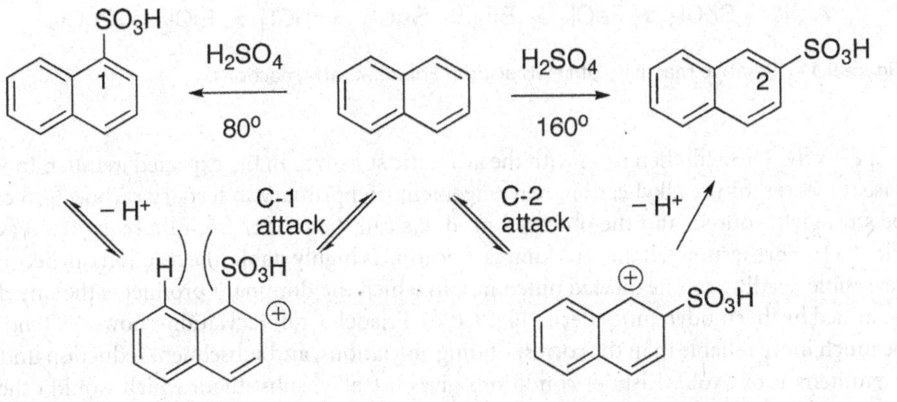

Figure 9.9 Kinetic versus thermodynamic control in sulfonation reactions of naphthalene.

since any resonance stabilisation immediately disrupts the aromatic ring, but nor does this carbocation suffer from destabilising steric interactions; it is therefore thermodynamically more stable product. Because sulfonation is reversible, equilibration of the C-1 product to the C-2 one is possible.

9.2.2.5 Diazo coupling

Diazonium salts, prepared by the reaction of aromatic amines with nitrous acid (itself generated from sodium nitrite and acid; Figs. 9.10a and 9.10b), are sufficiently electrophilic to attack an aromatic ring and give the corresponding diazo coupling product (Fig. 9.4; $X = H$, $E = ArN_2^+$). These compounds are of considerable importance, since they are intensely coloured and form the basis of many modern dyestuffs. Importantly, the UV–visible chromophore of these compounds is readily altered by changing substitution patterns on the aromatic rings; thus, electron-releasing groups tend to shift the wavelength of absorption to the blue end of the spectrum and electron-withdrawing groups to the red end.

9.2.2.6 Friedel–Crafts alkylation and acylation

The reaction of alkyl and acyl halides with aromatic substrates in the presence of a Lewis acid, normally aluminium halides, is called Friedel–Crafts alkylation and acylation respectively (Fig. 9.4; $X = H$, $E = R$ or $RC(O)$). However, a range of other Lewis acids can be used, and their order of reactivity is shown in Fig. 9.11. The reaction proceeds by initial coordination of the Lewis acid to the alkyl or acyl halide (Figs. 9.12a and 9.12b), which assists the removal of the leaving group from the halide substrate, to generate an alkyl or acylium cation,

$$HCl + NaNO_2 \longrightarrow HONO + NaCl \qquad \text{(a)}$$

$$ArNH_3 + HONO \longrightarrow Ar\overset{+}{N}\equiv N + H_2O \qquad \text{(b)}$$

Figure 9.10 Diazotisation of aromatic amines.

$$AlCl_3 > SbCl_5 > FeCl_3 > BF_3 > SnCl_4 > TiCl_4 > BiCl_3 > ZnCl_2$$

Figure 9.11 Relative reactivity of Lewis acids in Friedel–Crafts reactions.

respectively. This will then react with the aromatic substrate in the expected manner. In the case, however, of the alkyl cation, rearrangement of a primary to tertiary carbocation can be strongly favoured, and the observed products can then differ from those expected (see Fig. 4.11). Furthermore, if the attacking carbocation is highly stable, the reaction can become reversible, leading to unexpected outcomes, in which the dominant product is the one determined by thermodynamic effects (Fig. 9.12c). Friedel–Crafts acylations, however, tend to be much more reliable than the corresponding alkylations, and subsequent reduction under Clemmensen or Wolff–Kishner conditions gives the alkyl substituent which would otherwise not be available by Friedel–Crafts alkylation. One problem, however, can arise when loss of carbon monoxide (CO) from the attacking acylium ion $RC(O)^+$ if the corresponding carbocation R^+ is very stable.

Reactive electrophiles related to that of the Friedel–Crafts reaction can be generated under other conditions. For example, the Vilsmeier–Haack reaction generates the iminium electrophile $ClC(R)=NMe_2^+$ by treating an amide with phosphoryl chloride (Fig. 9.13b). This is typically dimethyl formamide, in which case $R = H$. This electrophile, however, being weaker than the carbocation of the Friedel–Crafts process, will react only with particularly activated aromatic systems, such as anilines and phenols, and heterocyclic systems, such as indoles. Another reaction related to the Friedel–Crafts set of processes includes the Gatterman–Koch reaction (Fig. 9.13a); this leads to formylation of the aromatic ring and is achieved by treating an aromatic ring with carbon monoxide and HCl in the presence of aluminium chloride, under which conditions the reactive electrophile is likely to be $[HC=O]^+$. In the Fries rearrangement, the electrophile is generated as an intermediate by the reaction of an ester with the Lewis acid catalyst (Fig. 9.13c). Hydroxyalkylations may be achieved

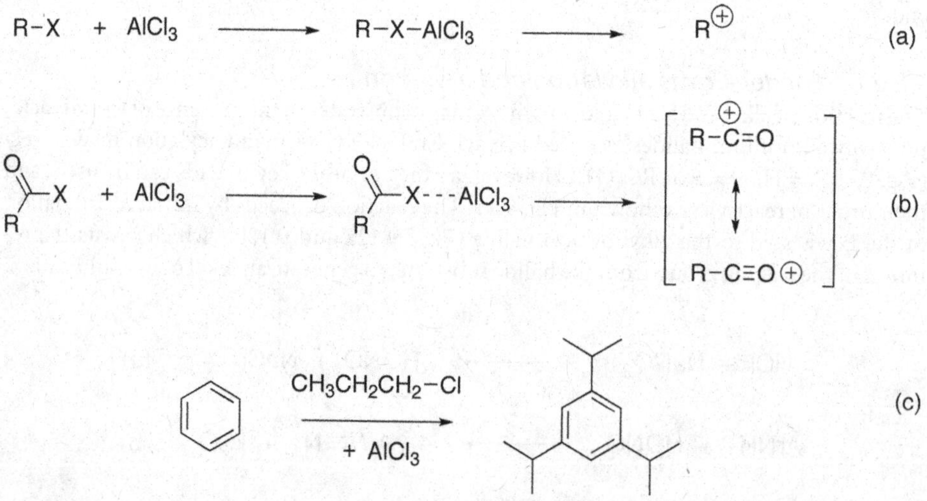

Figure 9.12 Generation and reactivity of Friedel–Crafts electrophiles.

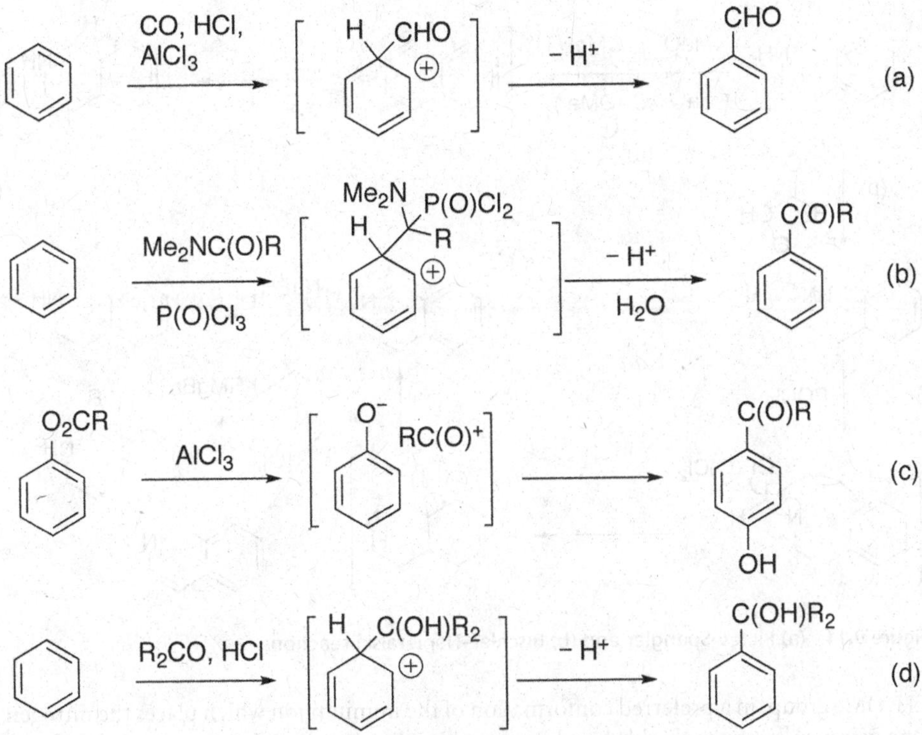

Figure 9.13 Electrophilic substitution of benzene: (a) Gatterman–Koch reaction, (b) Vilsmeier–Haack reaction, (c) Fries rearrangement and (d) hydroxyalkylation.

by treating an aromatic system with an aldehyde under acidic conditions; this generates an oxonium species which is readily attacked by the nucleophilic aromatic system (Fig. 9.13d).

It is possible for electrophilic aromatic substitutions to be conducted internally in such a way that ring closures occur; in this case, ring closure is assisted by the proximal relationship of nucleophilic and electrophilic centres. For example, the Pictet–Spengler reaction, in which a β-arylethylamine is reacted with an aldehyde or aldehyde equivalent, firstly generates an intermediate iminium species, which then cyclises by an intramolecular S_EAr reaction, leading to the formation of the tetrahydroisoquinoline product (Fig. 9.14a). An alternative approach, leading to a similar outcome, is the Bischler–Napieralski reaction (Fig. 9.14b), which starts from the same β-arylethylamine; this time, formylation, followed by treatment with phosphorus oxychloride, generates an iminium intermediate, which ring closes in an analogous manner by an S_EAr mechanism. The resulting imine can be reduced or reacted with an organometallic reagent to give either the C-2-unsubstituted or C-2-substituted product respectively. An asymmetric variant of this process has been developed by the inclusion of a chiral auxiliary (Fig. 9.15); thus, reaction of phenylacetyl chloride with a chiral hydrazone gives the expected amide. Reduction of the amide (lithium aluminium hydride, LAH), formylation and reaction with $POCl_3$ leads to a ring closure followed by elimination to generate an iminium ion, which upon reaction with an organometallic derivative gives the product tetrahydroisoquinoline with high diastereoselectivity; the stereochemistry of this addition is controlled by steric preference for addition of the nucleophile *anti-* to the

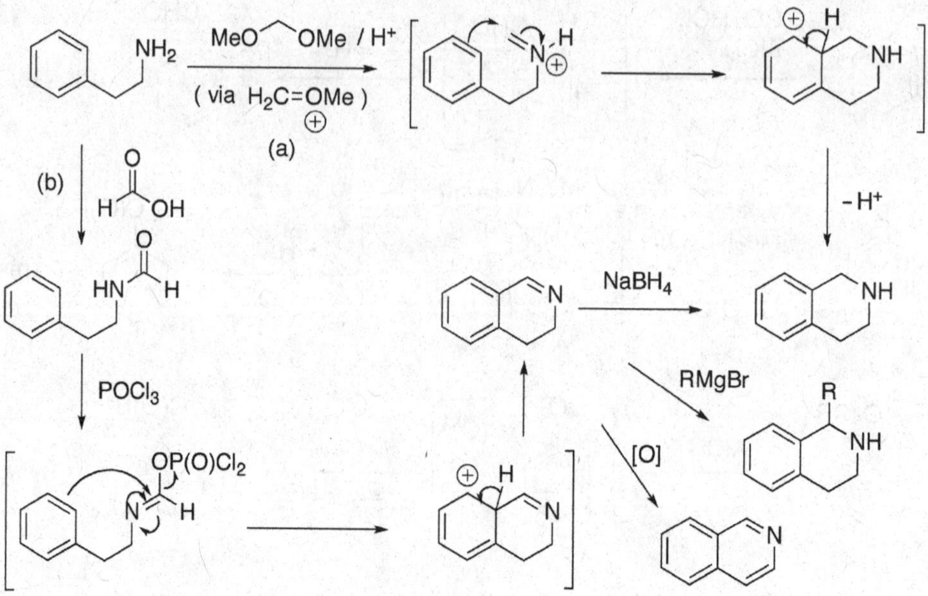

Figure 9.14 (a) Pictet–Spengler and (b) Bischler–Napieralski reactions.

CH_2OMe group, in a preferred conformation of the iminium ion which places the nitrogen lone pair parallel to the double bond so as to minimise $A_{1,3}$ strain.

Alternatively, the ring closure leading to tetrahydroisoquinolines can be done distal to the nitrogen (Fig. 9.16) in a reaction called the Pomeranz–Fritsch synthesis. In this case, ring closure of an acetal derived from benzylamine occurs under acidic conditions, leading to a dihydroisoquinoline; further reduction generates the product tetrahydroisoquinoline. This process can be conducted in a stereoselective sense (Fig. 9.17); chiral ketone reduction, followed by Mitsunobu reaction and displacement with a tosyl amine, gives the α-arylethylamine product. Acid-catalysed ring closure, hydrogenation and deprotection

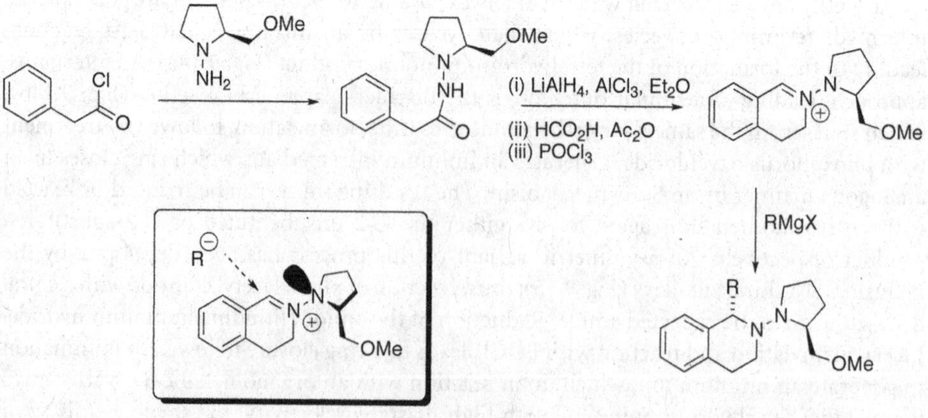

Figure 9.15 Asymmetric Bischler–Napieralski reaction.

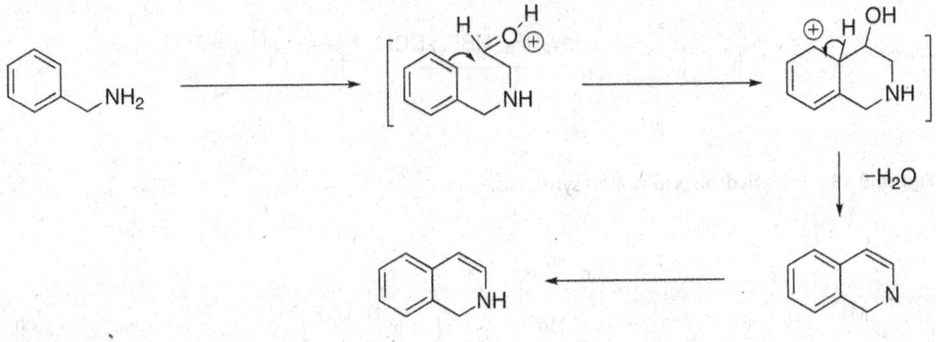

Figure 9.16 Pomeranz–Fritsch synthesis.

under reductive conditions led to the desired product in high enantioselectivity. Although this type of reaction is typically done using acid catalysis, other electrophiles will suffice; thus, conversion of a suitable alkene to the iodonium intermediate leads to ring closure via an aromatic electrophilic substitution (Fig. 9.18). This type of ring closure can even be done in such a way that more than two components can be coupled; thus, reaction of aniline with benzaldehyde gives the corresponding imine, and this can be coupled with an alkene under acidic conditions (Fig. 9.19) to give a tetrahydroquinoline.

Pyridines are more electron poor than a simple benzene due to the presence of the electron-withdrawing nitrogen atom, and the highest occupied molecular orbital (HOMO) energy is therefore lowered. They are therefore intrinsically less reactive to electrophilic attack than is benzene, and this reactivity is lowered further if the electrophile interacts with the nitrogen atom (Fig. 9.20a), placing a positive charge on the nitrogen atom; typical S_EAr reactions effective for benzene therefore proceed only very slowly with pyridine. If the reaction does occur, however, attack at C-3, rather than at C-2 or C-4, occurs, because this ensures that no resonance canonical form occurs in which the nitrogen atom carries a positive charge (Figs. 9.20b–9.20d). For this reason, the pyridine ring will be more readily

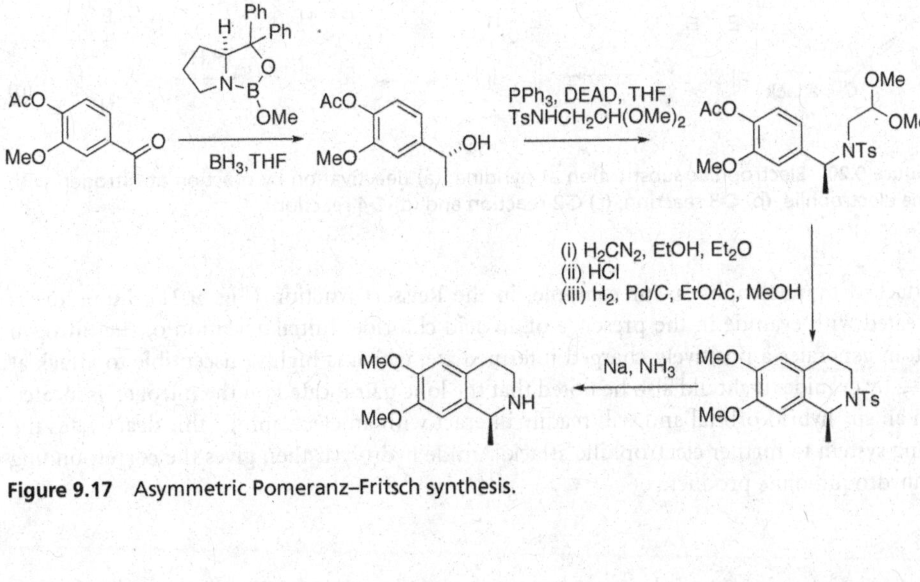

Figure 9.17 Asymmetric Pomeranz–Fritsch synthesis.

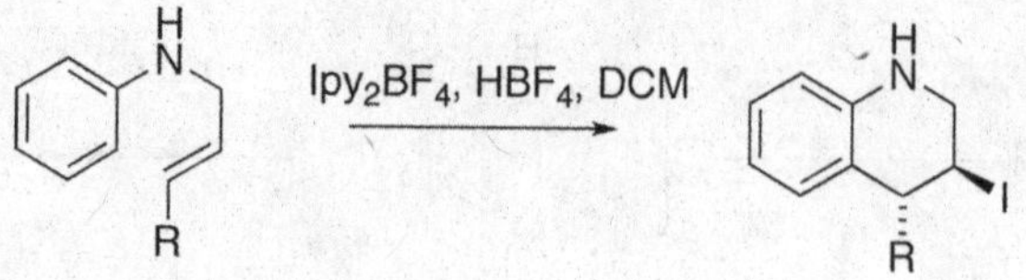

Figure 9.18 Tetrahydroisoquinoline synthesis.

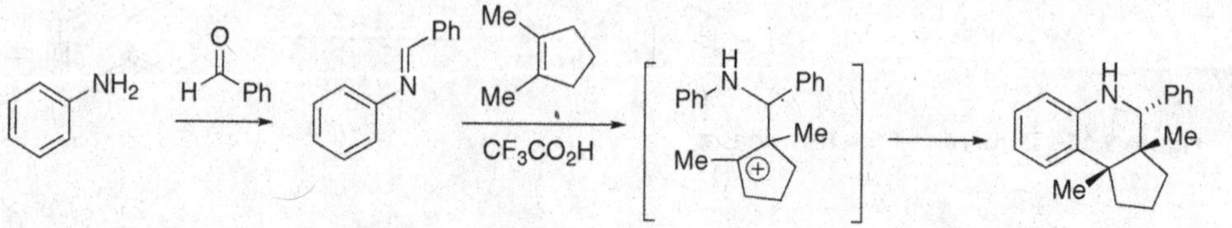

Figure 9.19 Tetrahydroquinoline synthesis.

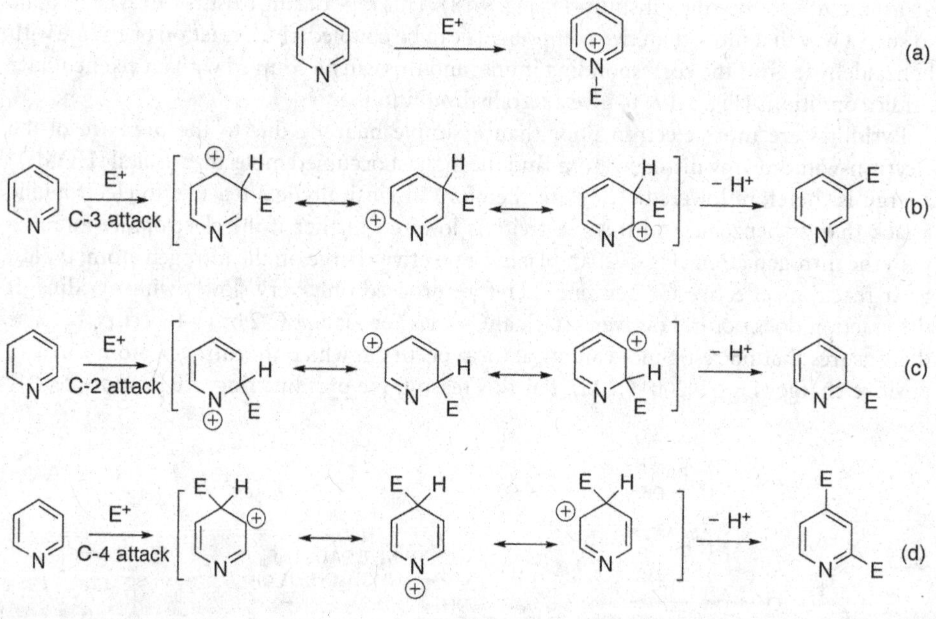

Figure 9.20 Electrophilic substitution at pyridine: (a) deactivation by reaction at nitrogen with the electrophile, (b) C-3 reaction, (c) C-2 reaction and (d) C-4 reaction.

attacked by nucleophiles; for example, in the Reissert reaction (Fig. 9.21), quinoline is treated with cyanide in the presence of an acid chloride. Initial acylation of the nitrogen atom generates a positively charged intermediate, which is highly susceptible to attack at C-2 by cyanide. It should also be noted that the lone pair located on the nitrogen is located in an sp^2 hybrid orbital and will readily interact with an electrophile; this deactivates the ring system to further electrophilic attack. Amide hydrolysis then gives the corresponding dihydroquinoline product.

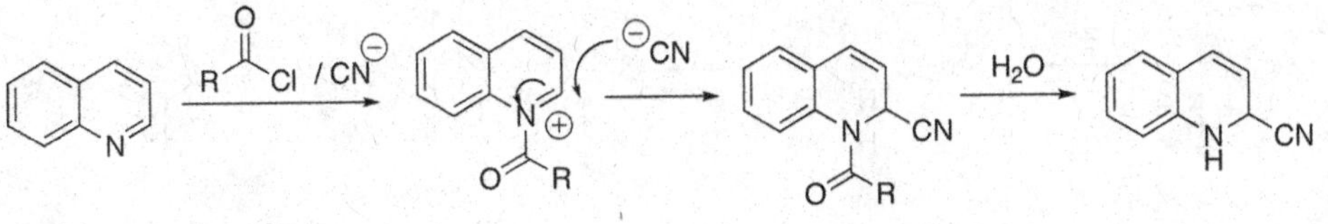

Figure 9.21 Reissert reaction.

The reaction of five-membered ring aromatic systems (X = O, N, S) is slightly different again; these systems do not possess the same level of aromatic stabilisation of benzene, and pyrrole, like pyridine, can be deactivated by interaction of the nitrogen lone pair with the electrophile (Fig. 9.22a). Pyrrole will nonetheless react effectively with electrophiles to substitute at either of the C-2 or C-3 positions; the former is marginally favoured because there is one more canonical form which stabilises the intermediate (Figs. 9.22b and 9.22c). However, this selectivity is reversed in indoles (Figs. 9.23a and 9.23b), in which C-3 selectivity is preferred because such a reaction maintains the aromaticity of the benzene system throughout the reaction, and the intermediate cation has two resonance structures which do not disrupt this aromaticity; in C-2 reaction, there is no resonance stabilisation unless the aromaticity of the carbocyclic ring is disrupted. However, if the C-3 position is blocked, then the substitution reaction to C-2 is enforced (Fig. 9.23c).

9.2.2.7 Metallation
It is possible to directly metallate suitable aromatic rings (those which are not deactivated by an electron-withdrawing group) with mercury(II), thallium(III) or lead(IV) acetates or halides; the reactive electrophile is the metal cation in these cases (Fig. 9.24). Although these reagents are toxic, and the reactions rarely used, the reaction has synthetic merit since it provides direct access to products which are otherwise difficult to obtain. The resulting organometallic derivatives are themselves of substantial value, since they can be further transmetallated and reacted on to generate a variety of functional groups (see Chapter 12).

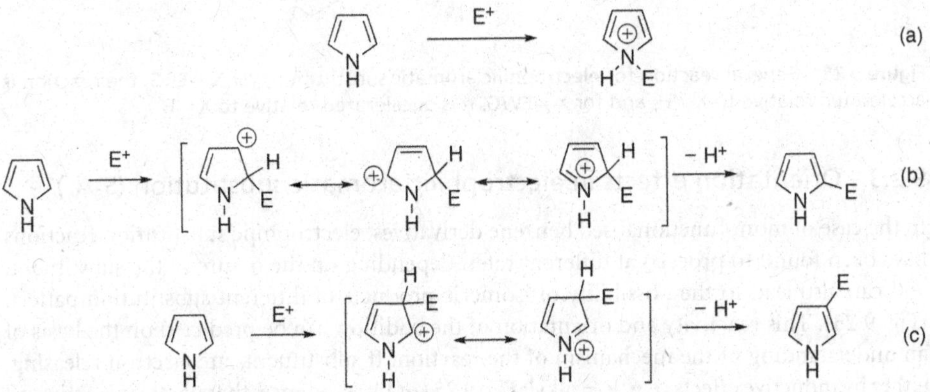

Figure 9.22 Electrophilic substitution at pyrrole: (a) deactivation by reaction at nitrogen with the electrophile, (b) C-2 reaction and (c) C-3 reaction.

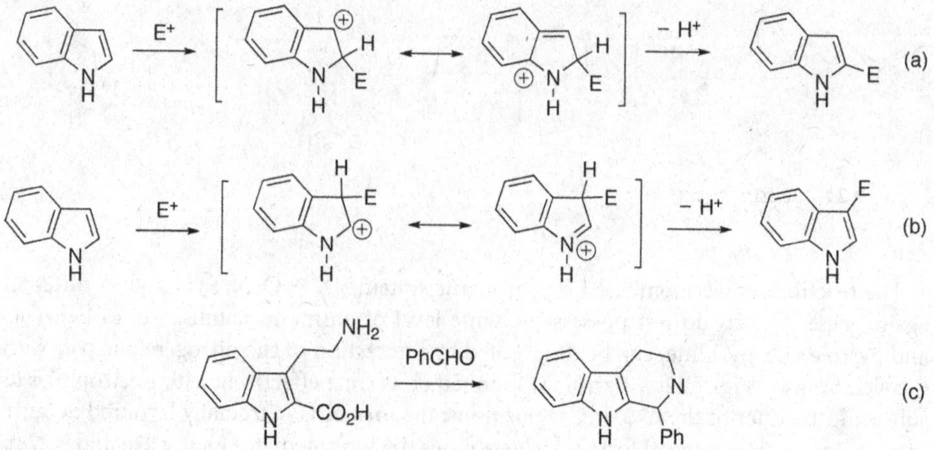

Figure 9.23 Electrophilic substitution at indole: (a) C-2 reaction, (b) C-3 reaction and (c) C-2 reaction with ring closure.

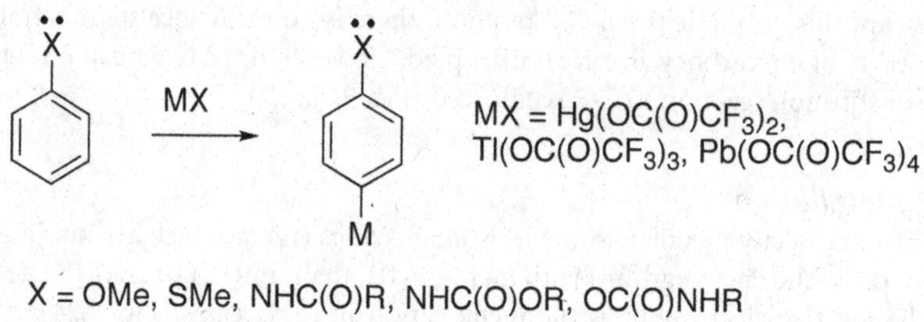

MX = Hg(OC(O)CF$_3$)$_2$,
Tl(OC(O)CF$_3$)$_3$, Pb(OC(O)CF$_3$)$_4$

X = OMe, SMe, NHC(O)R, NHC(O)OR, OC(O)NHR

Figure 9.24 Direct metallation of an aromatic ring.

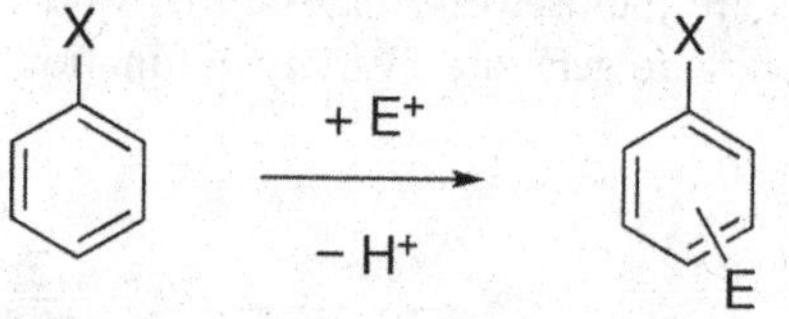

Figure 9.25 General reaction for electrophilic aromatic substitution; for X = ERG, the reaction is accelerated relative to X = H, and for X = EWG, it is decelerated relative to X = H.

9.2.3   Orientation effects in electrophilic aromatic substitution (S$_E$Ar)

In the case of monofunctionalised benzene derivatives, electrophilic substitution reactions have been found to proceed at different rates, depending on the nature of the substituent, and can also lead to the possibility of isomeric products of different substitution pattern (Fig. 9.25). This reactivity and orientation of the addition can be predicted on the basis of an understanding of the mechanism of the reaction. If substituents are electron releasing, either by inductive effects (e.g. R = alkyl; Fig. 9.25) or by resonance, the substrate is activated relative to benzene, and the reaction proceeds at a faster rate than benzene; unsurprisingly, if it is electron withdrawing, then it is deactivated relative to benzene, and the reaction proceeds

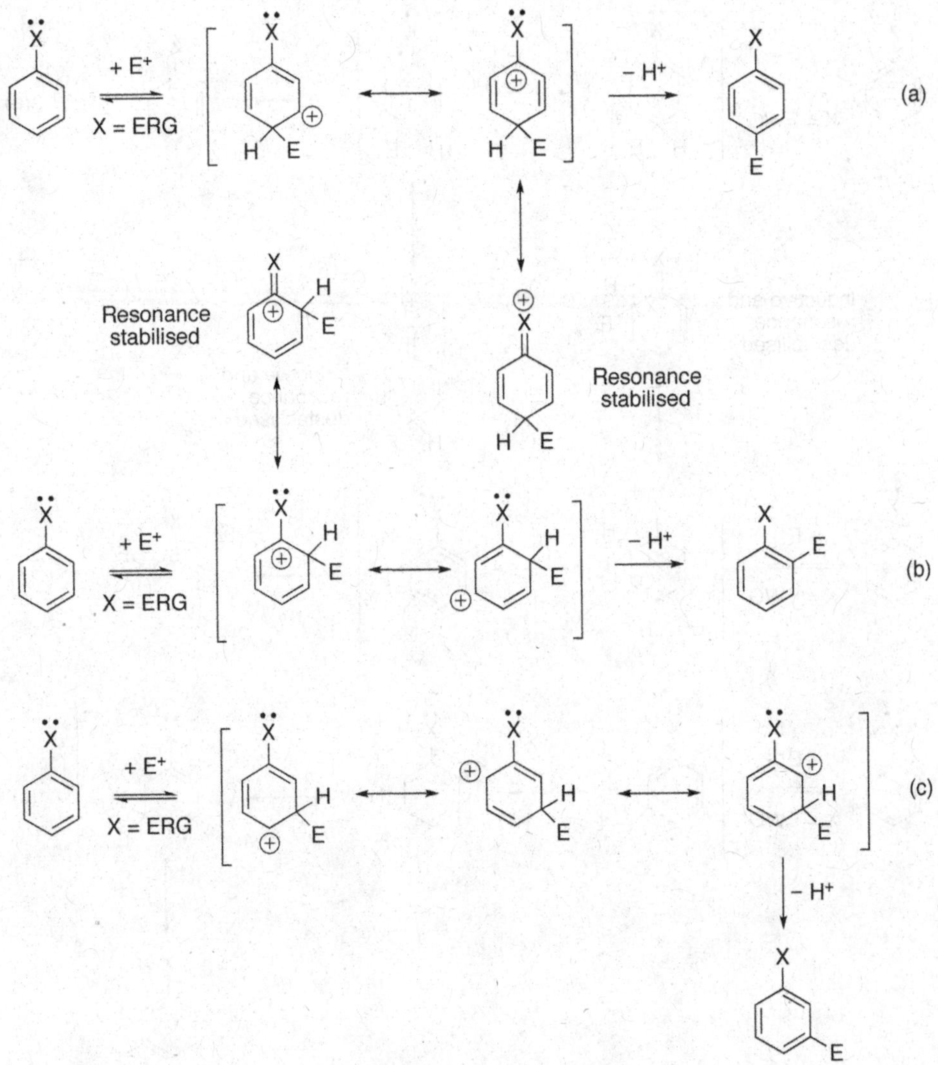

Figure 9.26 Orientation effects in electrophilic aromatic substitution for X=ERG (e.g. X=OR, SR, NR$_2$): (a) *p*-substitution, (b) *o*-substitution and (c) *m*-substitution.

more slowly than for benzene (e.g. R = NO$_2$; Fig. 9.25). However, it has been observed that in addition to altering the rate of reaction, electron-releasing groups (i.e. groups carrying a lone pair capable of resonance donation) favour reaction at the ortho and para positions, and this arises due to the intermediacy of a particularly favourable resonance stabilisation in which the positive charge can be delocalised onto the electron-releasing substituent; these groups are said to be ortho and para directing (Figs. 9.26a and 9.26b). This is in stark contrast to the meta position, where similar delocalisation is not possible (Fig. 9.26c). On the other hand, electron-withdrawing substituents are meta directing and deactivating; in this case, the positive charge needs to be kept as far away from the substituent as possible,

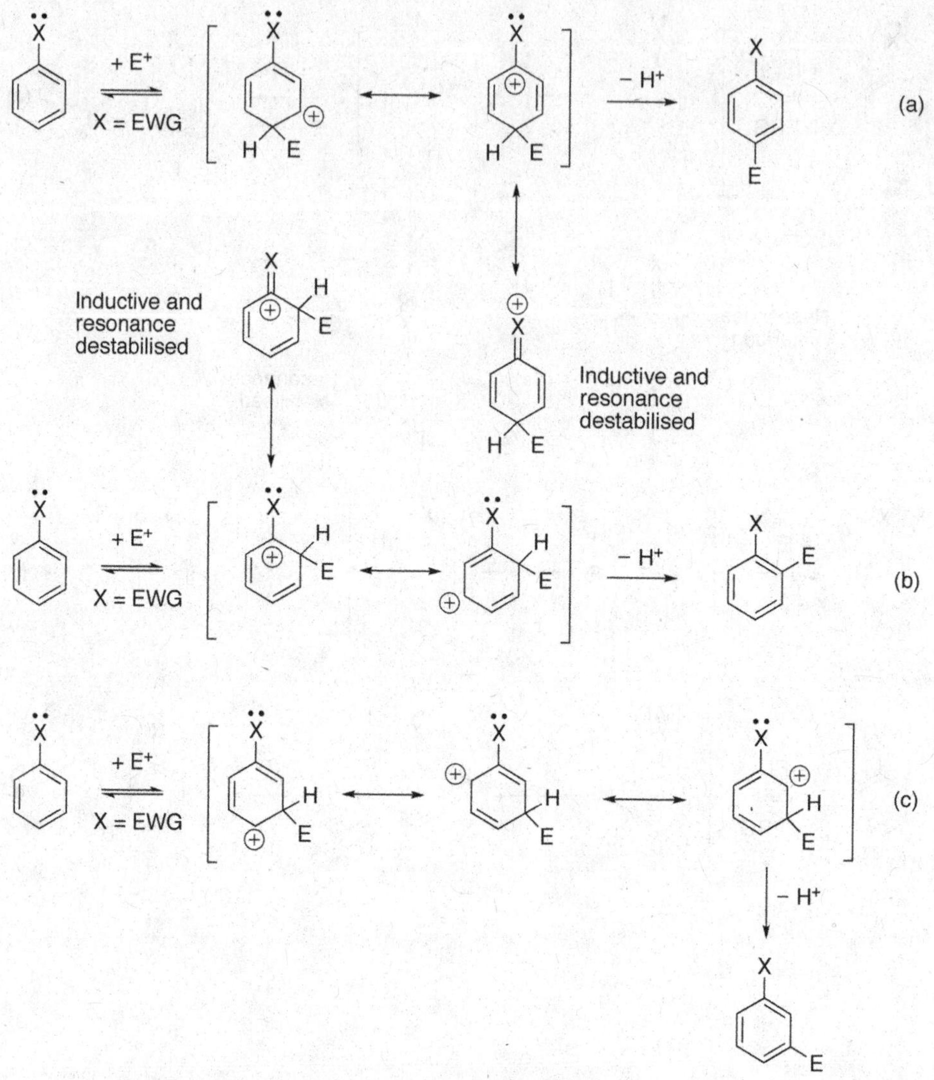

Figure 9.27 Orientation effects in electrophilic aromatic substitution for X=EWG (e.g. X=CO₂R, C(O)R, CN): (a) *p*-substitution, (b) *o*-substitution and (c) *m*-substitution.

and this is best achieved by attack at the meta position (Figs. 9.27a–9.27c). Halogens form an important class of substituent, since they are ortho and para directing by virtue of their lone pair, but are deactivating due to their strong inductive electron-withdrawing effects (Figs. 9.28a–9.28c). These outcomes are summarised in Table 9.1. Note that inductive effects are weaker than resonance effects, and the outcome of a reaction is always determined from a consideration of resonance effects. To a first approximation, it can be considered that multiple substitution is additive based on the groups initially present on the aromatic ring.

An important variant of this process is substitution in arylsilanes. In Chapter 4, the stabilisation of β-positive charges relative to a silicon atom was indicated; this results from

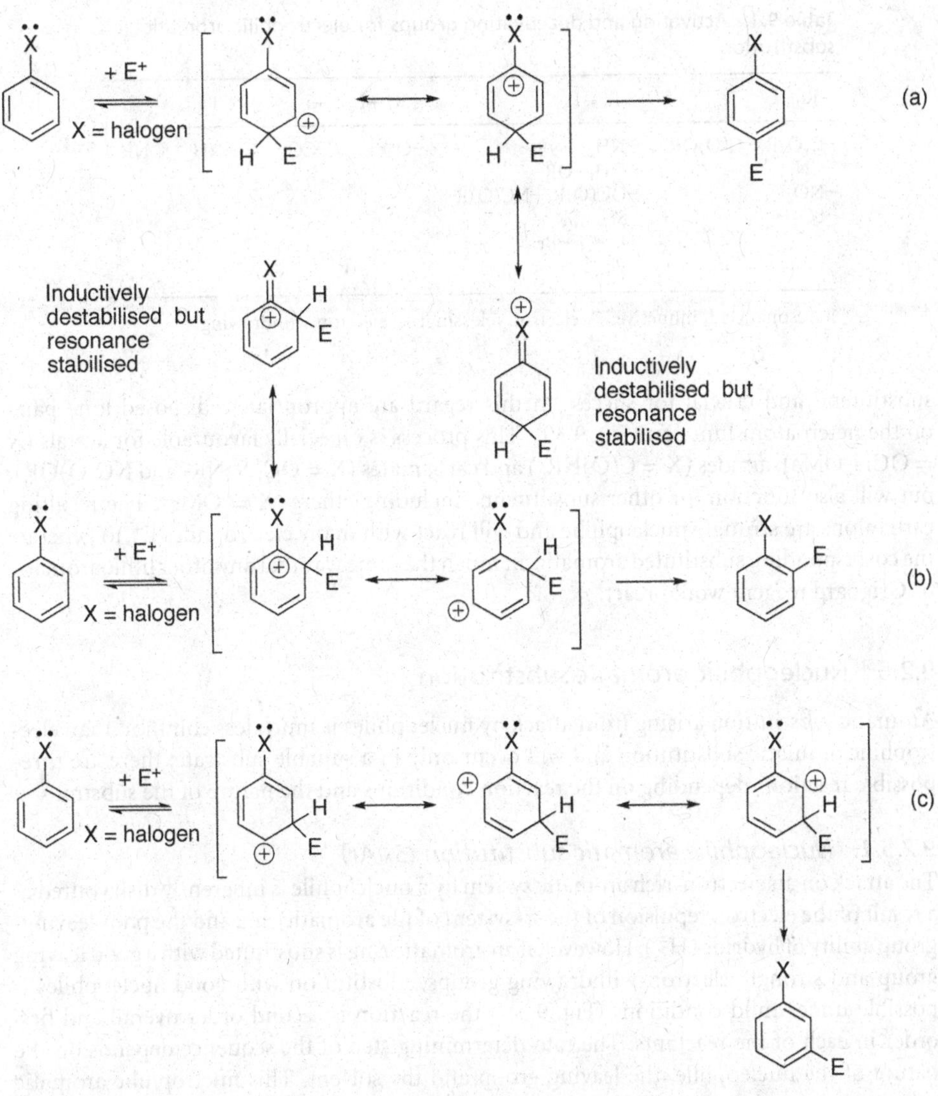

Figure 9.28 Orientation effects in electrophilic aromatic substitution for X = halogen (Cl, Br, I): (a) *p*-substitution, (b) *o*-substitution and (c) *m*-substitution.

favourable orbital overlap of the filled σ-orbital of the carbon–silicon bond with the proximal empty p orbital. In an aromatic system, a silyl substituent strongly supports ipso electrophilic attack, since the carbocation which is generated ends up in a similar β-relationship to the silicon substituent (Fig. 9.29). Nucleophilic attack on silicon then regenerates the aromatic system.

9.2.4 *o*-Lithiation

o-Lithiation of aromatic rings using BuLi or *s*-BuLi with heteroatom-containing substituents is favoured; this results from internal chelation of the metal cation with the adjacent

Table 9.1 Activating and deactivating groups for electrophilic aromatic substitution

$-R, -I$	$+R, -I$	$+R, +I$	$+I$	$-I$
$-C(O)R, -C(O)OR$	$-NH_2, -NR_2$	$-O^-$	$-CO_2^-$	$-NH_3^+, -NR_3^+$
$-CN$	$-OH, -OR$		$-R$	
$-NO_2$	$-OC(O)R, -NC(O)R$			
$-SO_3H$	$-SH, -SR$			
	$-I, -Cl, -Br$			
	$-Ph$			

R, resonance; I, inductive; +, electron releasing; −, electron withdrawing.

substituent, and crucial for success in this regard are appropriately disposed lone pairs on the heteroatom function (Fig. 9.30). This process is especially favourable for acetals (X = OCH_2OMe), amides (X = $C(O)NR_2$) and carbamates (X = $OC(O)NR_2$ and $NC(O)OR$), but will also function for other substituents, including ethers (X = OMe). The resulting carbanions are strongly nucleophilic and will react with many electrophiles E^+ to generate the corresponding substituted aromatic, in much the same way that any other lithioaromatic or Grignard reagent would react.

9.2.5 Nucleophilic aromatic substitution

Aromatic substitution arising from attack by nucleophiles is much less common than electrophilic aromatic substitution and will occur only in a suitable substrate; there are three possible reactions depending on the reaction conditions and the nature of the substrate.

9.2.5.1 Nucleophilic aromatic substitution (S_NAr)

The attack on an electron-rich aromatic system by a nucleophile is inherently disfavoured as a result of the electron repulsion of the π-system of the aromatic ring and the poor leaving-group ability of hydride (H^-). However, if an aromatic ring is substituted with a good leaving group and strongly electron-withdrawing groups, substitution with good nucleophiles is possible under mild conditions (Fig. 9.31); the reaction is second order overall and first order in each of the reactants. The rate-determining step of the sequence depends on the nature of the nucleophile, the leaving group and the solvent. This nucleophilic aromatic substitution reaction generally requires activation with at least two nitro groups or similarly electron-withdrawing groups (Fig. 9.32); these both make the aromatic ring susceptible to attack by the nucleophile and stabilise by resonance the anion (the Meisenheimer complex) once formed. The reaction is more favoured when it can occur in an intramolecular sense, such as the Smiles rearrangement. One reaction, however, of particular note is the Sanger reagent, which is widely used for the identification of the N-terminal groups of peptides

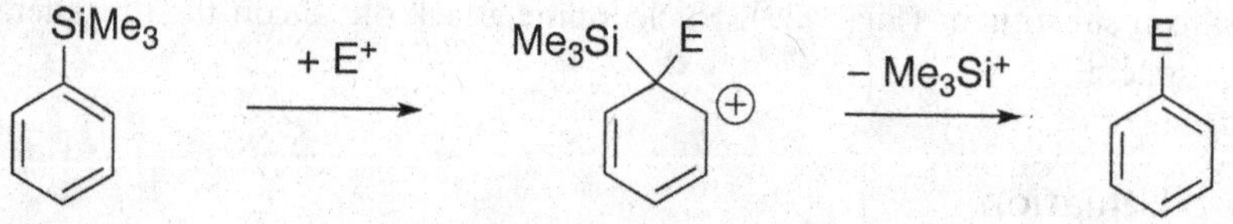

Figure 9.29 General reaction for ipso electrophilic aromatic substitution.

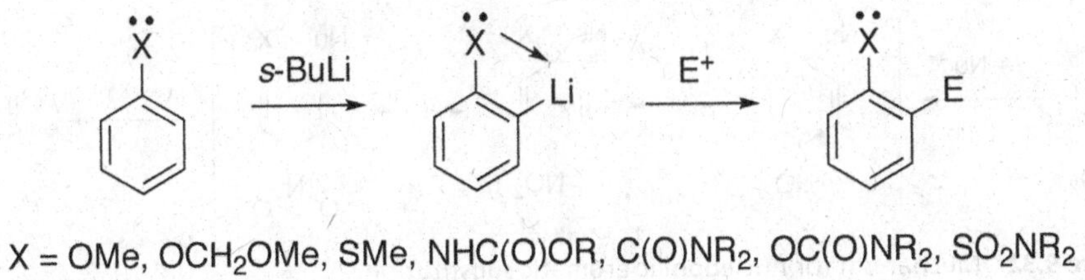

X = OMe, OCH₂OMe, SMe, NHC(O)OR, C(O)NR₂, OC(O)NR₂, SO₂NR₂

Figure 9.30 General process for *o*-lithiation.

(Fig. 9.33); reaction of the terminal amine with the Sanger reagent by addition–elimination with a highly electron-poor dinitroaromatic gives a strongly UV-active residue which can be readily identified after amino acid hydrolysis.

Nucleophilic substitution with loss of hydride does, however, occur in pyridines; in the Chichibabin reaction (Fig. 9.34), attack by amide anion on the electron-deficient ring of pyridine is followed by rearomatisation by loss of hydride. Evolution of hydrogen gas then leads to product formation.

9.2.5.2 *Reaction via diazonium salts (S_N1)*
Treatment of aniline with nitrous acid, generated from sodium nitrite under acidic conditions, leads to the formation of a diazonium salt (Fig. 9.35). These diazonium salts are unstable, and if treated with a variety of nucleophiles in the presence of a metal catalyst, normally Cu(I)X (X = Cl, Br, CN), these will undergo substitution with loss of nitrogen gas, thereby providing a powerful driving force for the reaction (Fig. 9.35). This is called the Sandmeyer reaction, and the reaction most likely proceeds via the intermediacy of aromatic radicals generated by reduction of the carbocations formed in the presence of the metal catalyst. The reaction is preparatively of value because firstly the generation of diazonium salts (from an aniline and nitrous acid) is very easy and secondly a variety of new substituents may be introduced into the ring, including chloride, bromide and cyanide.

9.2.5.3 *Benzyne reaction*
Aromatic halides will undergo elimination in the presence of a strong base, typically sodamide or potassium amide in ammonia solvent (Fig. 9.36). The elimination leads to an aryne or benzyne intermediate; a key piece of the evidence for such a species is that labelled chlorobenzene gives approximately a 1:1 mixture of direct substitution at the 1-position and adjacent substitution at the 2-position when treated with amide anion. In contrast to normal alkynes, benzyne possesses a highly electrophilic triple bond and will be readily attacked by nucleophiles such as alkoxides, amide anions (R_2N^-) or carbanions (Fig. 9.37a), or participate as a highly reactive dienophile in pericyclic cycloaddition processes

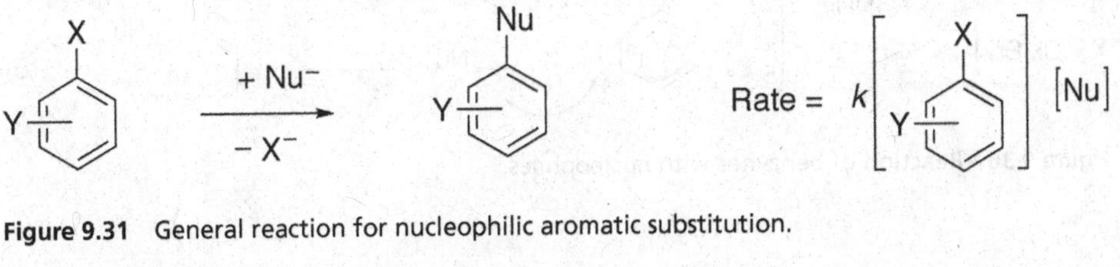

$$\text{Rate} = k\left[\begin{array}{c} X \\ Y \end{array} \right][\text{Nu}]$$

Figure 9.31 General reaction for nucleophilic aromatic substitution.

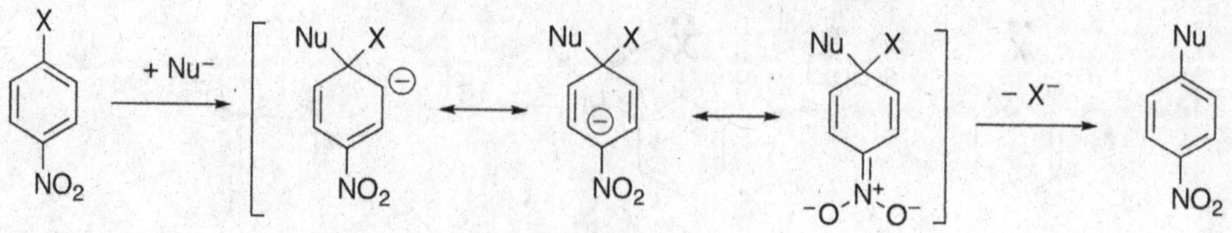

Figure 9.32 Mechanism for nucleophilic aromatic substitution.

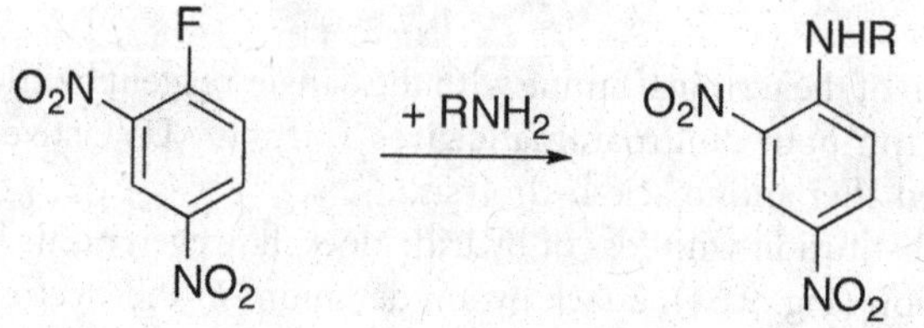

Figure 9.33 Sanger reagent.

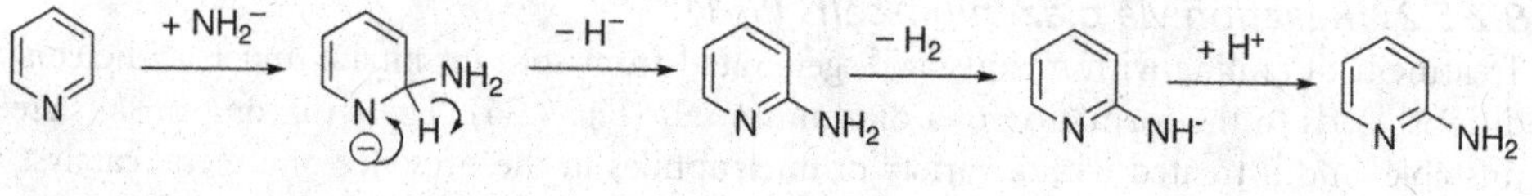

Figure 9.34 Chichibabin reaction.

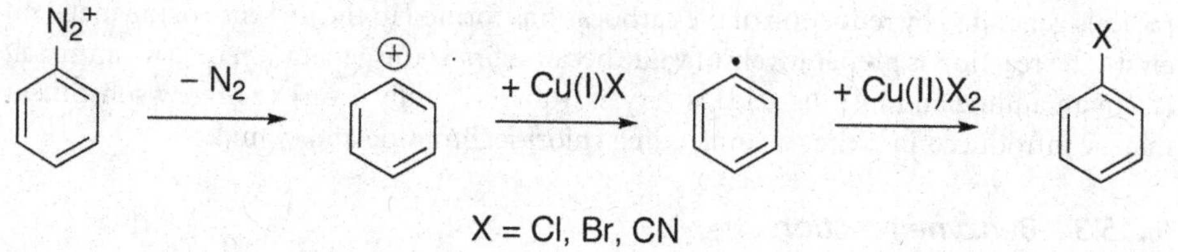

X = Cl, Br, CN

Figure 9.35 Sandmeyer reaction.

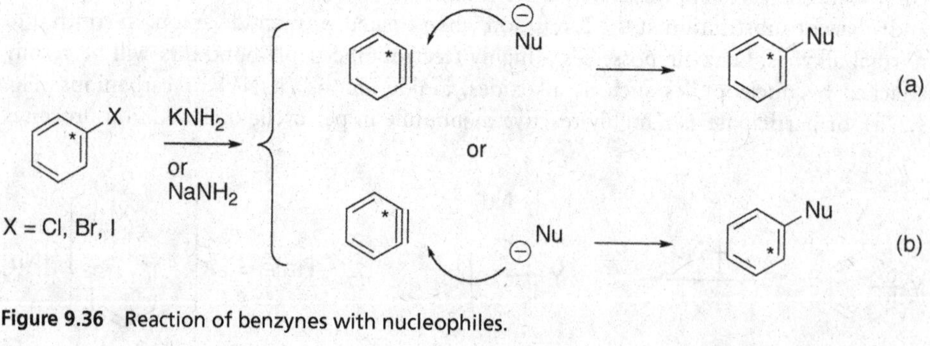

X = Cl, Br, I

Figure 9.36 Reaction of benzynes with nucleophiles.

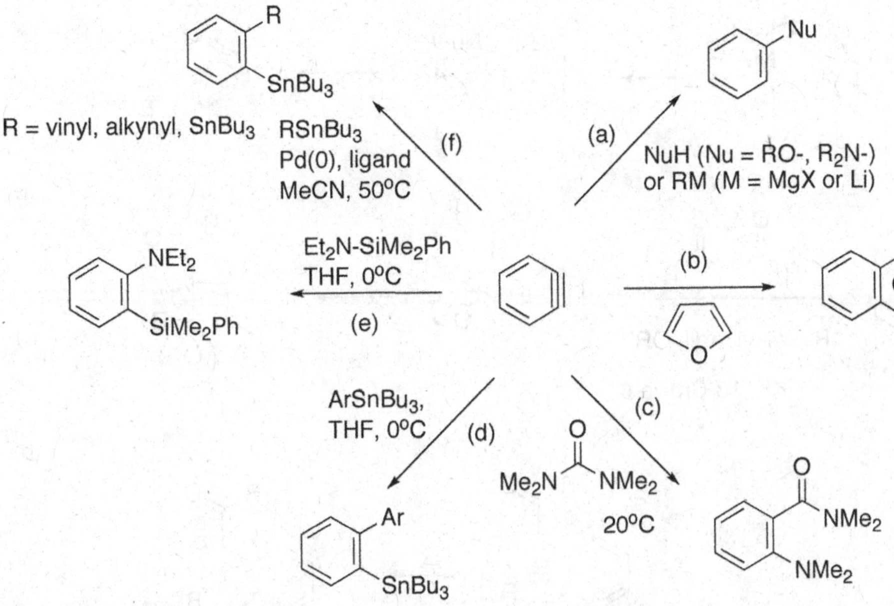

Figure 9.37 Trapping of the benzyne intermediate.

(Fig. 9.37b). The reaction is regioselective, and this addition in substituted aromatics occurs so as to give the more stabilised carbanion intermediate. Note that this anion is not resonance stabilised; its stability comes from the presence of one or more inductive effects which operate. More recent work has indicated that the benzyne intermediate can insert directly into a σ-bond which comprises nucleophilic and electrophilic components, in a scheme which may be generalised as shown in Fig. 9.38a, including carbonyl–nitrogen, sulfur–tin and silicon–nitrogen bonds (Figs. 9.37c–9.37e); the reactions proceed by initial attack of the nucleophilic component, leading to an aryl anion which is immediately trapped by the electrophilic component. Additionally, carbon–tin and tin–tin (Fig. 9.37f) will also react to give the expected insertion products, provided palladium(0) catalysis in the presence of suitable ligands is applied. Addition of β-dicarbonyls has also been found to lead to simultaneous formation of two carbon–carbon bonds (Fig. 9.38b): Initial nucleophilic addition of the enolate of the β-dicarbonyl substrate to give an aryl anion is followed by intramolecular attack at the remaining carbonyl group, leading to a cyclobutene which collapses to the product. These reactions tolerate substitution on the ring of the benzyne intermediate and provide a novel entry into 1,2-difunctional aromatic products of important synthetic value.

9.2.5.4 Nucleophilic aromatic substitution (radical) (S_RN1)

A mechanistically unusual process which results in the direct substitution of aromatic rings is the S_RN1 process (Fig. 9.39). The reaction is commonly used for aryl groups carrying leaving groups (such as halide, quaternary ammonium groups and phosphate) and for nucleophiles such as enolate, thiolate and amide anions, but not harder species such as alkoxides and aryloxides.

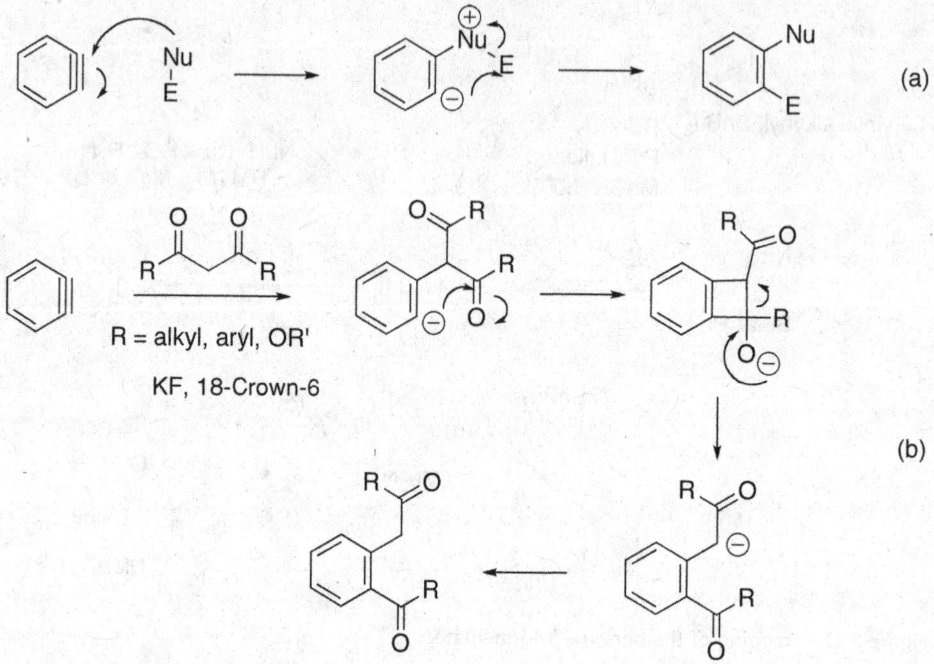

Figure 9.38 Internal trapping of the benzyne intermediate.

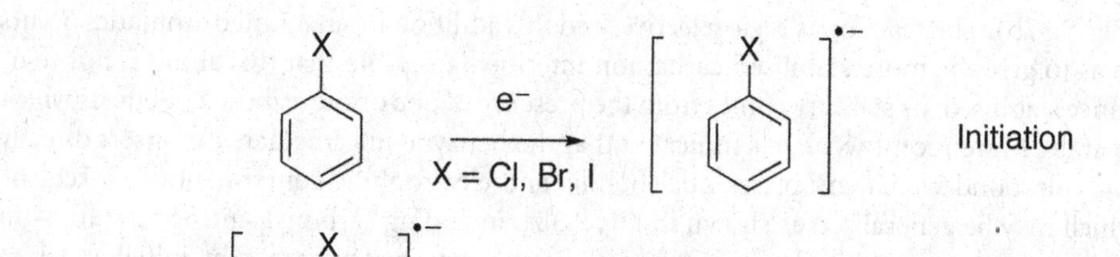

Figure 9.39 The S$_R$N1 reaction.

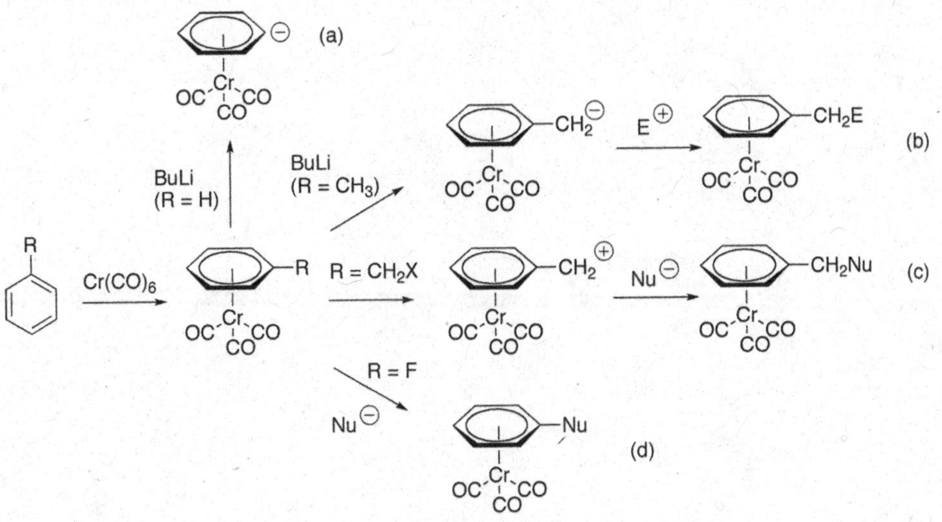

Figure 9.40 Arene metal complexes and their reactivity.

9.2.6 Arene chromium tricarbonyl complexes

In addition to their nucleophilicity, aromatic rings have the potential to be excellent ligands; they will chelate very effectively with transition metals, such as chromium, cobalt and iron, to generate organometallic derivatives which are valuable synthetic intermediates. Coordination to chromium hexacarbonyl, for example, results in acidification of the ring as well as the benzylic protons by electron withdrawal, making the corresponding carbanions more readily available by treatment with strong base such as butyllithium (Figs. 9.40a and 9.40b). The chromium residue also stabilises benzylic carbocations by electron release (Fig. 9.40c), as well as activates the ring to nucleophilic attack if a suitable leaving group is present (Fig. 9.40d). The chromium metal can be easily removed under oxidative conditions, typically by air, to regenerate the aromatic system.

Chapter 10
Sequential Addition and Elimination Reactions

In previous chapters, we have considered addition and elimination as individual and unrelated processes, but it is not uncommon for them to proceed in sequence. Reactions can be initiated either by addition or by elimination and then followed by elimination or addition.

10.1 Addition–elimination reactions

Addition–elimination by nucleophiles is particularly important at carbonyl groups which carry a leaving group (RC(O)X), such as X = halide, alkoxide or amide: Clearly, nucleophilic addition is favoured for the reasons that were considered in Chapter 7, but in this case the tetrahedral intermediate which is formed can now collapse with loss of the leaving group to regenerate a new carbonyl group. The order of reactivity is controlled by the nature of the leaving group and is shown in Fig. 10.1, with acyl halides being the most reactive and amides the least (carboxylic acids are not normally reactive to nucleophilic attack at the carbonyl group as a result of the presence of a highly acidic proton on the hydroxyl group).

This order of reactivity is explained by a consideration of the electronic effect and the leaving-group ability of the heteroatom; thus, for halogens, the electrophilicity of the carbonyl is substantially enhanced by their powerful inductive and only weak resonance effect, but this effect is successively diminished in the sequence S, O, N as inductive withdrawal falls and resonance release increases. Furthermore, leaving-group ability reduces in the same sequence X > S > O > N, making acid halides by far the most reactive in this type of reaction. Note that leaving-group ability correlates with the pK_a of the conjugate acid.

There are several mechanistic nuances in the reactions of carboxylic acid derivatives. For example, the hydrolysis of amides can be acid or base catalysed. For the latter, addition of hydroxide to give a tetrahedral intermediate is followed by the rate-limiting expulsion of amide anion (Fig. 10.2a); because this is not a good leaving group, reversion to starting materials can be favoured and the overall hydrolysis reaction becomes inefficient. For this reason, acid catalysis is normally used (Fig. 10.2b). On the other hand, ester hydrolysis is complicated by the fact that either alkyl–oxygen or acyl–oxygen cleavage is possible; both lead to the same outcome, but mechanistically, the processes are very different. The former is effectively an S$_N$2 process in which carboxylate behaves as the leaving group, and because it relies on the leaving-group ability of carboxylate, which is normally not very high, this type of cleavage is not very common. Much more commonplace is the latter acyl–oxygen cleavage; like amides, both acid- and base-catalysed hydrolyses are possible. However, there are a substantial number of mechanistic subtleties in this reaction, arising from different possibilities for the course of the reaction. Firstly, it should be noted that esters are normally formed under acidic conditions, since the equilibrium may be displaced to the right using an

$$RCOX \ (X = Cl, \ O_2CR) > RCOSR' > RCO_2R' > RCONR'_2 > RCO_2H$$

Figure 10.1   Order of reactivity of acyl derivatives.

excess of alcohol or by removal of water using a dehydrating agent, but hydrolysed under basic conditions, since the alkali shifts the equilibrium reaction to the right by deprotonating the carboxylic acid product (Figs. 10.3a and 10.3b respectively). Base-catalysed ester hydrolysis normally proceeds by $B_{AC}2$ cleavage (base-catalysed, acyl–oxygen cleavage, bimolecular), involving initial addition of hydroxide followed by expulsion of the alkoxide leaving group (Fig. 10.4a). Acid-catalysed hydrolysis typically proceeds by $A_{AC}2$ cleavage (acid-catalysed, acyl–oxygen cleavage, bimolecular), involving protonation, followed by initial addition of water followed by expulsion of the alcohol leaving group (Fig. 10.4b); in this case rapid protonation of the ester leads to an activated carbonyl which is then attacked by water in a slow rate-limiting step. The formation of tetrahedral intermediates features in the course of these reactions; for example, in the case of the hydrolysis of cyclohexane esters, the rate of hydrolysis of the equatorial ester is about 20 times faster than that of the axial, and this is a result of the 1,3-diaxial interactions which destabilise the intermediate in the axial case (Figs. 10.4c and 10.4d).

Depending on the nature of the alkoxy residue of the ester, important alternative mechanisms are possible; thus, in *t*-butyl esters initial protonation leads to an intermediate which can collapse by loss of a highly stabilised cation in a unimolecular process, by alkyl–oxygen cleavage (called $A_{AL}1$), and this process is common for all systems in which the generation of a stabilised carbocation is possible (Fig. 10.5a). The resulting carbocation can be intercepted by water to give *t*-butanol or can lose a proton to generate isobutylene. In the case of benzoates with ortho substituents which sterically impeded the attack of hydroxide at the carbonyl group, direct loss of the protonated alcohol generates an acylium species which is then attacked by water via an $A_{AC}1$ mechanism (Fig. 10.5b).

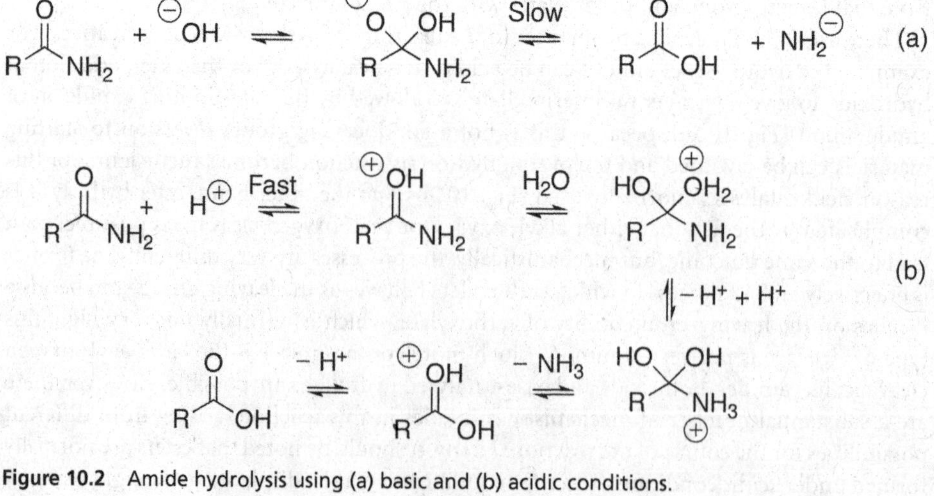

Figure 10.2   Amide hydrolysis using (a) basic and (b) acidic conditions.

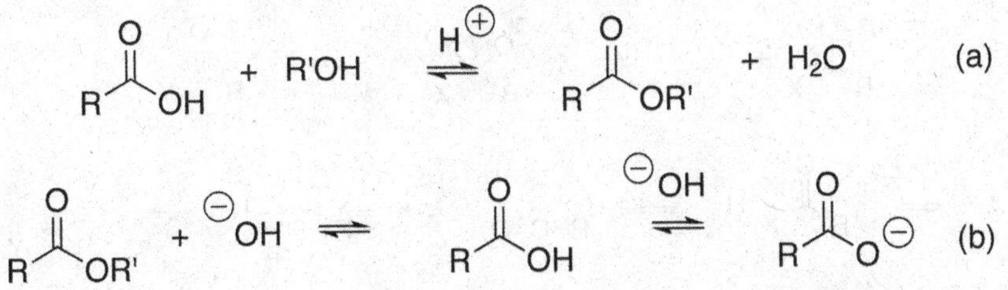

Figure 10.3 (a) Formation of esters under acidic conditions and (b) hydrolysis under alkaline conditions.

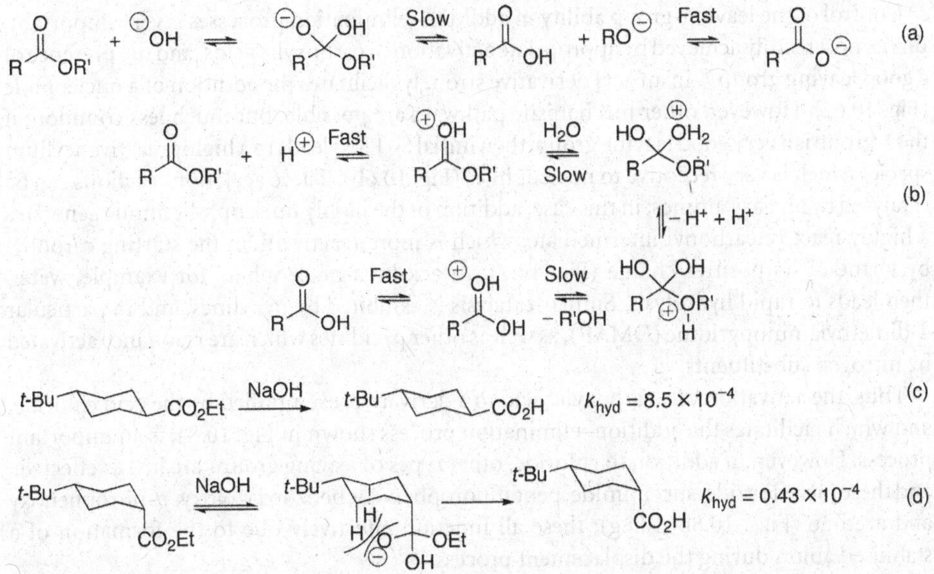

Figure 10.4 Ester hydrolysis using (a) $B_{AC}2$ mechanism under basic conditions and (b) $A_{AC}2$ under acidic conditions (c) of equatorial and (d) of axial esters.

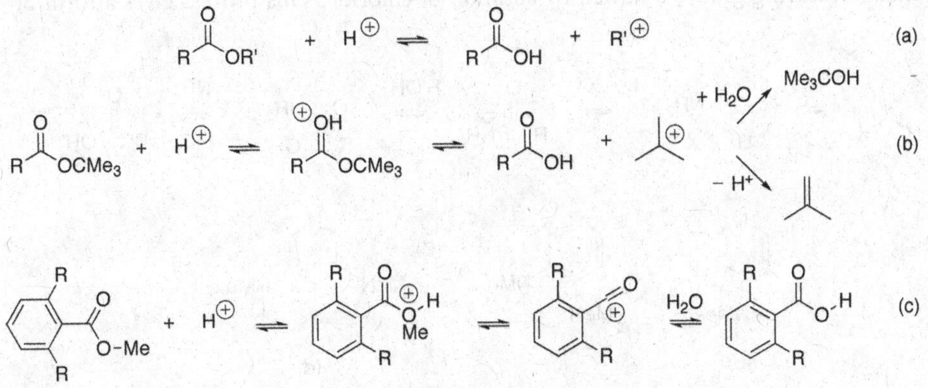

Figure 10.5 Ester hydrolysis by (a) $A_{AL}1$ and (b) $A_{AC}1$ conditions.

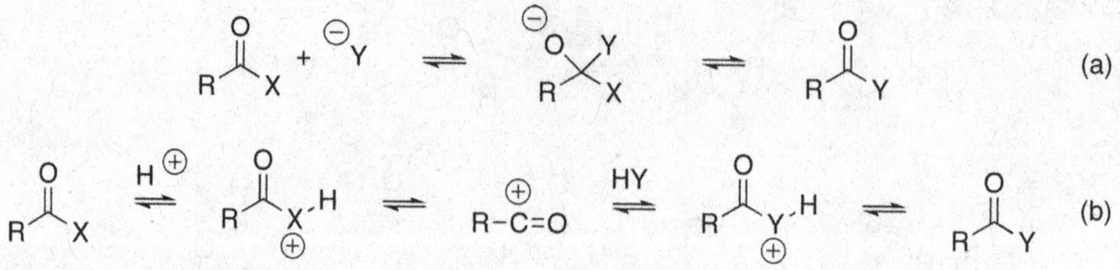

Figure 10.6   Addition of nucleophiles to carboxylic acid derivatives by (a) addition–elimination or (b) elimination–addition.

Control of the leaving-group ability in addition–elimination processes is very important; this is most readily achieved by appropriate activation of carboxylic acids, and the presence of a good leaving group X in an acyl derivative strongly facilitates the addition of a nucleophile (Fig. 10.6a). However, other mechanistic pathways are possible, but much less common; if the X group is a very good leaving group, then initial S_N1 loss leads to a highly reactive acylium species which is very receptive to nucleophiles (Fig. 10.6b). These acylation reactions can be catalysed by tertiary amines; in this case, addition of the highly nucleophilic amine generates a highly reactive carbonyl intermediate, which is more reactive than the starting carbonyl by virtue of its positive charge (Fig. 10.7). Attack by a nucleophile, for example, water, then leads to rapid hydrolysis. Similar catalysis is exhibited by pyridines, and in particular 4-dimethylaminopyridine (DMAP), as well as other pyridines which are resonance activated by nitrogen substituents.

Thus, the activation of a carboxylic acid to a derivative, exemplified by the acid chloride, and which facilitates the addition–elimination process shown in Fig. 10.8a, is an important process. However, in addition to chloride, other types of leaving groups are just as effective, and these include azide, succinimide, pentafluorophenoxy, benzotriazoloxy, p-nitrophenoxy and acetone (Figs. 10.8b–10.8g); these all function effectively due to the formation of a stabilised anion during the displacement process.

The activation of a carboxylic acid is most readily achieved by treatment with either of thionyl chloride, phosphoryl chloride or oxalyl chloride to give the corresponding acid chloride (Figs. 10.9a–10.9c); this reaction proceeds by initial transient formation of an excellent leaving group, followed by addition of chloride. This process gives a form of a

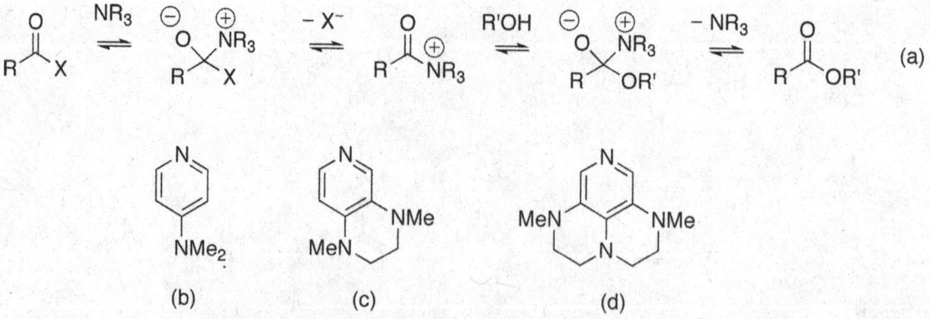

Figure 10.7   (a) Nucleophilic catalysis of acyl derivatives by tertiary amines in hydrolysis reactions, and (b–d) DMAP and its derivatives as nucleophilic catalysts.

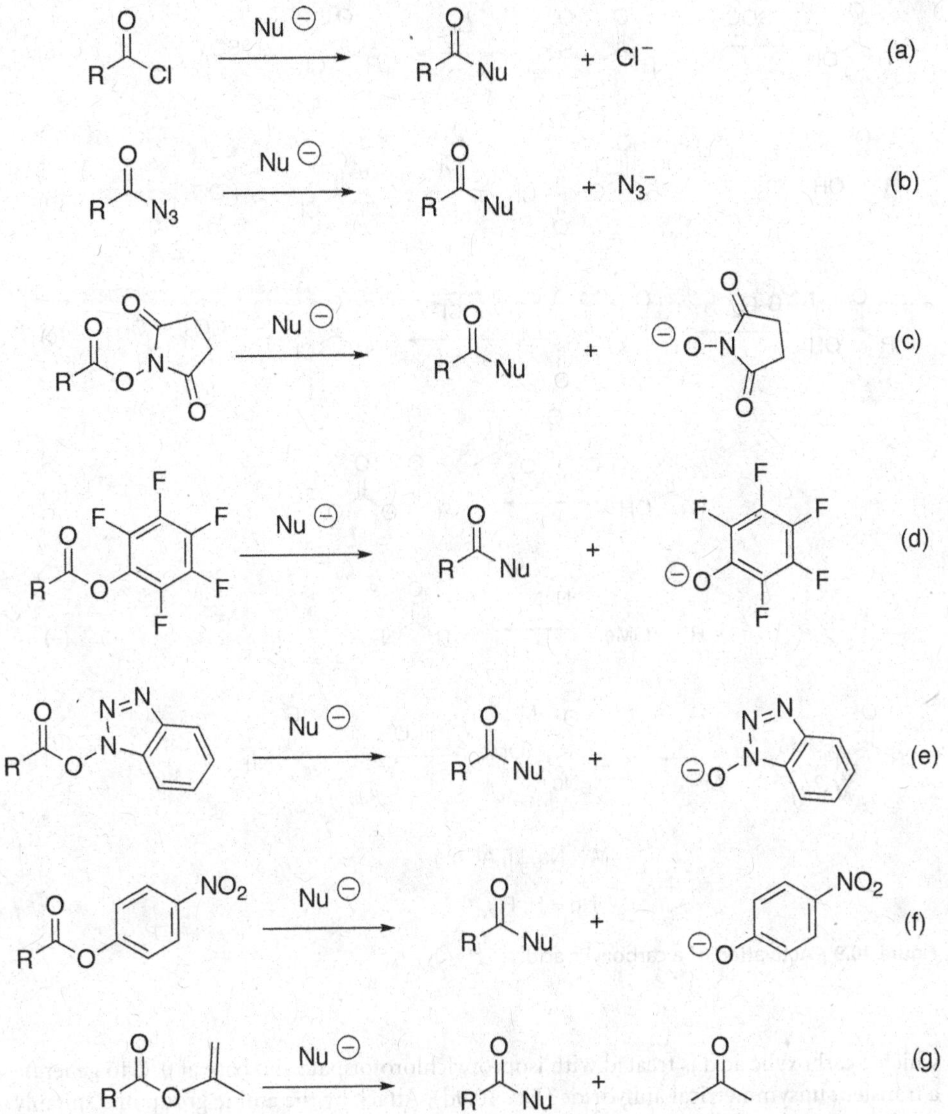

Figure 10.8 Alternative leaving groups in nucleophilic addition–elimination processes.

carbonyl group which is highly activated towards nucleophilic addition, but the process for generating the acid chloride requires highly vigorous conditions and is not always suitable for sensitive substrates. However, because such acylation reactions using acid chlorides generally require vigorous conditions, a large number of milder reagents and conditions for the activation of carboxylic acids have been developed; one way leaving-group ability is altered is by the use of groups which fragment on departure, thereby providing a powerful entropic driving force for the reaction. Such activation of the carboxyl component can be readily achieved using various reagents. These include the mixed anhydride method, in

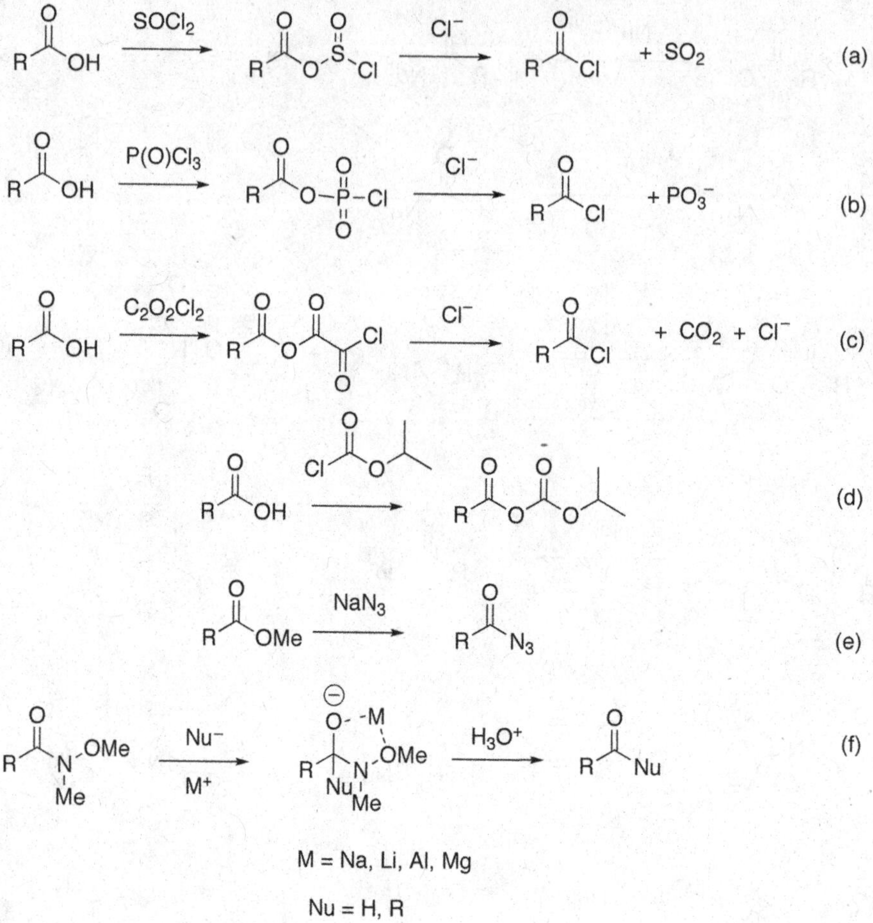

Figure 10.9 Activation of a carboxylic acid.

which a carboxylic acid is treated with isopropylchloroformate and base at 0°C to generate a transient unsymmetrical anhydride (Fig. 10.9d). Attack by the amine group of a suitably protected amino acid group then forms the desired amide bond, with release of carbon dioxide and isopropanol. The formation of these stable and gaseous products provide a clear thermodynamic and kinetic driving force and is a strategy which has been widely adopted to permit these reactions to be conducted under very mild conditions, and with sensitive substrates. The acyl azide method (Fig. 10.9c) is different, insofar as it allows a methyl ester to be directly converted to an activated intermediate by displacement with the powerful nucleophile, azide, to give an acyl azide intermediate, which is then coupled once again to a protected amino acid; this approach is especially convenient, since it saves a deprotection step and allows the direct conversion of an ester to an amide. On the other hand, sometimes it is of value to reduce the leaving-group ability of X in order to achieve greater selectivity in reactions; Weinreb amides are a case in point, and these stabilise the tetrahedral intermediate formed upon addition of the nucleophile by facilitating chelation between the

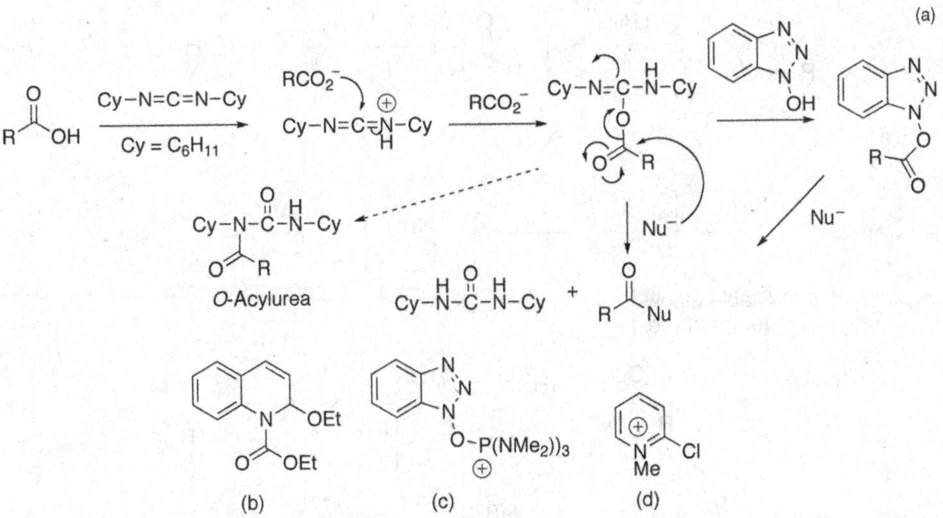

Figure 10.10 Reagents for transient activation of a carboxylic acid.

X group and the metal cation (Fig. 10.9f). Hydrolysis then generates the carbonyl product at the ketone oxidation level.

Another strategy has involved the development of alternative reagents for the coupling of acid and amine components but without requiring the isolation of an intermediate activated ester. One of the most important of these is dicyclohexylcarbodiimide, which functions by addition of the carboxylic acid component to give a transient activated ester intermediate (Fig. 10.10a), which is in turn attacked by an amine to generate the highly stable urea product. Problems with this approach, however, are that the active ester can rearrange to give an *O*-acylisourea product, which is inactive to further reaction, and removal of the dicyclohexylurea by-product can be difficult. This concept has spawned an array of reagents which behave in a mechanistically similar manner, and these include 2-ethoxy-1-ethoxycarbonyl-1,2-dihydroquinoline, bis(2-oxo-3-oxazolidinyl)phosphonic chloride and the Mukaiyama chloropyridine reagent (Figs. 10.10b–10.10d); the significance of these is that the reactions can be conducted under milder conditions, are not prone to the *O*-acylisourea-forming pathway and are suitable for sensitive systems, for example, which are sensitive to racemisation. Each of these reagents functions by transient formation of an activated intermediate, which is intercepted in solution to give the formation of the desired amide product. These mild coupling methods suitable for repeated application enabling the preparation of a peptide polymer from its amino acid components have been especially important.

A further level of sophistication in the chemistry of these coupling reactions comes by application of nucleophilic catalysts. Since a coupling reaction of this type is bimolecular, during the course of the reaction, the overall rate of reaction decreases rapidly as both starting materials are consumed; this means that active ester intermediates have longer residence times and are more likely to engage in undesired side-reaction pathways (rearrangement or racemisation; Fig. 10.10a). If a powerful nucleophile is added, such as imidazole, pyridine, DMAP or *N*-hydroxybenzotriazole, this intermediate will be trapped to form

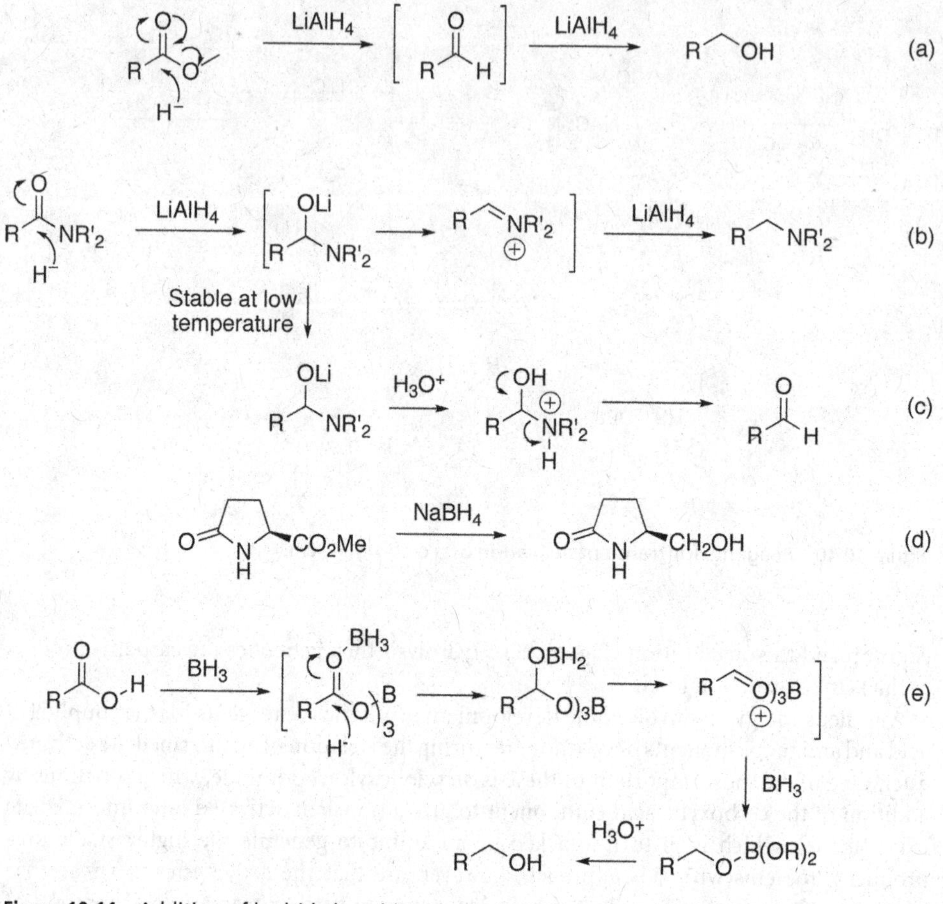

Figure 10.11 Addition of hydride by addition–elimination.

another transient intermediate faster than rearrangement of the first one can proceed, and the overall rate of reaction will be accelerated.

Many types of nucleophiles will add by addition, and these reactions are considered in more detail below.

10.1.1 Addition of hydride

The addition of hydride leads to reduction reactions for which additional selectivity can be achieved by control of the reaction temperature, reduction reagent or nature of the carbonyl group. Lithium aluminium hydride (LAH) is the most reactive of this class of reagents and will reduce an ester or acid halide to an alcohol, or an amide to an amine (Figs. 10.11a and 10.11b). Note that the mechanistic pathways in these two processes are quite different and reflect the differing leaving-group abilities of alcohols versus amines; in the case of esters, elimination of alkoxide during the course of the reaction generates the aldehyde in situ, which is directly further reduced to the alcohol product. In the case of amides, however, the better leaving group is lithium oxide, and this generates an iminium ion in situ, which is directly further reduced to the amine product. However, if this reaction is done at low temperature, the intermediate hemiaminal ether is stable and after quenching with acid it collapses to the aldehyde product (Fig. 10.11c). Sodium borohydride is normally not reactive enough to reduce an ester, unless it is activated by an adjacent electron-withdrawing substituent (Fig. 10.11d). The reactions are thought to proceed by initial coordination of the lithium or sodium salt to the carbonyl group, thereby activating it, followed by addition of the nucleophilic hydride. The reducing activity of these hydride donors can be controlled using a combination of steric and electronic effects; thus, sodium *s*-butylborohydride is a particularly hindered source of hydride, and lithium salts tend to be more reactive than sodium salts and aluminium hydrides more than borohydrides, on account of the greater Lewis acidity of the former in each case. Furthermore, introduction of electron-withdrawing substituents attenuates reactivity, so that, for example, lithium triethoxyborohydride is substantially less reactive than lithium borohydride, but the electron-releasing groups in lithium triethoxyborohydride make it substantially more reactive than lithium borohydride. On the whole, this type of reducing reagent is ineffective for the reduction of carboxylic acids, since their high basicity leads to initial deprotonation and this significantly deactivates the system towards further nucleophilic attack.

Diborane and Dibal-H, much less powerful reducing reagents, belong to a class of reducing reagents which might be described as electrophilic reducing reagents (Fig. 10.11e). As a result of their Lewis acidic character (they are both electron deficient), they function by initial coordination of the metal to the carbonyl group; this increases the nucleophilicity of the hydride and addition then occurs. Elimination of borate then generates another oxonium-like species that is able to accept a further equivalent of hydride. Acidic work-up quenches the reducing agent and liberates the product.

10.1.2   Addition of heteroatom nucleophiles

The addition of heteroatom nucleophiles, principally water, but also alcohols and amines, gives hydrolysis (or alcoholysis or aminolysis, respectively), as we have seen in Section 10.1 above. These reactions typically proceed by nucleophilic addition to generate a tetrahedral intermediate, which collapses to give the expected product (see Figs. 10.2 and 10.4). These reactions can be acid or base catalysed, and in the case of esters this gives rise to several specific hydrolytic mechanisms; particularly important are the $B_{AC}2$ and $A_{AC}2$ mechanisms, since they are very common, but other less common mechanisms have been identified (see Table 10.1).

These mechanistic differences can be exploited in the design of appropriate protecting groups for alcohols (Fig. 10.12); conversion of alcohols to acetate, trifluoroacetate, benzoate and pivaloate esters by reaction with the corresponding acid chloride or anhydride under basic conditions (such as Et_3N or pyridine) enables protection of the hydroxyl group as a result of steric effects and resonance delocalisation of the lone pairs on the oxygen. Such groups are easily and selectively removed by appropriate application of the alkaline hydrolysis conditions; the better and the less hindered the carboxylate leaving group, the milder the conditions for hydrolysis that can be employed. Thus, trifluoroacetate esters are the most easily cleaved, using bicarbonate as base, and pivaloates are the most difficult, requiring more forcing conditions. A similar approach can be used for amines which have been protected as carbamates (Fig. 10.13a). Conversion of amines to carbamates, normally done by reacting an amine with chloroformate or similar reactive system, generates a protecting group which can

Table 10.1   Mechanisms of ester hydrolysis[a]

Name	Type of reaction	Conditions
Acid-catalysed ester hydrolysis		
$A_{AC}1$	S_N1	Only if substrate is sterically hindered
$A_{AC}2$	Via tetrahedral intermediate	Standard hydrolysis
$A_{AL}1$	S_N1	Only if R'^+ is stable (see Fig. 10.5a)
$A_{AL}2$	S_N2	Not observed
Base-catalysed ester hydrolysis		
$B_{AC}1$	S_N1	Not observed
$B_{AC}2$	Via tetrahedral intermediate	Standard hydrolysis
$B_{AL}1$	S_N1	Only if R'^+ is stable and in weakly basic conditions (see Fig. 10.5a)
$B_{AL}2$	S_N2	Only in α-lactones

[a] See also Figs. 10.4–10.6.

be readily removed under alkaline conditions; here the reactivity pattern is again dictated by the R′ group in exactly the same way as that for esters. The phthalimide group is an example of an amine protecting group cleavable by hydrazine in an addition–elimination sequence (Fig. 10.13b).

On the other hand, protection of carboxylic acids, in order to remove their acidic hydrogen and to provide a steric blocking effect at the carbonyl group, as esters is readily possible, and these groups can be removed by hydrolysis as required. Bulky esters cannot generally be easily removed under basic hydrolysis conditions, since this requires attack at the carbonyl group by a nucleophile, so methyl or ethyl esters are usually used for deprotection under alkaline conditions.

10.1.3   Addition of carbon nucleophiles

The addition of carbon nucleophiles, which can be organometallic or enolates, to carbonyls is very important.

10.1.3.1   Organometallic derivatives

The addition of Grignards and organolithiums to carbonyl groups is very straightforward, as a result of their high nucleophilicity, and has been very widely used, as a result of their ready availability. Esters react with 2 equivalents of such organometallics to give a tertiary alcohol (Fig. 10.14a). This reaction proceeds by initial addition of one organometallic residue to

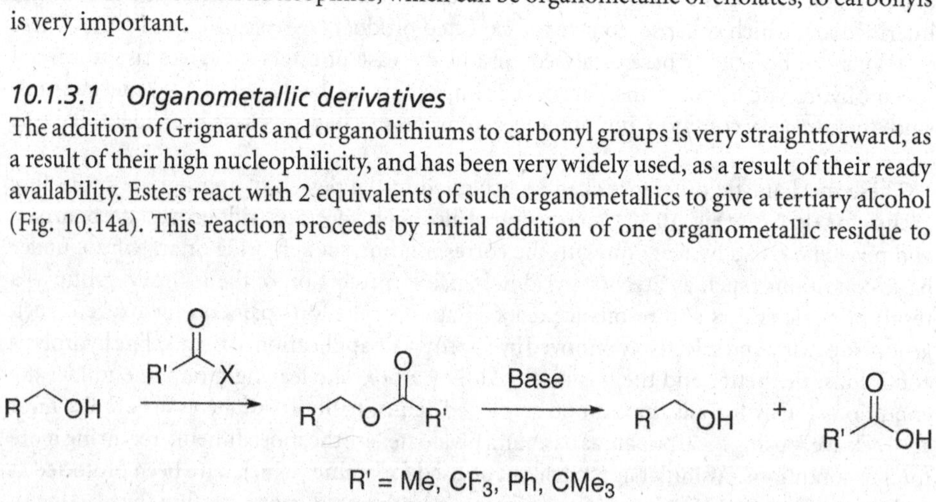

R′ = Me, CF$_3$, Ph, CMe$_3$

Figure 10.12   Protection of alcohols as esters.

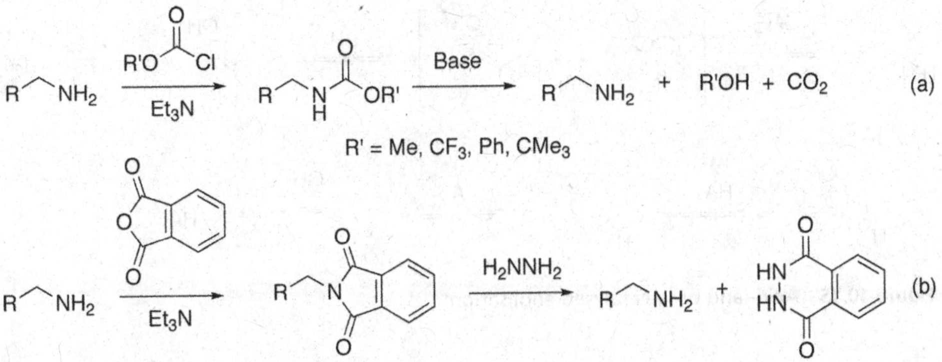

Figure 10.13 Protection of amines as groups cleavable by addition–elimination.

generate an intermediate ketone; however, this is more reactive than the starting ester and rapidly reacts with a second equivalent of the organometallic derivative to give a tertiary alcohol product. Amides are susceptible to a similar addition, but in this case the lower leaving-group ability of the amine residue means that the ketone is usually the observed product, which is generated under acidic work-up (Fig. 10.14b). The addition of these organometallics to carboxylic acids proceeds differently (Fig. 10.14c); the first equivalent leads to deprotonation of the carboxylic acid, since the organolithium is a very good base. However, it is also such a good nucleophile that the carboxylate anion is further attacked by another equivalent of the organolithium, and this leads to a tetrahedral intermediate. Under the reaction conditions, this is stable, but the addition of aqueous acid both quenches any unreacted organolithium and protonates the intermediate. This collapses to the product

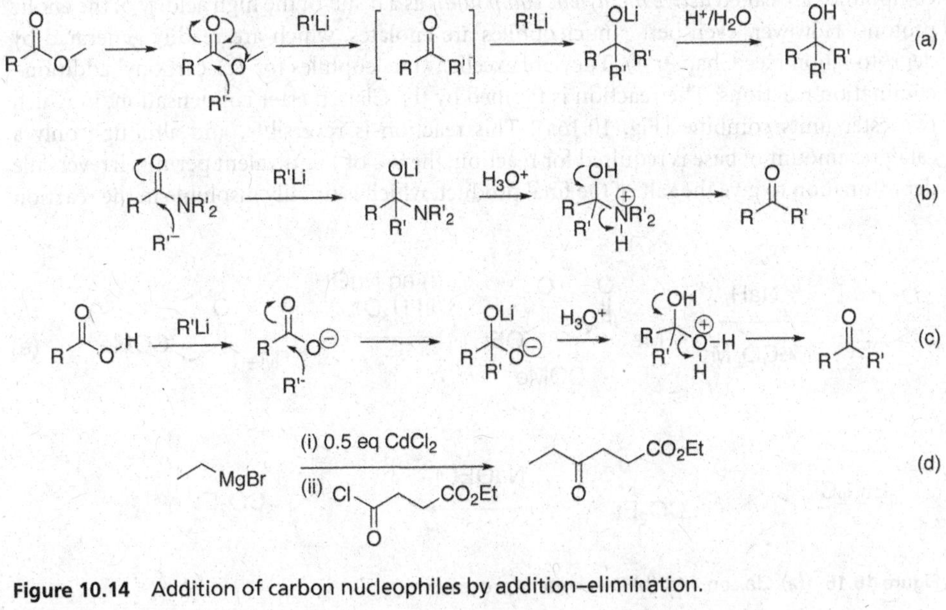

Figure 10.14 Addition of carbon nucleophiles by addition–elimination.

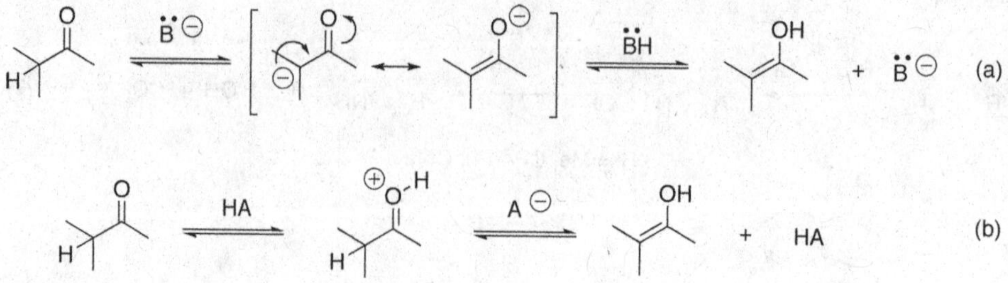

Figure 10.15 Acid- and base-catalysed enolisation.

ketone. One difficulty with these reactions can be caused by the high basicity of Grignards and organolithiums; if the carbonyl group has acidic α-hydrogens, deprotonation can compete effectively with addition. If this is a problem, the addition of an oxophilic lanthanide salt, usually cerium chloride, has been found to enhance addition reactions, presumably since it effectively coordinates to the carbonyl oxygen. It is possible to use reagents with a more attenuated reactivity, and organocadmium reagents are a useful example, although have fallen into disuse as a result of toxicity concerns (Fig. 10.14d). The required organocadmium reagent is easily generated from a Grignard and cadmium chloride, and this will react selectively with an acid chloride to give a ketone product; this is not susceptible to a further addition of another equivalent of the organocadmium reagent. Furthermore, this reaction is very selective for the acid chloride group and can be done in the presence of esters.

10.1.3.2   Enols and enolates

Enols are important tautomeric forms of carbonyl-containing compounds as a result of the enhanced acidity of the α-hydrogens; enolisation can be facilitated under basic or acidic conditions (Figs. 10.15a or 10.15b), and the resulting enol form is a good nucleophile. Such compounds are called *active methylene compounds* as a result of the high acidity of the enolic protons. However, even better nucleophiles are enolates, which are readily generated by deprotonation (see Chapter 4). They are excellent nucleophiles for 1,2-carbonyl addition–elimination reactions: The reaction is typified by the Claisen ester condensation, in which two ester units combine (Fig. 10.16a). This reaction is reversible, and although only a catalytic amount of base is required for reaction, the use of 1 equivalent permits irreversible deprotonation to give the salt of the final product, which is usually insoluble in the reaction

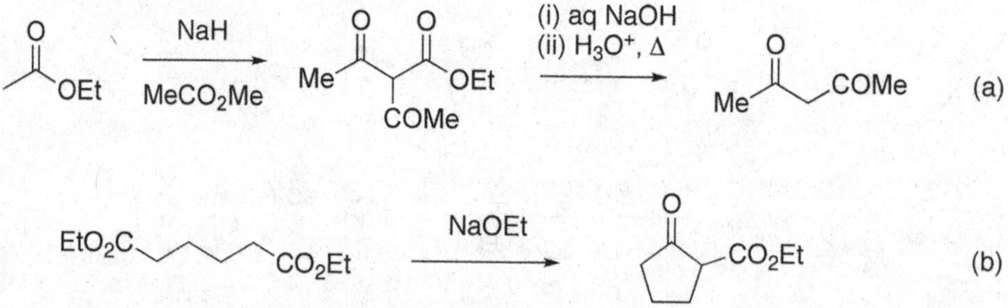

Figure 10.16 (a) Claisen and (b) Dieckmann processes.

medium; this drives the reaction to completion. The product thus formed may be hydrolysed under alkaline conditions and then decarboxylated by thermolysis under acidic conditions, in order to generate a 1,3-diketone. If the process occurs in an intramolecular sense, the reaction is called the Dieckmann cyclisation (Fig. 10.16b). There are a wide variety of similar highly effective reactions, all of which proceed by an initial nucleophilic addition; the subsequent process varies according to the nature of the reactant. These include the Knoevenagel, Perkin, Stobbe, Reformatski and Darzens condensations (Figs. 10.17a, 10.17b, 10.17c, 10.17d and 10.17e respectively), amongst many other variants. These reactions can be used for the formation of rings, and a particularly important example is the Robinson ring annelation (Fig. 10.17f); a β-dicarbonyl enolate, generated under standard conditions, reacts with an α,β-unsaturated carbonyl by Michael addition. The resulting enolate then equilibrates to an alternative one, which then undergoes an aldol reaction with the adjacent carbonyl group and gives ring closure. Dehydration then leads to the observed decalin product. Equilibration processes in these reactions are very important; for example, in the process shown in Fig. 10.17g, deprotonation of the starting material leads to a retro-aldol reaction, followed by equilibration of the enolates and then readdition to the aldehyde. The reaction proceeds in the forward direction because the final addition–elimination sequence is irreversible, giving the observed product.

10.1.4 Addition of phosphorus nucleophiles

The addition of phosphorus ylides to carbonyl compounds can be used to prepare alkenes. The reaction is simple to execute and generally high yielding, and provides a synthetically very valuable entry to alkenes, since a wide range of substituted components can be accommodated. In its simplest form (Fig. 10.18), the reaction involves the treatment of a carbonyl compound, either an aldehyde or a ketone, with a phosphorus ylide readily generated by the deprotonation of a phosphonium salt (itself obtained from the alkylation of a triorganophosphorane); addition of the nucleophilic ylide to the carbonyl group is the first and rate-limiting step and generates a betaine intermediate, and this is followed by cyclisation to an oxaphosphetane intermediate. This intermediate then collapses with loss of triphenylphosphine oxide to give the alkene product by a concerted *syn*-elimination; the formation of the P=O system provides the thermodynamic driving force for the reaction. Stabilisation of the ylide by incorporating electron-withdrawing groups makes them very easy to handle since they are isolable, but also slows the rate of this addition, and the reaction proceeds only under more forcing conditions, usually at reflux, to give predominantly the *trans*- or *E*-product. On the other hand, if the ylide is not substituted with such electron-withdrawing groups, it is called non-stabilised, cannot be isolated and is more reactive towards the addition reaction but tends to give a mixture of *cis*- and *trans*-products, but one in which the *cis*-isomer predominates. If this reaction is conducted under 'salt-free' conditions, in which lithium ions are avoided, the proportion of the *cis*-alkene increases substantially and can in fact be the sole product. The simplest explanation for this outcome is that the initial addition reaction controls the stereochemical outcome. In the case of the stabilised ylide, this is reversible, and the addition proceeds so as to minimise steric interactions between R and R′ in the oxaphosphetane intermediate; formation of the most stable oxaphosphetane is then followed by elimination, leading to the *trans*-product. However, for the non-stabilised ylide, the initial addition is not reversible, and the reaction outcome

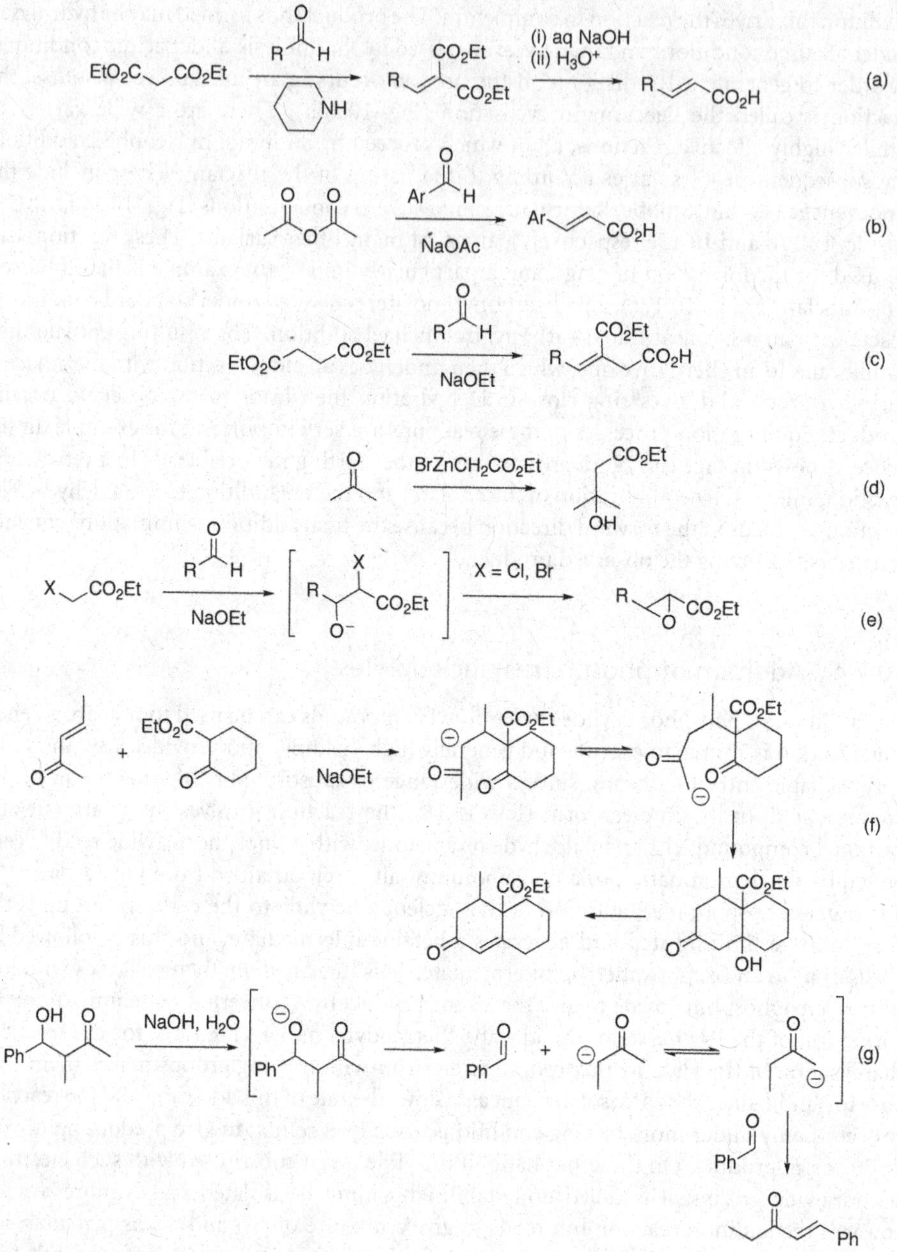

Figure 10.17 Enolate condensation processes.

therefore results from the fastest reaction; this kinetic control depends on the minimisation of steric interactions between R and R' in the transition state.

Variants of the basic reaction are of particular synthetic value; for example, in the Schlosser modification, a non-stabilised triphenylphosphonium ylide is generated in the presence of an excess of lithium bromide, in which case the *trans*-alkene is favoured (Fig. 10.19). Here,

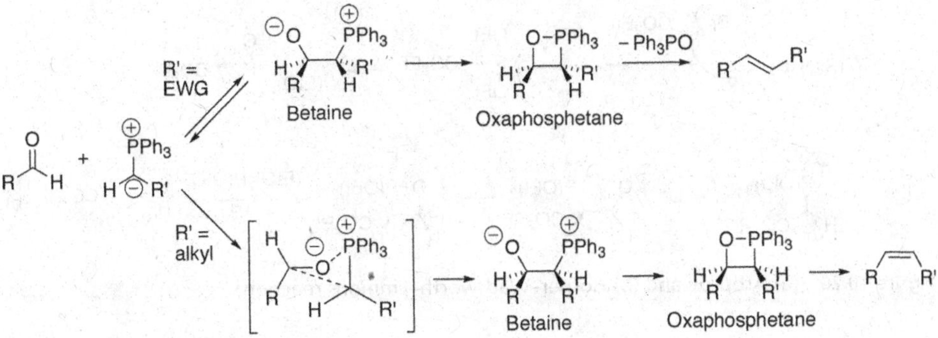

Figure 10.18 Wittig reaction.

the lithium bromide catalyses opening of the normally formed oxaphosphetane intermediate into a lithio betaine, which can easily stereochemically equilibrate by generation of the corresponding ylide. Reclosure to the corresponding oxaphosphetane and elimination of the phosphine oxide then leads to the more stable *trans*-alkene product. The Horner–Wadsworth–Emmons reaction uses phosphonate esters, prepared according to the Arbusov reaction (Fig. 10.20a). Deprotonation of such species using BuLi, sodium hydride or potassium *t*-butoxide, for example, generates a highly nucleophilic phosphonate anion, which reacts with carbonyls analogously to the basic Wittig reaction, selectively giving the *trans*-alkene product (Fig. 10.20b); a key advantage in this case, however, is that the phosphonate by-product is easily removed because of its high water solubility. Of even potentially greater use is the possibility of controlling the stereoselectivity of the reaction outcome, so that by using bis(trifluoroethyl) or diaryl phosphonate esters, it is possible to obtain exclusive formation of the *cis*-isomer; these are called the Still–Gennari and Ando modifications respectively (Fig. 10.21). This neatly complements the *trans*-selectivity of the classical Wittig process. Alternatively, the Horner–Wittig reaction uses collapse of β-hydroxyphosphine oxides (Fig. 10.22); these are generated from the lithio derivatives of alkyldiphenylphosphine oxides and aldehydes. This addition usually proceeds diastereoselectively to give the *anti*-product, and upon *syn*-elimination of the phosphorus oxide component, the *cis*-alkene product is generated. The *trans*-product can be accessed using a variation; reaction of the lithio species

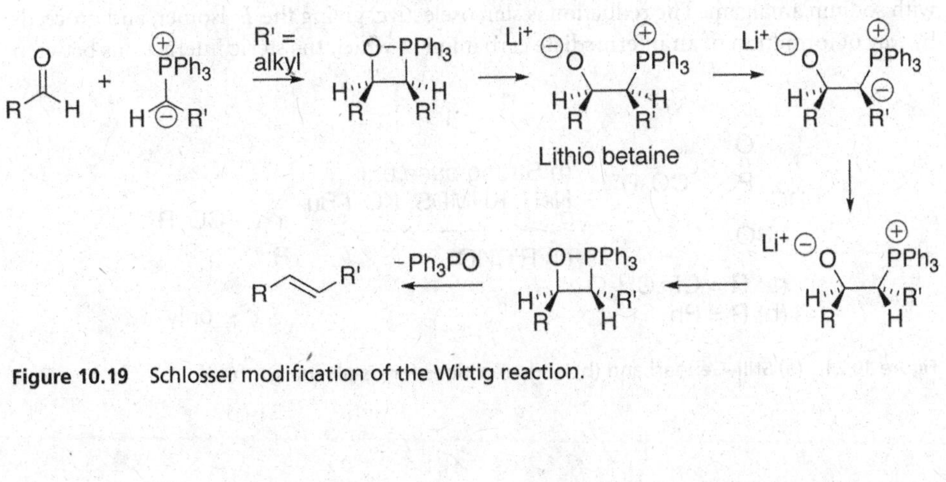

Figure 10.19 Schlosser modification of the Wittig reaction.

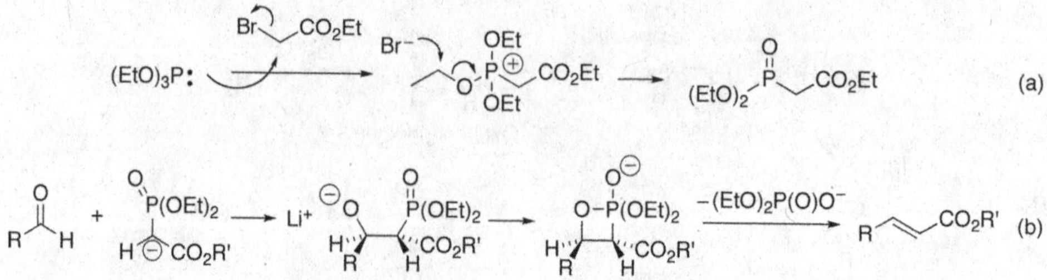

Figure 10.20   (a) Arbusov and (b) Horner–Wadsworth–Emmons reaction.

with an ester gives the corresponding ketone, and this can be diastereoselectively reduced to give the *syn*-product. Elimination in this case then proceeds to the *trans*-alkene.

Worthy of mention is the Corey–Fuchs alkynylation reaction, which is analogous to the Wittig reaction, but leads to the formation of an alkyne (Fig. 10.23). In this case, the required phosphonium ylide is generated by treating triphenylphosphine with carbon tetrabromide and zinc, and this readily reacts with aldehydes to make the 1,1-dibromoalkene. This can be further treated with excess butyllithium in order to force a double elimination, generating the corresponding alkyne.

There are other processes which are analogous to the Wittig reaction, but which use other heteroatoms to stabilise the intermediate carbanion intermediate. The Peterson reaction is a particularly valuable variant which uses organosilanes (Fig. 10.24); in this case, the addition of an α-silylcarbanion to the carbonyl group generates a β-hydroxysilane, and its subsequent elimination can be controlled such that *cis*- or *trans*-alkenes may be obtained stereospecifically. Thus, under acidic or Lewis acidic ($BF_3 \cdot Et_2O$) conditions, an E_2-type elimination proceeds via the normal *anti*-transition state, but if the reaction is conducted in the presence of base, normally potassium hydride, initial intramolecular cyclisation to the corresponding siloxane occurs followed by a synperiplanar elimination. By appropriate choice of starting *syn*- or *anti*-diastereomer of the β-hydroxysilane (Fig. 10.24) and elimination conditions, it is possible to access *cis*- or *trans*-alkenes selectively.

The sulfur-stabilised equivalent, the Julia olefination reaction, proceeds by way of addition of an α-lithiosulfone (generated by deprotonation of a sulfone) to a carbonyl compound; in this case, the reaction is quenched with an acid halide, such as acetyl chloride, to generate a β-acyloxysulfone, which can be reductively eliminated to give the alkene product, usually with sodium amalgam. The reduction is stereoselective, giving the *E*-isomer, and proceeds by way of formation of an intermediate carbanion in which the steric interactions between

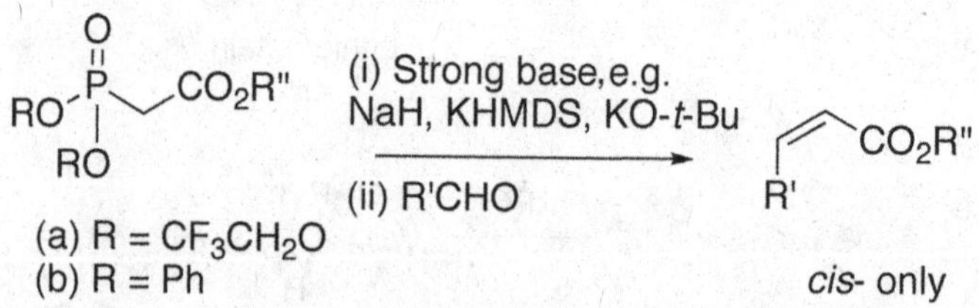

Figure 10.21   (a) Still–Gennari and (b) Ando reactions for *cis*-alkene synthesis.

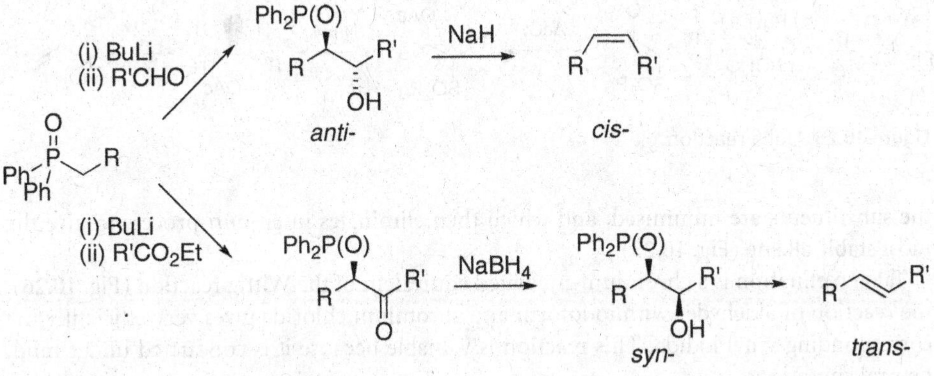

Figure 10.22 Horner–Wittig reaction.

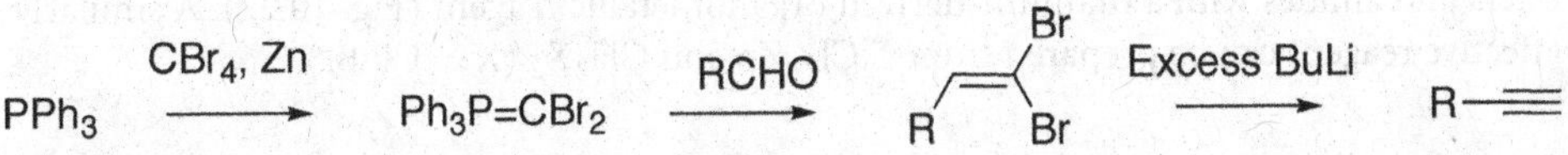

Figure 10.23 Corey–Fuchs reaction.

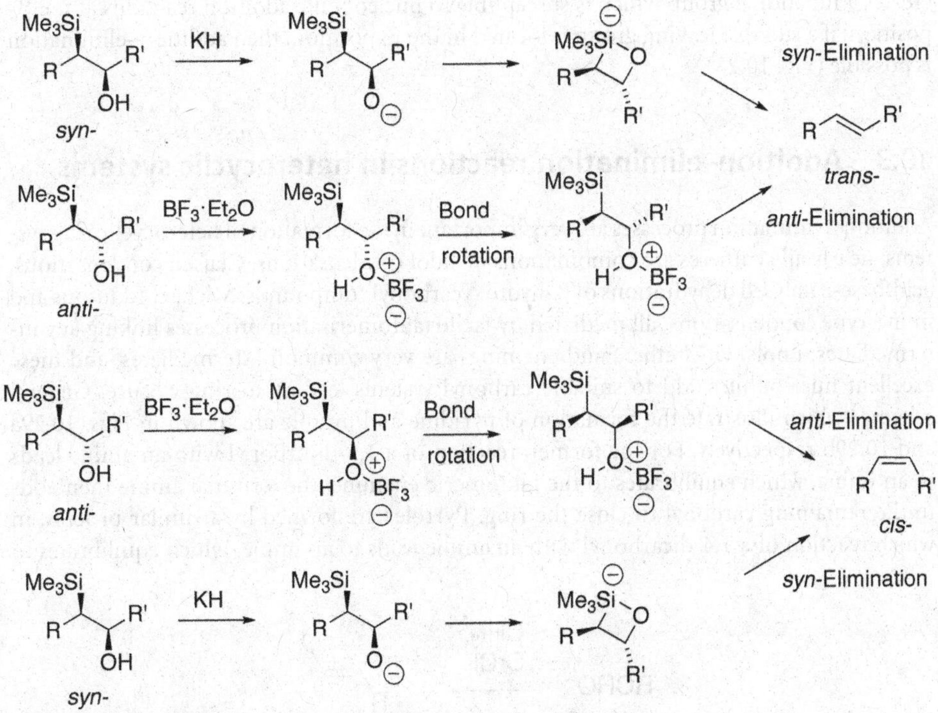

Figure 10.24 Peterson reaction.

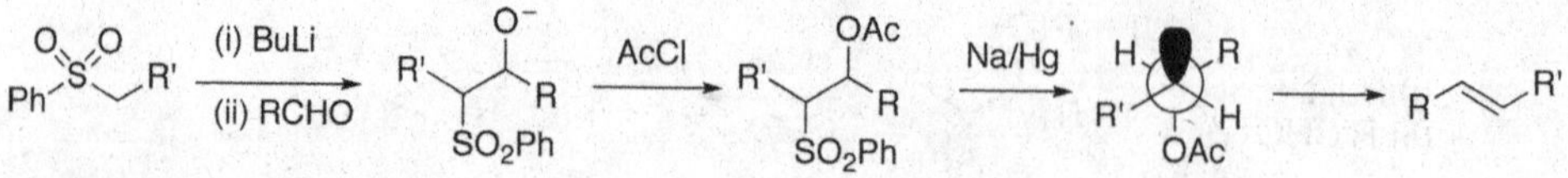

Figure 10.25   Julia reaction.

the substituents are minimised, and which then eliminates in an *anti*-process to give the most stable alkene (Fig. 10.25).

Takai olefination is a chromium-mediated equivalent of the Wittig reaction (Fig. 10.26); the reaction of aldehydes with iodoform and chromium chloride gives very efficiently the corresponding vinyl iodide. This reaction is valuable because it is conducted under mild, neutral conditions.

An important limitation to Wittig-type chemistry is that normally only aldehydes and ketones are appropriate substrates; a useful method for the homologation of other types of carbonyl groups is the Tebbe reaction, which involves the treatment of carbonyls, including esters and amides with a titanium-derived organometallic reagent (Fig. 10.27). A similarly effective reagent can be prepared from $TiCl_4$, Zn and CH_2X_2 (X = Cl, Br).

10.2   Addition–elimination reactions in conjugated systems

Conjugation of a carbon–carbon double bond to a carbonyl function (R″CH=CHC(O)R) creates a functional group which is susceptible to nucleophilic addition reactions at the β-position; if a suitable leaving group is located in the β-position, then addition–elimination is possible (Fig. 10.28).

10.3   Addition–elimination reactions in heterocyclic systems

Addition–elimination processes are very important in the formation of heterocyclic ring systems; nearly all syntheses are combinations of aldol condensations, Claisen condensations, acid/base-catalysed dehydrations of β-hydroxycarbonyl compounds, Michael additions and imine-type condensations, all mediated by facile tautomerisation processes linking key intermediates. Enols, vinyl ethers and enamines are very common intermediates, and these excellent nucleophiles add to suitable carbonyl systems leading to ring closure. General schemes which illustrate the formation of pyridine and pyrrole are shown in Figs. 10.29a and 10.29b respectively. For the former, reaction of a 1,5-dicarbonyl with an amine leads to an imine, which equilibrates to the tautomeric enamine; the terminal amine then adds to the remaining carbonyl to close the ring. Pyrroles are formed by a similar process, in which reaction of a 1,4-dicarbonyl with an amine leads to an imine, which equilibrates to

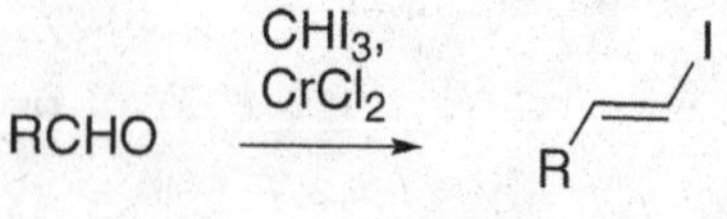

Figure 10.26   Takai reaction.

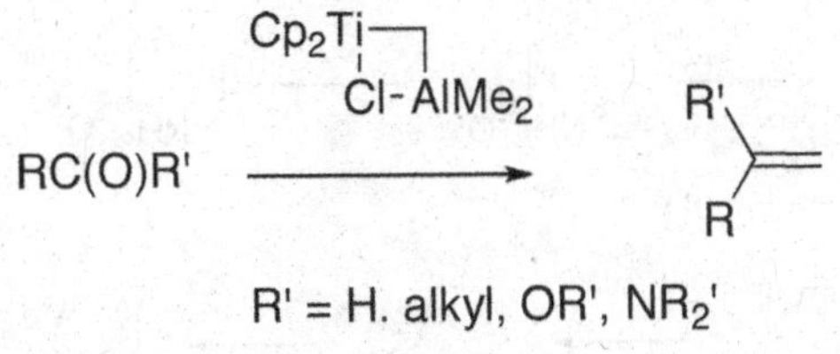

R' = H. alkyl, OR', NR$_2$'

Figure 10.27 Tebbe reaction.

its tautomeric enamine; the terminal amine then adds to the remaining carbonyl to close the ring. Typical examples of the types of heterocyclic systems which are accessed in this way are pyrroles (by the Knorr reaction; Fig. 10.29c), pyridines (by the Hantsch reaction; Fig. 10.29d), quinolines (by the Skraup and Friedlander reaction; Figs. 10.29e and 10.29f), isoquinolines (by the Bischler–Napieralski reaction; Fig. 10.29g) and tetrahydroisoquinolines (by the Pictet–Spengler synthesis; Fig. 10.29h).

10.4 Addition–elimination reactions in ring-closing metathesis

Alkene metathesis is the metal-catalysed redistribution of carbon–carbon bonds to form new ring systems and is most generally represented as shown in Fig. 10.30. The Grubbs first-generation catalyst reported in 1993 was the first readily available catalyst for this process (Fig. 10.31) and was later followed by the Grubbs second-generation catalyst with enhanced catalytic activity, functional group tolerance and stability; it is suitable for ring-closing metathesis, cross metathesis and ring-opening metathesis. The Chauvin mechanism is the generally accepted mechanism for this process, involving a series of [2 + 2]-cycloaddition–cycloreversions: intial cycloaddition of the alkene substrate and the ruthenium catalyst generates a metallacyclobutane intermediate, which collapses with loss of an alkene to give a metalloalkene, which in turn adds to the pendant alkene to generate another metallacyclobutane which with loss of the ruthenium metalloalkene gives the overall product (Fig. 10.32). The equilibrium is driven in favour of the product by the entropic gain from the loss of the gaseous alkene by-product. Olefin metathesis using molybdenum, tungsten and ruthenium catalysts has been shown to be effective. Careful optimisation of the catalysts has provided access to a synthetically valuable process, and this reaction now provides access to a wide range of substituted alkenes. There are several versions of this process, including ring-closing metathesis (RCM; Fig. 10.33), which is used to create rings systems from acyclic precursors, acyclic di-olefin methathesis (ADMET), alkyne methathesis and polymerisation, exemplified by ring-opening metathesis polymerisation (ROMP).

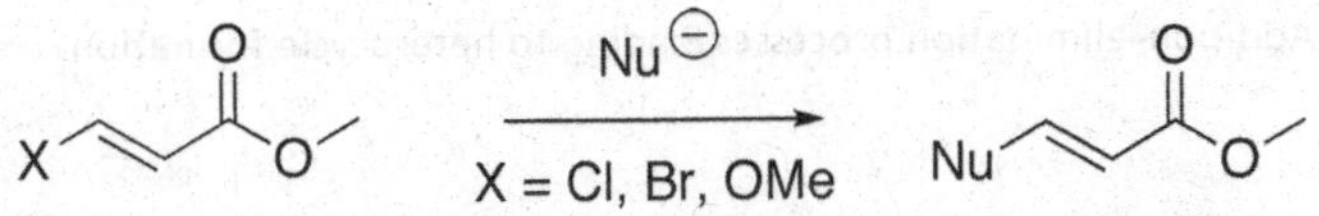

Figure 10.28 Conjugate addition–elimination processes.

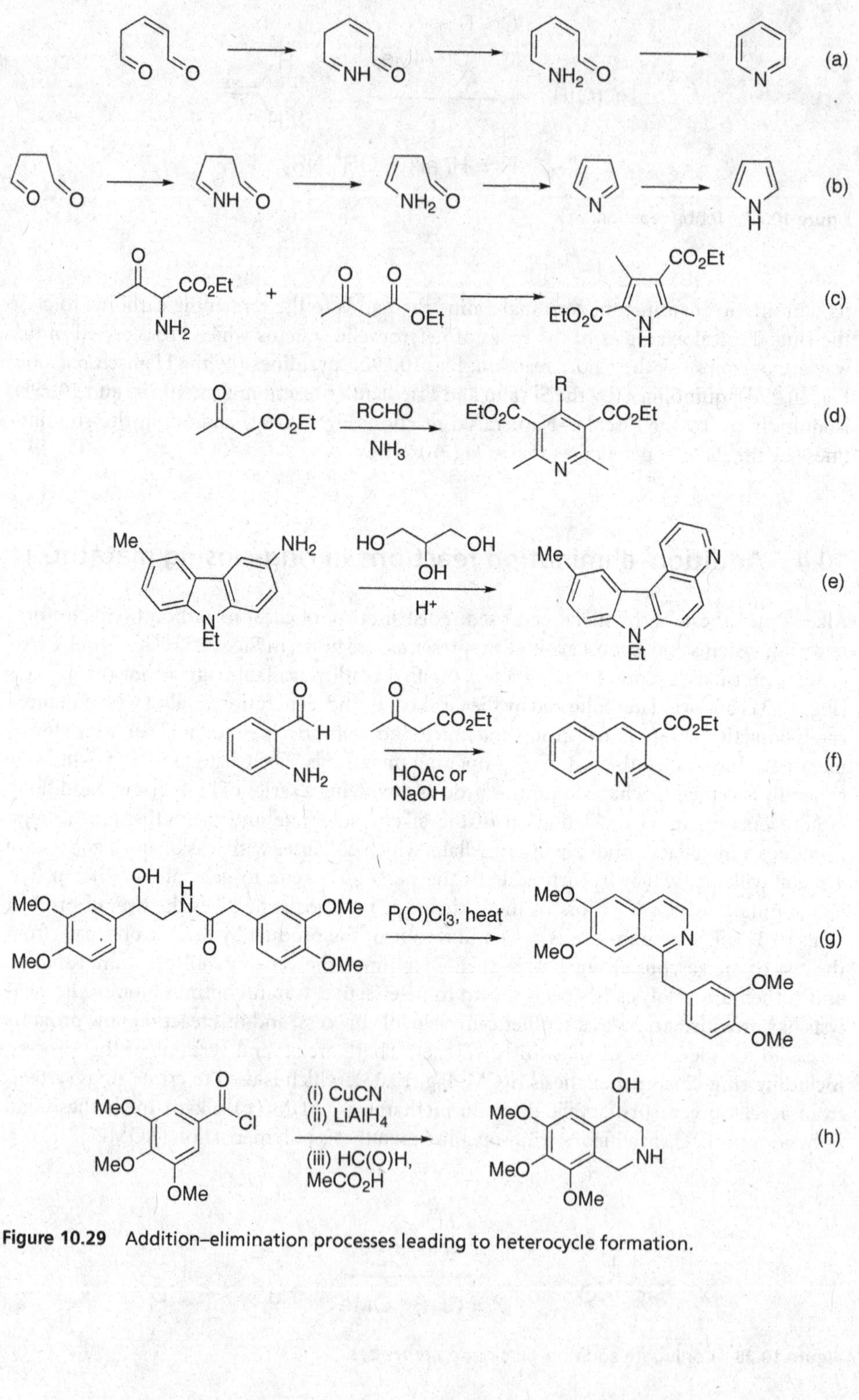

Figure 10.29   Addition–elimination processes leading to heterocycle formation.

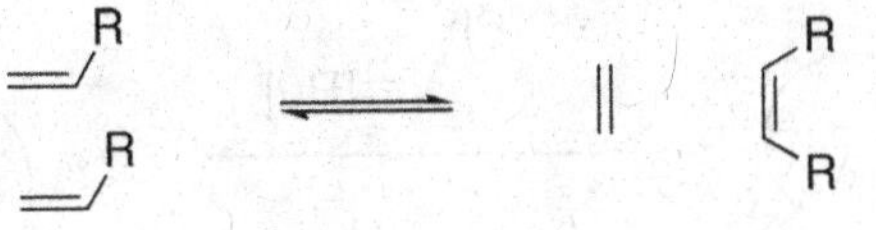

Figure 10.30 Generalised metathesis procedure.

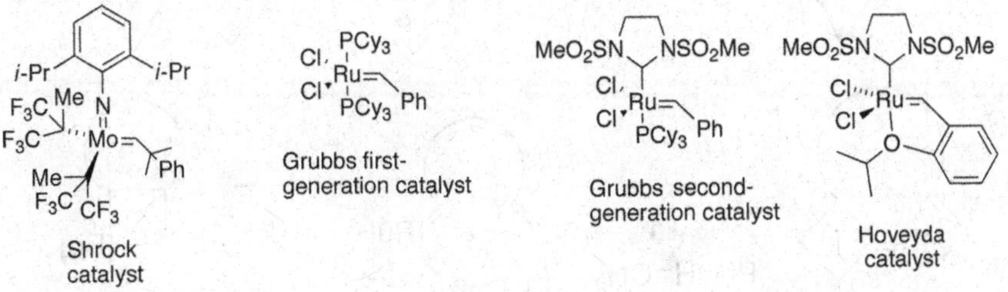

Figure 10.31 Catalysts for metathesis.

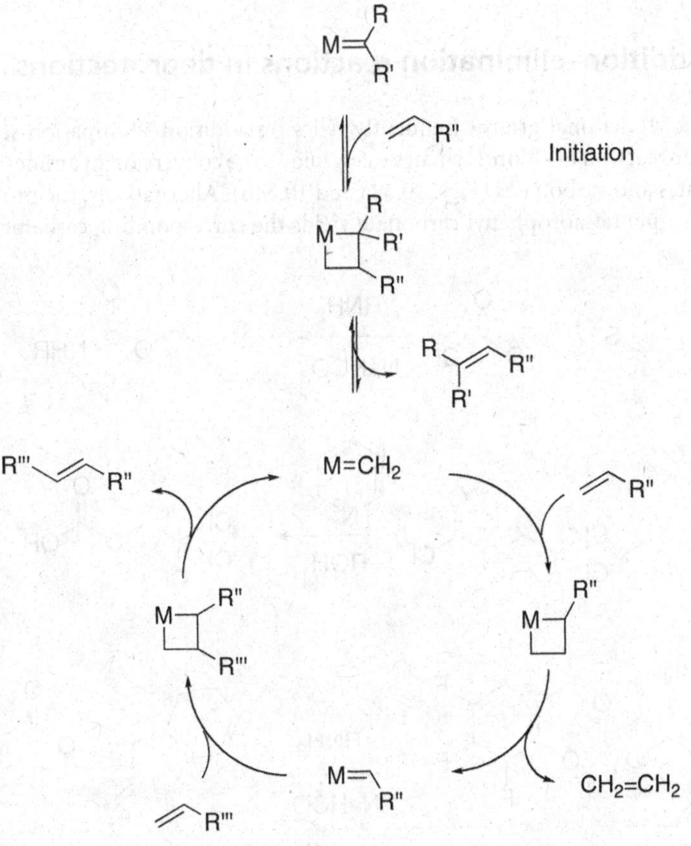

Figure 10.32 Chauvin mechanism for metathesis.

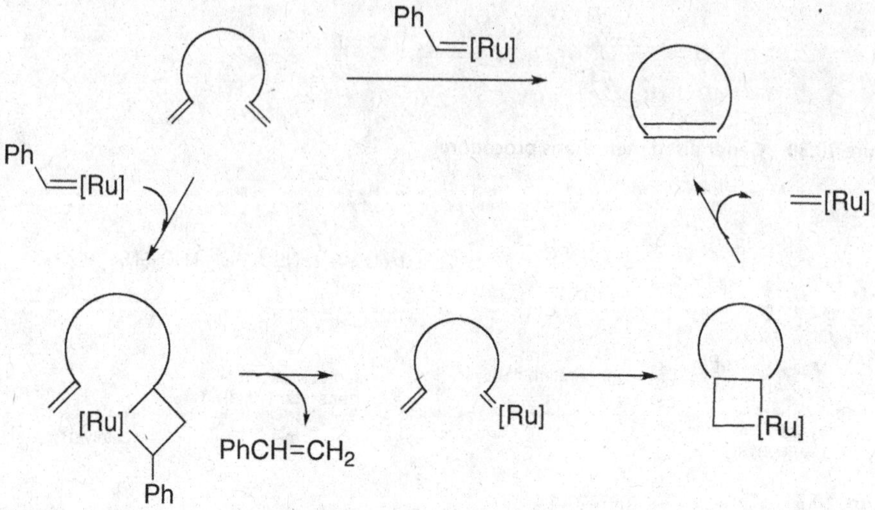

Figure 10.33 Ring closure using ring-closing metathesis.

10.5 Addition–elimination reactions in deprotections

Protection of functional groups frequently relies on addition–elimination sequences; for example, the reaction of chloroformates easily allows the conversion of amines and alcohols to carbamates and carbonates (Figs. 10.34a and 10.34b). Alternatively, the protection of an amine with a pentafluorophenyl carbonate yields the corresponding carbamate derivative

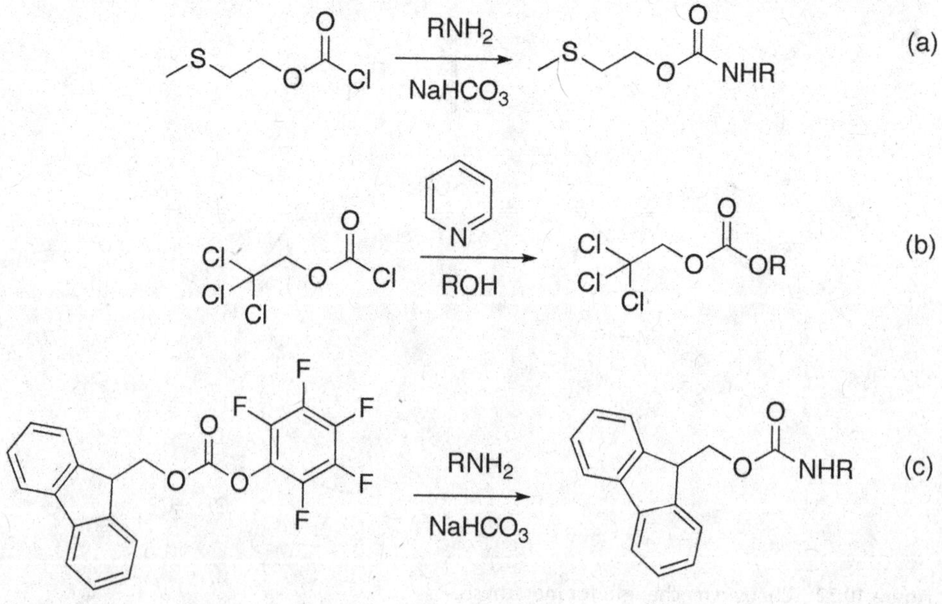

Figure 10.34 Protections as a result of addition–elimination.

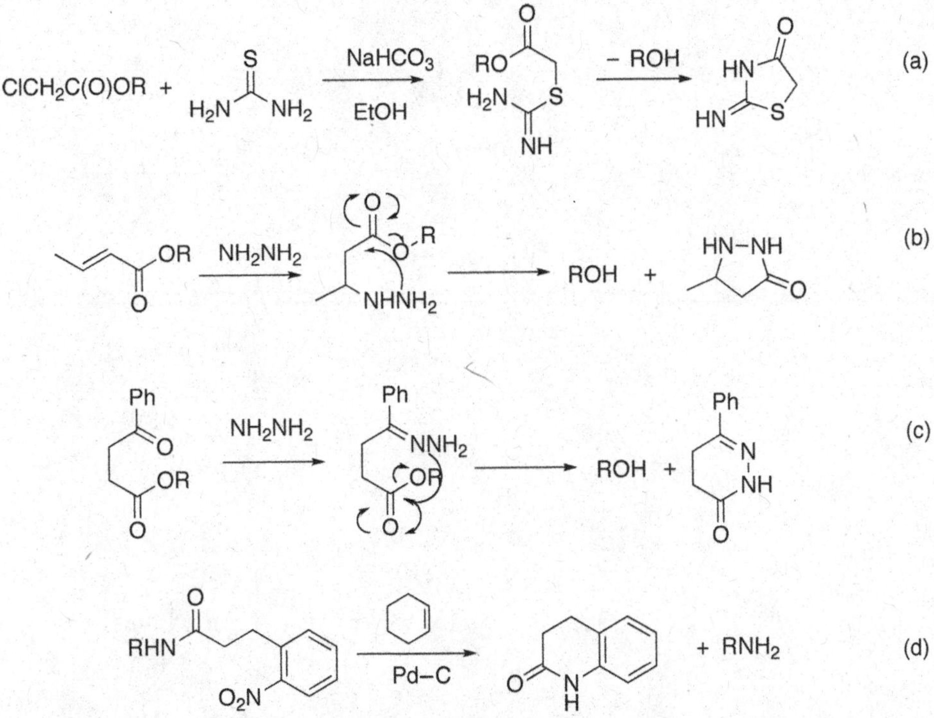

Figure 10.35 Deprotections as a result of addition–elimination.

in which the amine lone pair is stabilised by resonance (Fig. 10.34c); the advantage of this approach is that such carbonates are reasonably stable and easy to handle.

The application of nucleophilic addition–elimination sequences in deprotection reactions is neatly exemplified by the examples of Fig. 10.35. Chloroacetates may be cleaved by treatment with thiourea; initial alkylation on sulfur leads to an intermediate which efficiently collapses to a heterocyclic product, along with release of the alcohol (Fig. 10.35a). Of course, such esters may also be cleaved by basic hydrolysis. Cleavage of crotonate esters by initial conjugate addition of hydrazine is again followed by internal addition–elimination to release the alcohol along with a heterocyclic product (Fig. 10.35b). Another example relates to the deprotection of 4-oxoacyl esters, which are hydrolysed using hydrazine (Fig. 10.35c); initial addition–elimination at the ketone to generate the hydrazone is followed by an internal addition–elimination releasing the alcohol and again generating a heterocyclic product. Finally, reduction of an aromatic nitro group generates a highly nucleophilic amine, which is suitably disposed to an amide so that internal addition–elimination occurs, and this releases the amine with simultaneous formation of a lactam (Fig. 10.35d).

Chapter 11
Radical Reactions

Another important class of reaction which involves the formation of intermediates is radical reactions; although traditionally perhaps not as important as the polar reactions described earlier, much progress in radical chemistry has been made in the last 40 years, and reactions involving radical intermediates are now known to be highly selective and are therefore widely used in contemporary organic synthesis. It has emerged that these processes are particularly valuable, since a radical, unlike a cation or anion, is not crowded by the presence of an immediately adjacent solvation shell; radicals therefore tend to both react with sterically congested substrates and generate sterically congested products much more effectively than ionic processes.

11.1 Generation

Radicals may be generated in three main ways, as we have seen earlier in Chapter 4, and the most direct process normally involves the homolysis of weak heteroatom–heteroatom bonds, as illustrated in Fig. 11.1. Most commonly, the required weak bond is deliberately set up for collapse, for example, by the thermolysis of peroxides and azo compounds (Figs. 11.2a and 11.2b) and by the photolysis of halogens, aldehydes and ketones (Figs. 11.3a–11.3c).

Another important method for the generation of radicals is by the single-electron oxidation of carbanions or the reduction of carbocations (Fig. 11.4a). Two important examples are the radical formation from hydroperoxides using ferrous salts under alkaline conditions, and the dissolving metal reduction of ketones to generate ketyl radicals (Figs. 11.4b and 11.4c).

11.2 Reactions

Because radicals are so unstable (i.e. reactions in which they participate have very low E_{act} values), they undergo rapid reactions with themselves which terminate a radical chain (Fig. 11.5a), with other neutral species in reactions whose rate is largely controlled by the rate at which they are able to diffuse in solution (Figs. 11.5b and 11.5c) or by intramolecular processes which are kinetically highly favourable and for which reaction with another molecule is unnecessary (Figs. 11.5d–11.5f).

11.2.1 Termination

The reaction of two radicals gives a single neutral species, thereby destroying radicals and therefore terminating a sequence in which they are intermediates (Fig. 11.5a). Such combination reactions proceed with a very large rate constant, typically of the order of

$$R{-}X \longrightarrow R^{\bullet} + X^{\bullet}$$

Figure 11.1 Radical formation by homolysis.

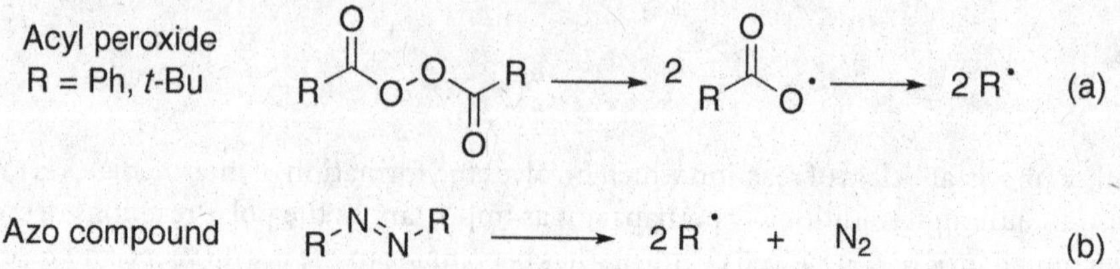

Acyl peroxide
R = Ph, *t*-Bu

(a)

Azo compound

$$\longrightarrow 2\,R^{\bullet} + N_2$$

(b)

Figure 11.2 Thermal formation of radicals.

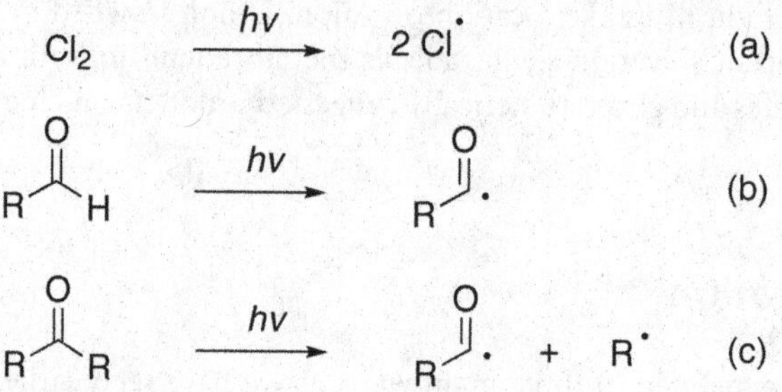

$$Cl_2 \xrightarrow{\;h\nu\;} 2\,Cl^{\bullet}$$

(a)

(b)

(c)

Figure 11.3 Photolytic formation of radicals.

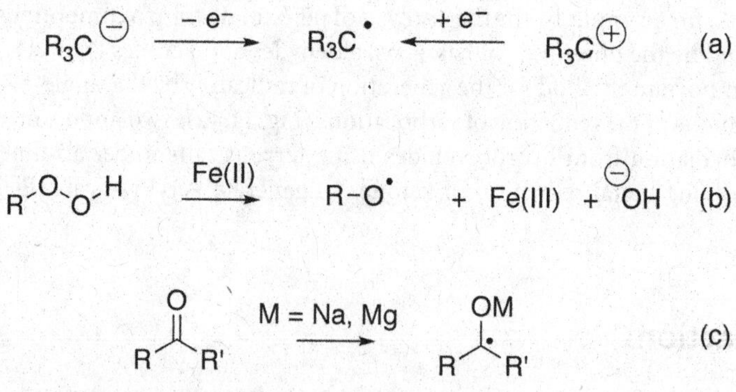

$$R_3C^{\ominus} \xrightarrow{\;-\,e^-\;} R_3C^{\bullet} \xleftarrow{\;+\,e^-\;} R_3C^{\oplus}$$

(a)

$$R{-}O{-}O{-}H \xrightarrow{\;Fe(II)\;} R{-}O^{\bullet} + Fe(III) + {}^{\ominus}OH$$

(b)

M = Na, Mg

(c)

Figure 11.4 Redox generation of radicals.

10^9–10^{10} M^{-1} s^{-1}, although *dimerisations*, in which R = R′, are generally kinetically very unfavourable because of the low probability of two identical radicals encountering each other.

11.2.2 Propagation

Before the termination of a radical process, a radical may react with another neutral molecule, in which case the reaction must lead to another radical, or it must rearrange to generate another radical. There are four possible processes.

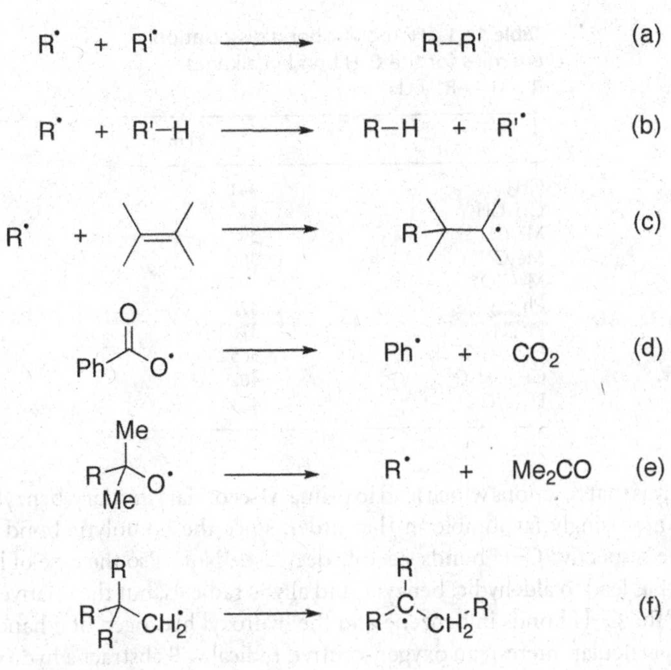

$$R^\bullet \; + \; R'^\bullet \longrightarrow R{-}R' \qquad \text{(a)}$$

$$R^\bullet \; + \; R'{-}H \longrightarrow R{-}H \; + \; R'^\bullet \qquad \text{(b)}$$

$$R^\bullet \; + \; \text{(alkene)} \longrightarrow R{-}\text{(radical)} \qquad \text{(c)}$$

$$Ph{-}C(O){-}O^\bullet \longrightarrow Ph^\bullet \; + \; CO_2 \qquad \text{(d)}$$

$$R{-}C(Me)_2{-}O^\bullet \longrightarrow R^\bullet \; + \; Me_2CO \qquad \text{(e)}$$

$$R_3C{-}CH_2^\bullet \longrightarrow R_2\overset{\bullet}{C}{-}CH_2{-}R \qquad \text{(f)}$$

Figure 11.5 Types of reactions of radicals.

11.2.3 Substitution

Radicals readily participate in the S_N2 reaction, which is the homolytic substitution of a suitable group in a bimolecular process. The reaction may be generalised as shown in Fig. 11.6, and overall reaction proceeds in a radical chain process involving a sequence of homolytic steps. Thus, homolysis of the starting material R—X forms the radical R$^\bullet$, which reacts with the other component A—B by homolytic abstraction of A to generate the desired product R—A. The by-product of this step, the radical B$^\bullet$, reacts with more starting material to abstract X, thereby generating another radical R$^\bullet$, along with the by-product B—X. The radical B$^\bullet$ is the so-called chain carrier, since it leads to overall propagation of the reaction by maintaining the homolytic dissociation steps.

Reaction of a radical with a hydrocarbon can lead to transfer of a hydrogen atom (Fig. 11.6; A = H). However, not all hydrocarbons are equally reactive, and in order to predict the likely hydrogen abstraction by a radical, it is necessary to consider bond dissociation energies (D) for characteristic chemical functions. Some typical values are given in Table 11.1 for the homolytic dissociation of a C—H bond, and this is graphically illustrated in Fig. 11.7.

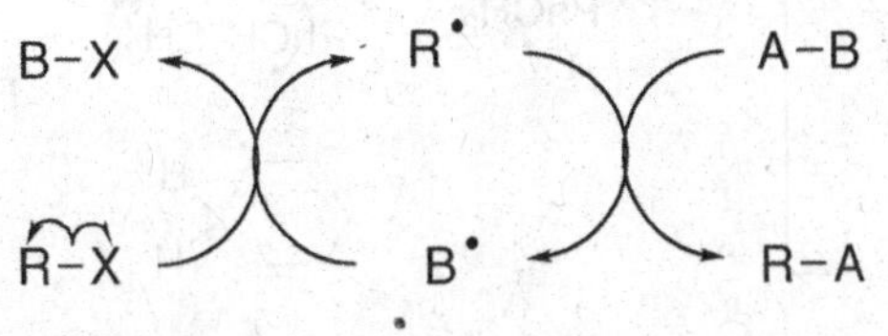

$$B{-}X \longleftarrow \quad R^\bullet \quad \longrightarrow A{-}B$$
$$R{-}X \longrightarrow \quad B^\bullet \quad \longleftarrow R{-}A$$

Figure 11.6 Free-radical substitution.

Table 11.1 Homolytic bond dissociation energies for the C–H bond in alkanes
$R - H \rightarrow R^{\bullet} + H^{\bullet}$

R	D_{298} (kJ mol^{-1})
CH_3	441
CH_3CH_2	412
Me_2CH	399
Me_3C	386
$MeC(O)$	365
Ph	470
$PhCH_2$	370
$H_2C=CHCH_2$	365
CH_3CH_2O	462
$H_2C=CH$	454

Noteworthy is that reactions which lead to primary, secondary, tertiary, benzylic and allylic radicals are increasingly favourable in that order, since the homolytic bond dissociation energy for the respective C–H bonds steadily decreases. Note also the ease of homolysis of C–H bonds that lead to aldehydic, benzylic and allylic radicals, but the relative difficulty of homolysis of the C–H bonds in benzene and the hydroxyl hydrogen of ethanol. The latter process is of particular interest; an oxygen-centred radical will abstract a hydrogen from an activated C–H bond (Fig. 11.8). In this case, the $D(R–H)$ value of 370 kJ mol^{-1} for the C–H bond is amply compensated by the formation of the O–H bond ($D(R–H) = 462$ kJ mol^{-1}). Note that there is no equivalent of this process in anionic chemistry, and it means that O–H groups do not normally require protection in radical reactions (although of course do often require it in basic reactions).

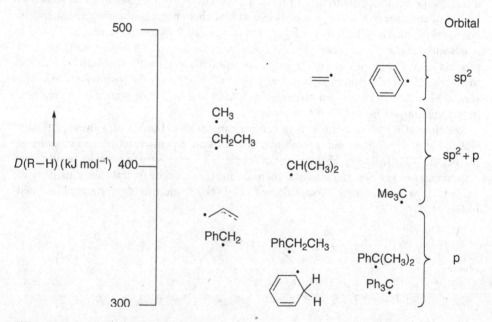

Figure 11.7 Relative stability of radicals.

$$RO^\bullet \ + \ R'-H \quad \longrightarrow \quad RO-H \ + \ R'^\bullet$$

$$D(R-H) = 370 \ kJ\,mol^{-1} \quad D(R-H) = 462 \ kJ\,mol^{-1}$$

Figure 11.8 Abstraction of hydrogen by alkoxy radicals.

Autoxidation is a radical process which leads to the oxidation of organic compounds (Fig. 11.9). The process is initiated by radical abstraction leading to a stabilised radical which then intercepts oxygen. The resulting hydroperoxy radical then abstracts a hydrogen and continues the chain. Autoxidation is one process which leads to oxidative degradation and is particularly important in living systems; metabolic defence mechanisms have been developed to intercept and remove these reactive radical intermediates.

The selectivity of this abstraction process is further affected by several factors, including the strength of the bond which is formed in the reaction and the strength of the bond which is broken. The former is illustrated by reaction of methyl radicals with various isomeric hydrocarbons as shown in Fig. 11.10. In this sequence, it has been shown that the logarithm of the rate of reaction and the bond dissociation energies correlate inversely; thus, the lower the bond energy of the bond being broken, the higher the rate of reaction. The latter is illustrated by the abstraction of hydrogen from methane by halogen radicals (Fig. 11.11). In this case, the logarithm of the rate of reaction correlates linearly with the homolytic dissociation energy of the bond which is being formed for all of the halogens except fluorine. This is accounted for in the case of fluorine, by the fact that the reaction is highly exothermic, and so the transition state is very early, with very little R–H bond breaking, in the reaction. On the other hand, in the case of bromination or iodination, the reaction is quite endothermic, and there is significant C–H bond breaking in the transition state, and so any structural effects which operate to stabilise the developing radical centre will be very important. For this reason, the selectivity of the halogen radicals decreases in the order:

$$I > Br > Cl > F$$

By way of exemplification, the abstraction of hydrogens from cumene by chlorine or bromine radicals leads to the outcome shown in Fig. 11.12, in which chlorine radicals will remove the

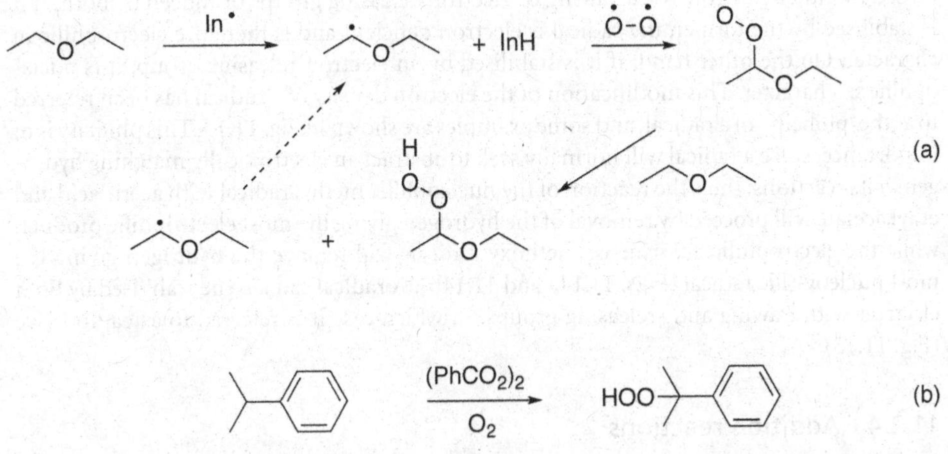

Figure 11.9 Autoxidation of (a) ether and (b) cumene.

Me$^\bullet$ ⌒ H–R ⟶ Me–H + R$^\bullet$

R	log k (M^{-1} s^{-1})	D(R—H) (kJ mol^{-1})
Me	1.65	435
Et	2.96	410
i-Pr	3.75	395
t-Bu	4.36	381

Figure 11.10 Rate control by bond breaking.

X$^\bullet$ ⌒ H–Me ⟶ X–H + Me$^\bullet$

X	log k (M^{-1} s^{-1})	D(X—H) (kJ mol^{-1})
F	10.22	569
Cl	7.54	431
Br	−2.64	364
I	−13.15	296

Figure 11.11 Rate control by bond making.

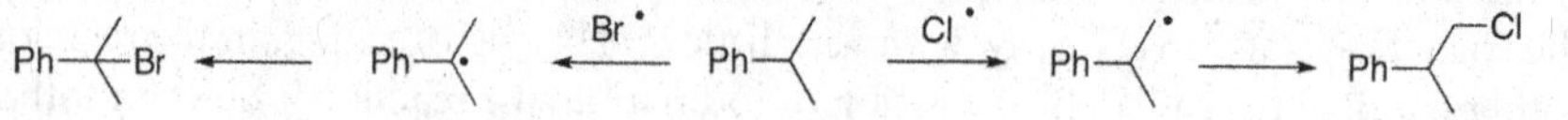

Ph—⟨—Br ⟵ Ph—⟨· $\overset{Br^\bullet}{\longleftarrow}$ Ph—⟨ $\overset{Cl^\bullet}{\longrightarrow}$ Ph—⟨· ⟶ Ph—⟨—Cl

Figure 11.12 Selectivity in halogen substitutions.

most kinetically available hydrogens (i.e. the most numerous and sterically most accessible methyl hydrogens), while bromine radicals will remove the hydrogen which leads to the most stable product radical.

A further parameter of critical importance for radical reactions relates to polarity effects; this arises because a radical centre, possessing as it does a half-filled orbital, can be stabilised by either electron-withdrawing or electron-releasing groups or indeed by both. If it is stabilised by the former, the radical is electron deficient and is therefore electrophilic in character. On the other hand, if it is stabilised by an electron-releasing group, it is nucleophilic in character. This modification of the electron density of a radical has been referred to as the 'philicity' of a radical, and some examples are shown in Fig. 11.13. This philicity is of importance, since a radical will normally seek to abstract an electronically matching hydrogen in its reactions; thus, the reaction of the nucleophilic methyl radical with acetic acid and ethyl acetate will proceed by removal of the hydrogen giving the most electrophilic product, while the electrophilic chlorine or methoxy radicals will remove the hydrogen giving the most nucleophilic radical (Figs. 11.14a and 11.14b). A radical can also be stabilised by both electron-withdrawing and -releasing groups, in which case it is referred to as captodative (Fig. 11.15).

11.2.4 Addition reactions

The addition of radicals to double bonds, carbon–carbon, carbon–oxygen or carbon–nitrogen, as shown in Fig. 11.5c, is very effective. Of course, the overall outcome is

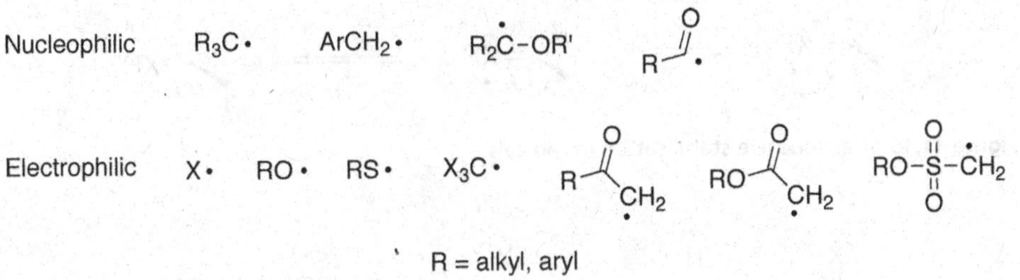

Figure 11.13 Philicity in radicals.

the same as that of addition of nucleophiles to these systems, but in this case the thermodynamics of the process can vary between π-systems. Thus, although the addition of radicals to carbon–carbon double bonds is strongly exothermic, in the case of carbon–oxygen double bonds it is endothermic and therefore reversible (Fig. 11.16), and so the equilibrium position lies to the side of starting materials. This arises because in the case of additions to alkenes, the loss of the carbon–carbon π-bond (280 kJ mol^{-1}) is more than compensated for by the formation of the carbon–carbon σ-bond (340 kJ mol^{-1}); however, in the case of carbonyl compounds, the loss of the carbon–oxygen π-bond (370 kJ mol^{-1}) is not compensated for by the formation of the carbon–carbon σ-bond (340 kJ mol^{-1}).

There are a number of effects which govern the addition of radicals to alkenes, and these include the nature of any substituents on the double bond. Such substituents are capable of modifying the electronics as well as the sterics of the reacting system. Thus, for a nucleophilic radical, electron-withdrawing groups at the β-position on the double bond will significantly enhance addition reactions, and some relative reaction rates are given in Figs. 11.17 and 11.18, which illustrate the magnitude of these effects. Unsurprisingly, resonance-withdrawing groups have the biggest effect on the rates of additions, since these enhance the electrophilicity at the β-position. This is seen further in the reaction of either electrophilic or nucleophilic radicals to differently substituted alkenes; thus, reaction of the electron-rich (nucleophilic) cyclohexyl radical and the electron-poor (electrophilic) malonate radical is accelerated by electron-withdrawing or electron-releasing groups respectively (Fig. 11.19). Overall, then, electron-withdrawing groups in the β-position of a double bond accelerate the addition of nucleophilic radicals and slow the addition of electrophilic radicals, and electron-releasing groups in the β-position of a double bond accelerate the addition of electrophilic radicals and slow the addition of nucleophilic radicals.

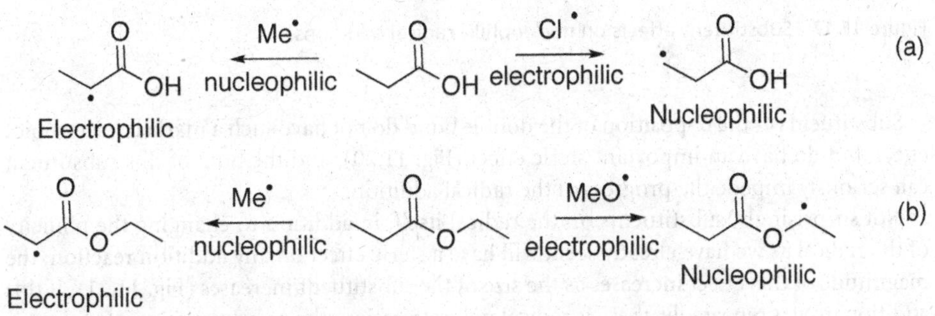

Figure 11.14 Selectivity in radical reactions.

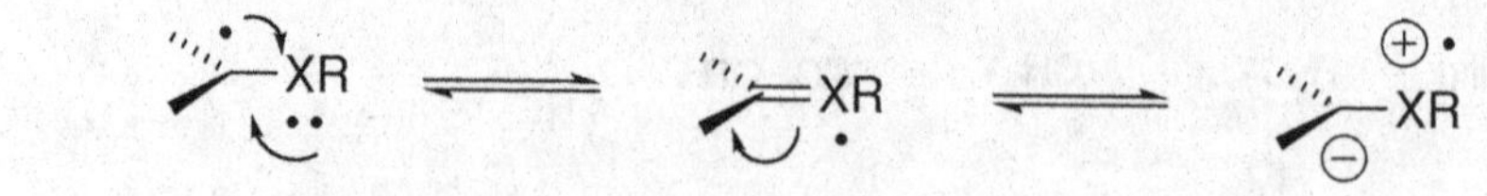

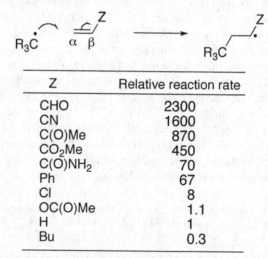

Figure 11.15 Captodative stabilisation in radicals.

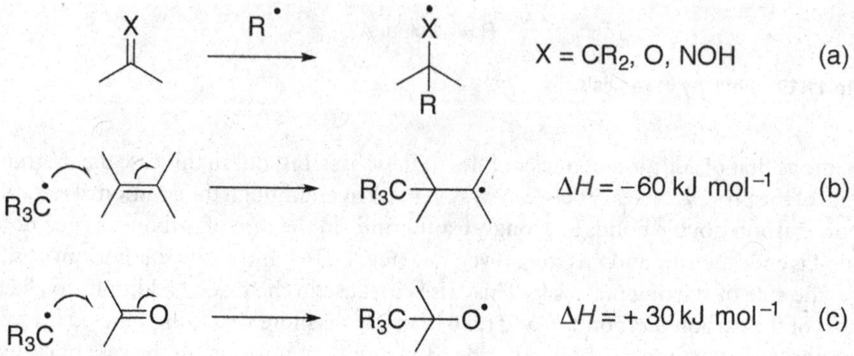

$X = CR_2, O, NOH$ (a)

$\Delta H = -60 \ \text{kJ mol}^{-1}$ (b)

$\Delta H = +30 \ \text{kJ mol}^{-1}$ (c)

Figure 11.16 Radical additions.

Z	Relative reaction rate
CHO	2300
CN	1600
C(O)Me	870
CO_2Me	450
C(O)NH$_2$	70
Ph	67
Cl	8
OC(O)Me	1.1
H	1
Bu	0.3

Figure 11.17 Substituent effects on nucleophilic radical additions.

Substituents at the α-position of the double bond do not have such a marked (electronic) effect, but do have an important steric effect (Fig. 11.20), and the bulk of this substituent can seriously impede the progress of the radical addition.

Not surprisingly, substituents on the radical itself, in addition to changing the philicity of the radical as we have already seen, will have a steric effect on any addition reaction; the magnitude of this effect increases as the size of the substituent increases (Fig. 11.21). If this addition occurs repeatedly, there is radical polymerisation; the polymerisation of alkenes is

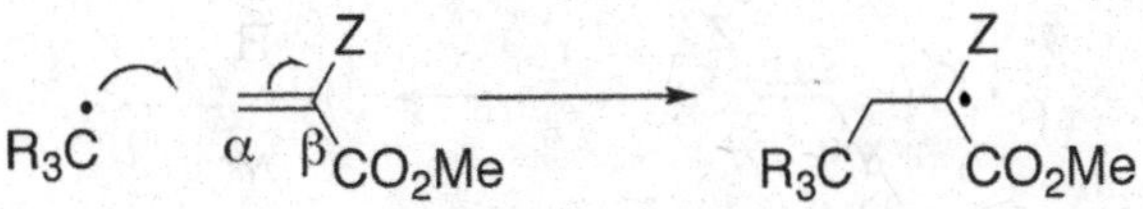

Z	Relative reaction rate
CN	310
CO_2Me	150
CF_3	40
Cl	12
SePh	10
CH_2Cl	8
Ph	6
CH_2CO_2Me	1.4
H	1
Me	0.75
CMe_3	0.20
OMe	0.16

Figure 11.18 Substituent effects on nucleophilic radical additions.

an important method for the formation of macromolecules based on a single or alternating repeating units (Fig. 11.22).

An example will illustrate the importance of the balance of each of these parameters, as shown in Fig. 11.23a. In this reaction, azobisisobutyronitrile (AIBN) homolyses to generate the initiating free radicals, which abstract hydrogen from tributyltin hydride to generate

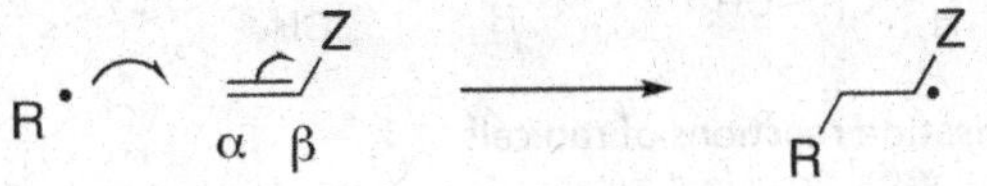

Z	Relative reaction rate R =	
	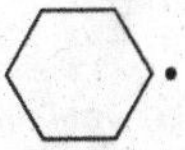 ⬡•	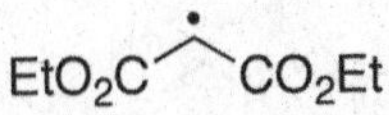 $EtO_2C\overset{\bullet}{\frown}CO_2Et$
CN	470	–
C(O)Ph	54	0.2
CO_2Me	46	0.3
Ph	4	1.2
Me	1	1.0
OMe	0.1	2.1
NEt_2	–	6.1

Figure 11.19 β-Substituent effects on radical additions.

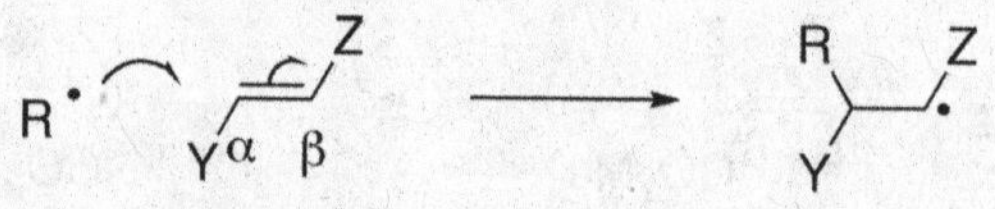

Y	Relative reaction rate
CN	6
CO$_2$Me	5
H	1
Cl	0.1
Me	0.01
Ph	0.009

Figure 11.20   α-Substituent effects on radical additions.

R	Relative reaction rate
Me	1000
i-Pr	250
t-Bu	50

Figure 11.21   Radical substituent effects on radical additions.

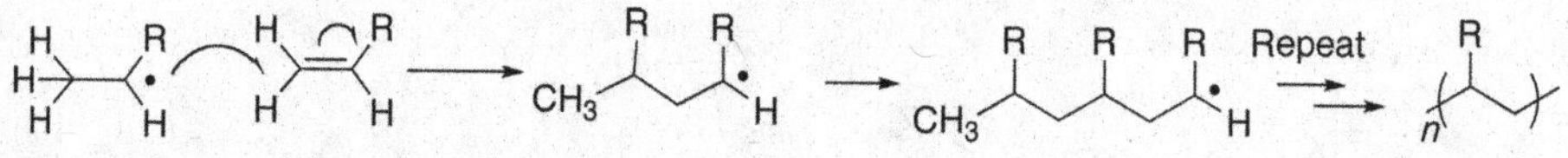

Figure 11.22   Polymerisation reactions of radicals.

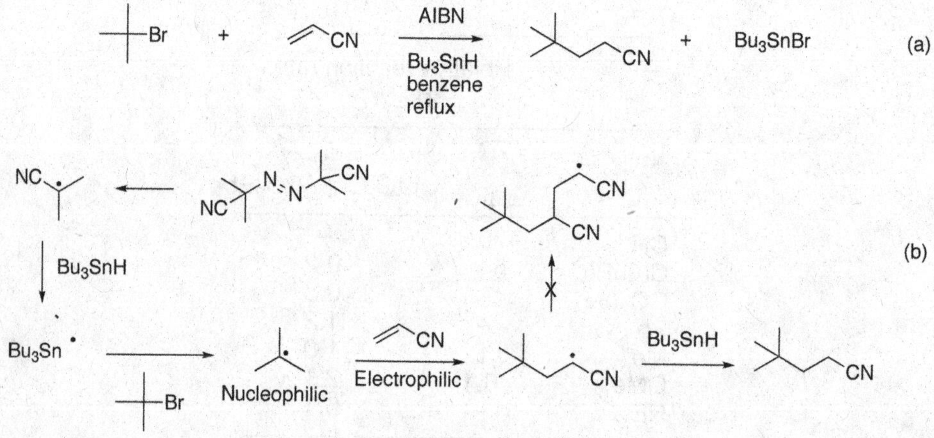

Figure 11.23   A radical addition reaction.

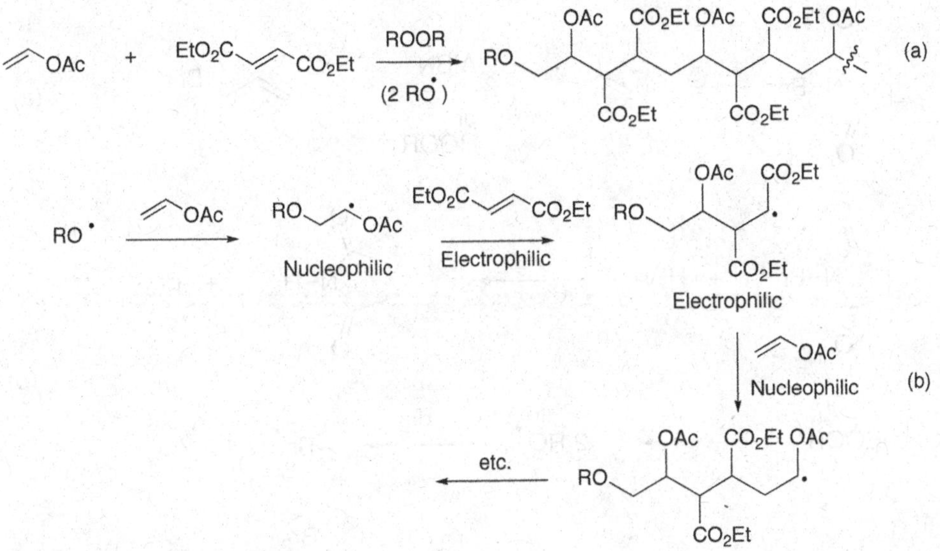

Figure 11.24 A radical polymerisation.

Bu$_3$Sn$^•$ radicals. These, abstract bromine radicals from *t*-butyl bromide by homolysis of the weakest bond. The resulting electron-rich *t*-butyl radicals effectively add to the electron-deficient acrylonitrile double bond (Fig. 11.23b) to generate an electrophilic radical which is reduced by the tributyltin hydride. This regenerates Bu$_3$Sn$^•$ radicals, which then carry the chain forward. Of interest in this sequence is that the rate of *t*-butyl radical addition to the acrylonitrile needs to be faster than its reduction by tin hydride; since the intrinsic rate of this process is very fast (rate typically at about 300 000 M^{-1} s^{-1}), it is necessary to keep the concentration of tributyltin hydride as low as possible to slow this undesired reduction down. However, it is also necessary that the final rate of reduction of the α-acrylonitrile radical is faster than its reaction with another molecule of acrylonitrile, which would otherwise lead to polymerisation. Fortunately, this is assured, since the polymerisation reaction is slowed by the fact that it represents an unfavourable matching of electrophilic radical and electrophilic double bond.

Similar control may be observed in polymerisation reactions which give alternating polymers (Fig. 11.24a). Initial addition of a *t*-butoxy radical to vinyl acetate generates an electron-rich radical which preferentially attacks the electron-poor diethyl fumarate (Fig. 11.24b); this then generates an electron-rich radical which selectively attacks the electron-poor vinyl acetate which once again generates an electron-rich radical. This guarantees the formation of polymer product based on a strictly alternating sequence.

The relative rates of radical processes can have a big impact on the outcome of a reaction as well; thus, Wohl–Ziegler allylic bromination with *N*-bromosuccinimide in the presence of a catalytic amount of hydrogen bromide proceeds via the so-called Goldfinger mechanism (Fig. 11.25). In this reaction, generation of a low concentration of bromine, as shown, ultimately leads to the formation of a low concentration of bromine radicals; these may add to the double bond, but in the absence of sufficient bromine to quench the resulting radical, this process is reversible and the bromine then abstracts a hydrogen atom from the allylic position to generate a resonance-stabilised allylic radical.

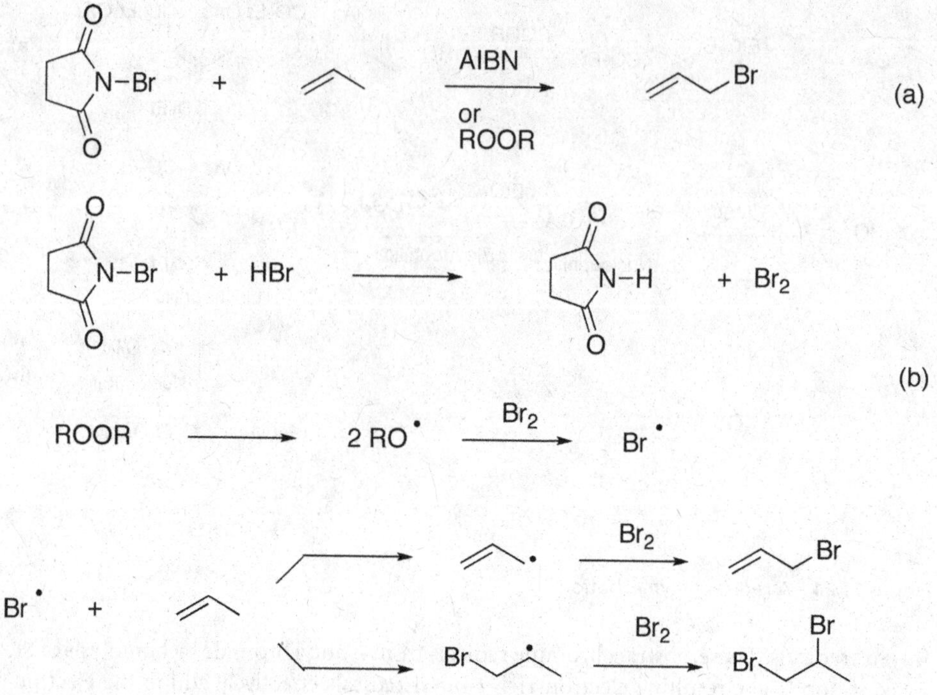

Figure 11.25 Wohl–Ziegler allylic halogenation.

The importance of initiation and propagation is illustrated by the peroxide-catalysed free-radical addition of hydrogen bromide to alkenes in Fig. 11.26. Overall, the reaction proceeds with anti-Markovnikov addition, since the regioselectivity is opposite to that observed for electrophilic addition (see Section 7.1.2). In this case, the regioselectivity is given by the stability of the first-formed radical from attack by Br$^\bullet$ (as opposed to H$^+$ for electrophilic addition).

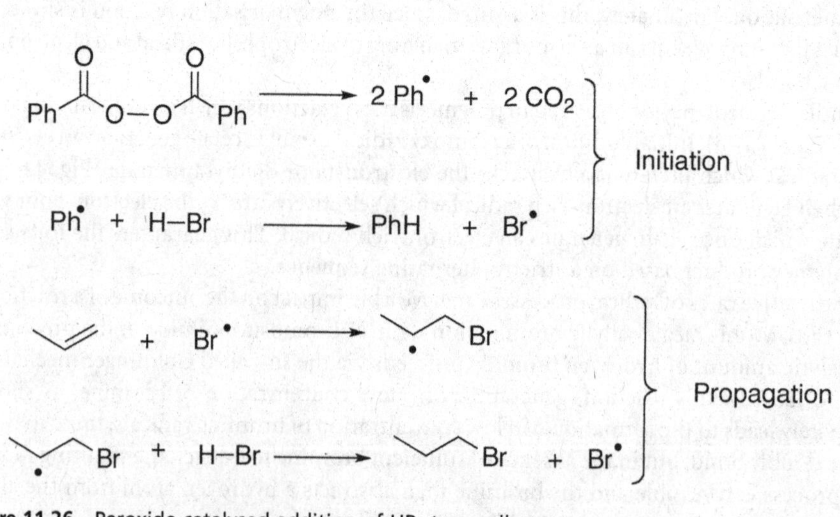

Figure 11.26 Peroxide-catalysed addition of HBr to an alkene.

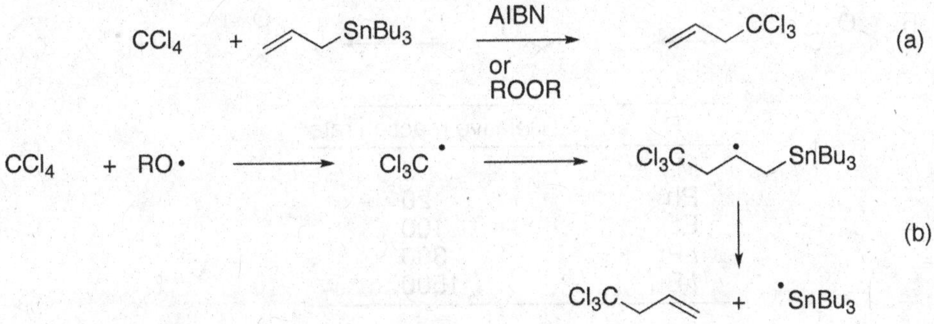

Figure 11.27 Allylic transfer from allylstannanes.

Finally, the application of allylic systems carrying stannyl residues can be very useful for allyl transfer reactions under mild conditions; for example, the reaction of carbon tetra-chloride with allyl tributyltin under radical-generating conditions leads to the very effective formation of the product indicated in Fig. 11.27a. This outcome arises because the $Cl_3C^•$ radical generated under these conditions effectively adds to the carbon–carbon double bond, and the intermediate radical reacts on to eject the chain-carrying tributylstannyl radical (Fig. 11.27b). This process can be elaborated even further, as shown in Fig. 11.28a, in which a three-component coupling can be achieved. This reaction proceeds by initial addition of methyl radical to an electron-deficient alkene, which is followed by trapping with allyl tributylstannane to give the product, along with the chain-carrying $Bu_3Sn^•$ radical (Fig. 11.28b).

11.2.5 Fragmentation

If a radical does not react with another entity as described in Sections 11.2.2 and 11.2.3 above, it may decompose to give other species, especially if a stable molecule, such as CO_2 or acetone, may be lost (Figs. 11.5d and 11.5e). This process is sometimes called β-scission

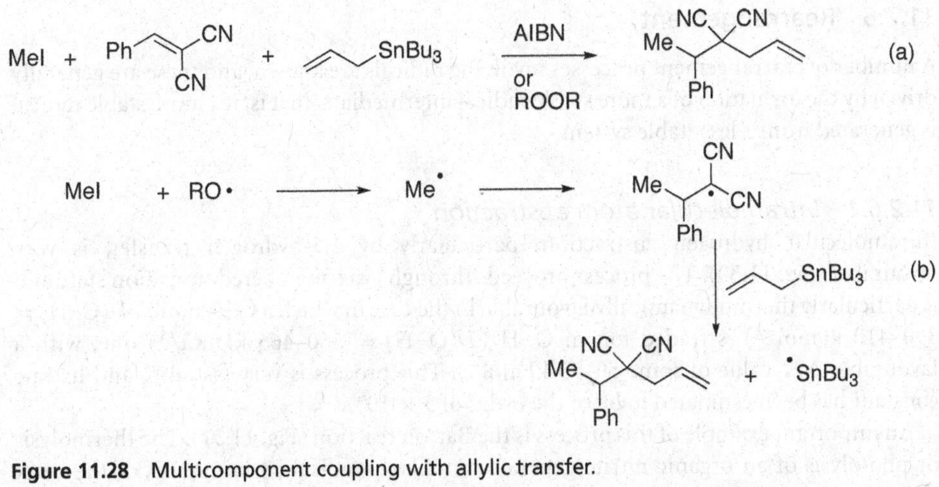

Figure 11.28 Multicomponent coupling with allylic transfer.

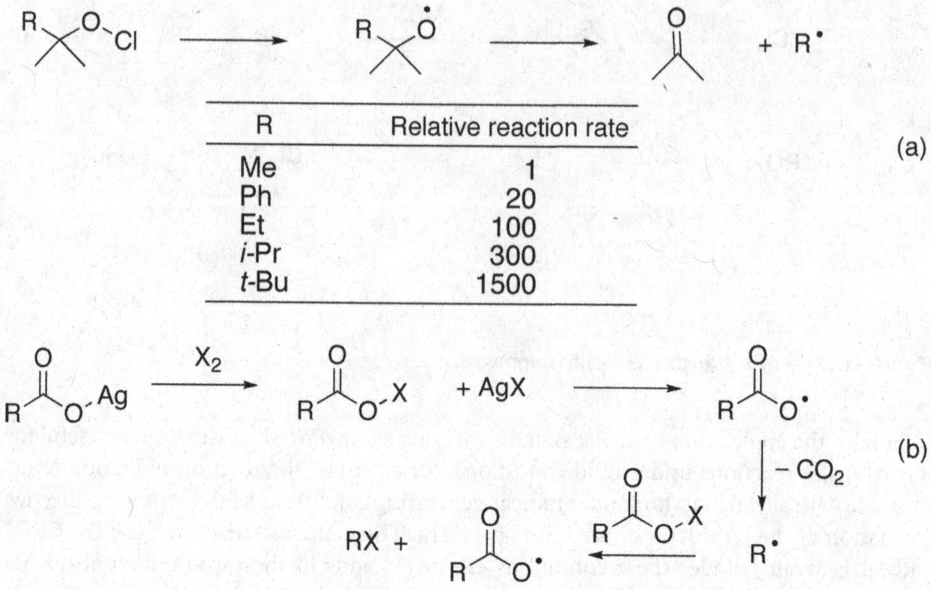

R	Relative reaction rate
Me	1
Ph	20
Et	100
i-Pr	300
t-Bu	1500

(a)

(b)

Figure 11.29 (a) Substituent effects on radical fragmentations and (b) Hunsdiecker reaction.

and occurs when the fragmentation of a radical leads to more stable components. Not surprisingly, the rate of this process depends on the bulk around the alkoxy radical centre and increases markedly with increasing bulk (Fig. 11.29a).

Another important example of the fragmentation process is the Hunsdiecker reaction; in this procedure, a silver salt of a carboxylic acid is treated with a halogen (normally bromine or iodine) to generate an intermediate acyl hypohalite. This readily homolyses to give the acyloxy radical, which in turn fragments and then the resulting alkyl radical is halogenated in the chain-carrying step (Fig. 11.29b).

11.2.6   Rearrangement

A number of rearrangement processes involving radicals are known, and these are generally driven by the formation of a more stable radical intermediate, that is, if a more stable radical is generated from a less stable system.

11.2.6.1   Intramolecular atom abstraction

Intramolecular hydrogen abstraction, particularly by 1,5-hydrogen transfer, is very favourable (Fig. 11.30). The process proceeds through a six-membered transition state and is particularly thermodynamically favourable in the case in which a C–H bond ($D(C–H) = 390–410$ kJ mol^{-1}) is traded for an O–H ($D(O–H) = 460–465$ kJ mol^{-1}) one, with a favourable ΔH value of some 50–75 kJ mol^{-1}. This process is very fast also, and its rate constant has been estimated to be of the order of 5×10^6 s^{-1}.

An important example of this process is the Barton reaction (Fig. 11.31). The thermolysis or photolysis of an organic nitrite generates an alkoxy radical by homolysis of the weak

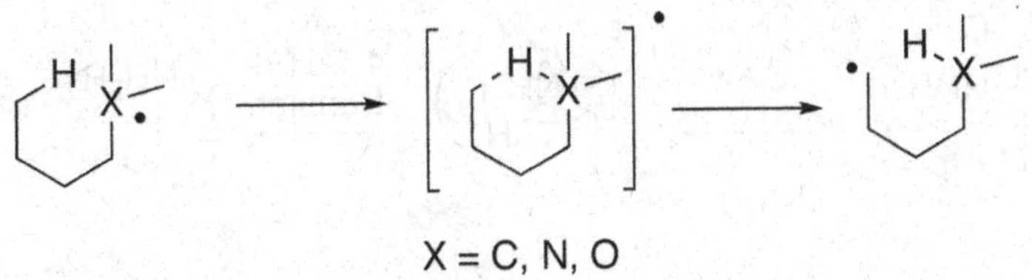

$$X = C, N, O$$

Figure 11.30 1,5-Hydrogen transfer.

oxygen—nitrogen bond; this will abstract an adjacent hydrogen from the proximal methyl group via a six-membered transition state to generate a methyl radical, which is terminated by the original nitroxyl radical, held in the solvent cage. The resulting nitroso product tautomerises and gives the observed product, the oxime. This may be hydrolysed in the usual way to generate the corresponding carbonyl product. This reaction is of significance, since the angular methyl group, normally considered not to be reactive, becomes functionalised in the course of the reaction; more generally, this reaction can be used for remote functionalisation of hydrocarbon substituents.

Another example is the Hofmann–Löffler–Freytag reaction (Fig. 11.32): Homolysis of a weak N—Cl bond generates a radical cation, which undergoes a 1,5-hydrogen shift and radical termination with chlorine radical to give a chloroalkane product. Intramolecular S_N2 reaction under basic conditions then generates the cyclic product.

11.2.6.2 1,2-Phenyl shifts

1,2-Phenyl shifts, which proceed via phenyl-bridged radicals, are common as much as they are for carbocations (Fig. 11.33a). These rearrangements occur because they lead to the formation of more stable radical systems, and the rearranging radical is partially stabilised by

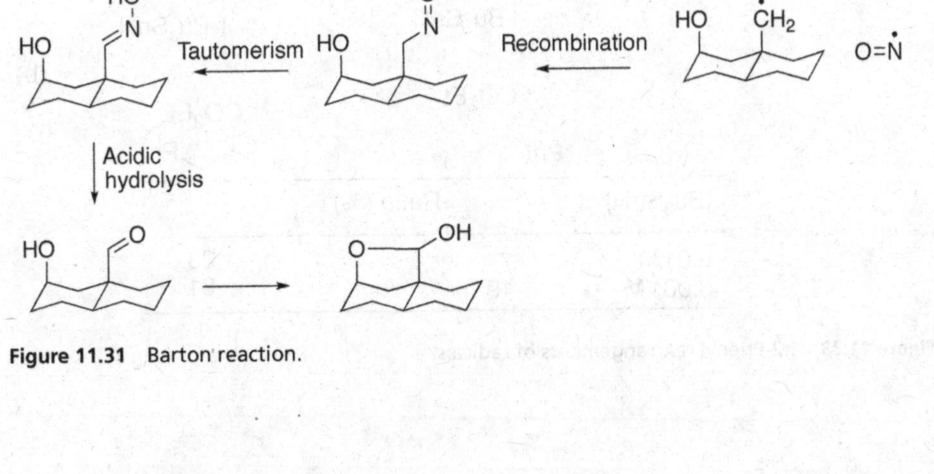

Figure 11.31 Barton reaction.

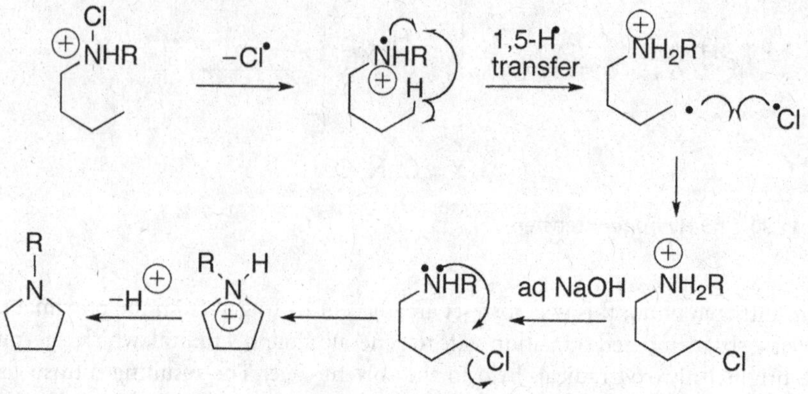

Figure 11.32 Hofmann–Löffler–Freytag reaction.

resonance delocalisation. The observed product in reactions proceeding by rearrangement can, however, depend on the reaction conditions; for example, as shown in Fig. 11.33b, generation of the radical in the standard way leads to a primary radical which may rearrange to a more stabilised resonance-stabilised system before reduction by tributyltin hydride. However, if the concentration of hydride is high enough, the primary radical will be intercepted before rearrangement can take place, giving the product of direct reduction.

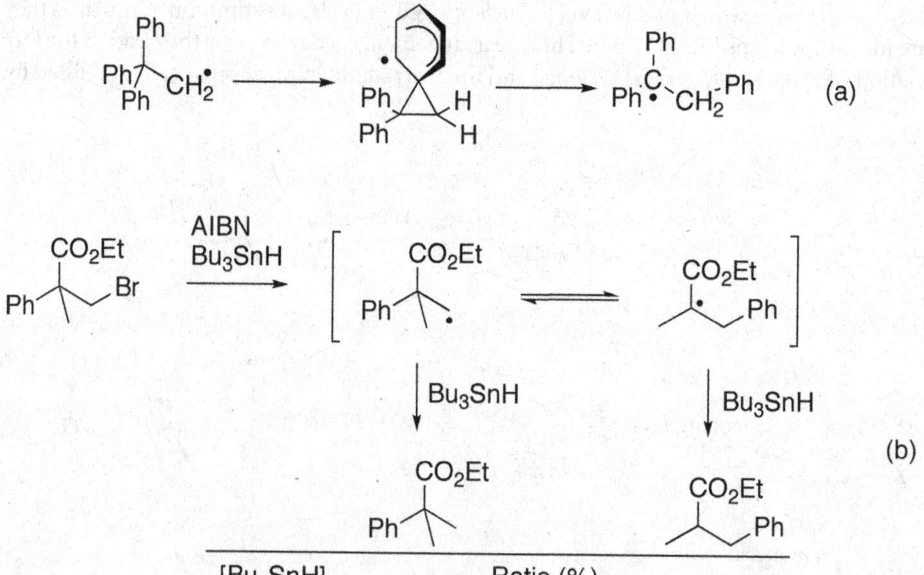

[Bu₃SnH]	Ratio (%)	
0.01 M	77	23
0.001 M	19	81

Figure 11.33 1,2-Phenyl rearrangements of radicals.

Table 11.2 The Baldwin rules for kinetically controlled ring closures

Favoured	Disfavoured
3–7 *exo-tet*	5–6 *endo-tet*
3–7 *exo-trig*	3–5 *endo-trig*
3–7 *endo-dig*	3–4 *exo-dig*
6–7 *endo-trig*	
5–7 *exo-dig*	

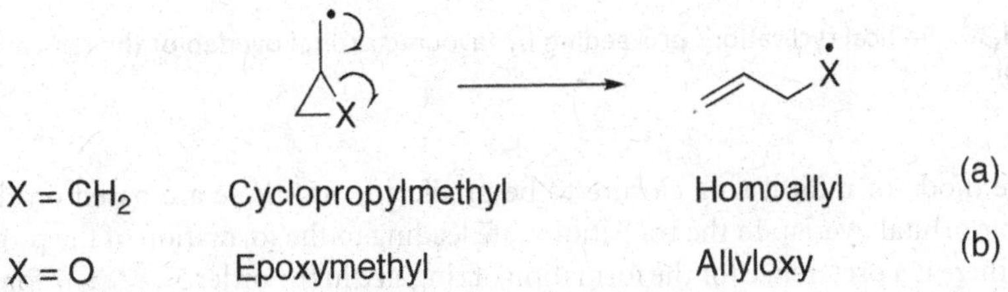

X = CH₂   Cyclopropylmethyl   Homoallyl   (a)

X = O   Epoxymethyl   Allyloxy   (b)

Figure 11.34 Homoallyl/cyclopropylmethyl type rearrangements of radicals.

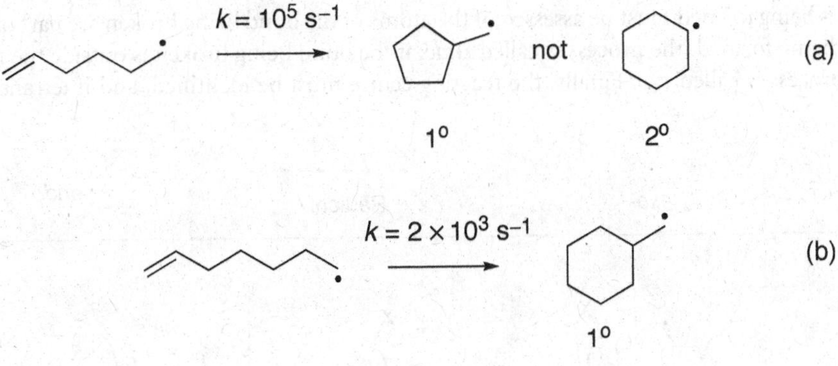

$k = 10^5 \text{ s}^{-1}$

not   1⁰   2⁰   (a)

$k = 2 \times 10^3 \text{ s}^{-1}$

1⁰   (b)

Figure 11.35 Radical cyclisations.

11.2.6.3 Intramolecular fragmentation

Radicals will fragment in an intramolecular sense, particularly if this leads to more stable species; a particularly important example is the homoallyl/cyclopropylmethyl rearrangement, which is known to proceed with a very high rate constant of 10^6 s^{-1} (Fig. 11.34a). A similar rearrangement is possible for the corresponding epoxymethyl/allyloxy radical system (Fig. 11.34b).

11.2.6.4 Intramolecular cyclisation reactions

Radicals will effectively cyclise, provided that there is a suitably disposed radical acceptor, which would normally be a double or triple bond; being intramolecular, such processes are normally kinetically highly favoured (Fig. 11.35). However, the process is favoured to different extents, depending on the length of the tether linking the radical and the π-acceptor function and therefore the size of the ring which is formed. Baldwin's rules (Table 11.2)

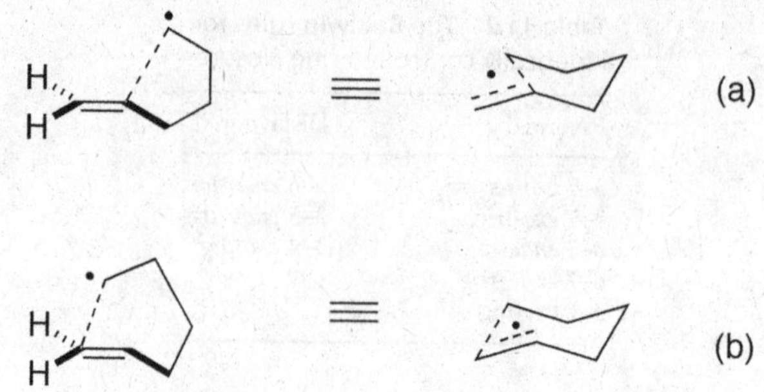

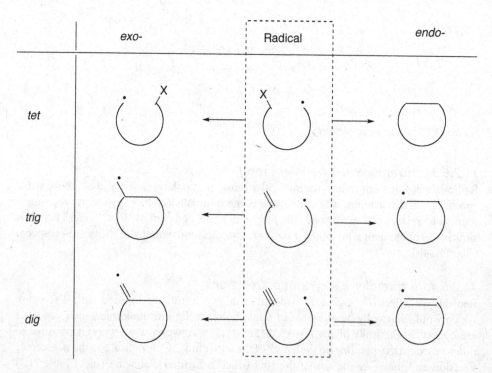

Figure 11.36 Radical cyclisations proceeding by favoured orbital overlap of the radical with the π^*-orbital.

allow the mode of radical ring closure to be predicted, and these are based on the most favourable orbital overlap in the transition state leading to the formation of the product. In general, there is a preference for the formation of ring size in the order $5 > 6 > 7$ for allowed processes, since in these cases, maximum orbital overlap is maintained during the reaction (Fig. 11.36). In order to use the rules, three parameters must be determined: Firstly, the ring size (N) being formed must be identified (Fig. 11.37). Secondly, the manner by which the ring is being formed must be assessed; if the atoms of the bond being broken are part of the ring being formed, the process is called *endo*; if the bond being broken is outside the ring, the process is called *exo*. Finally, the reacting centre must be identified, and if tetrahedral

Figure 11.37 Baldwin's rules for radical cyclisations.

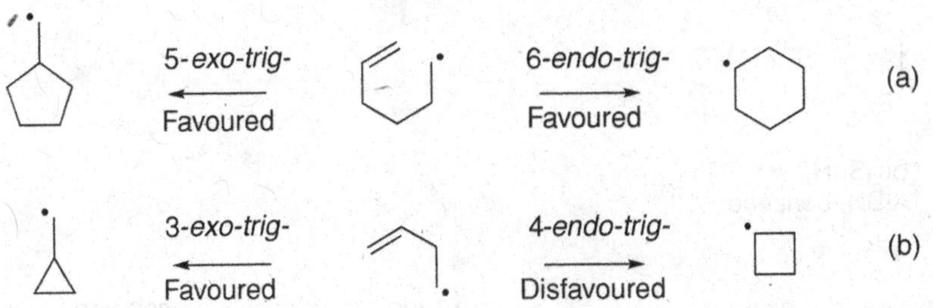

Figure 11.38 Some examples of the use of the Baldwin rules for radical cyclisations.

(sp^3), the process is designated *tet*; if trigonal (sp^2), the process is designated *trig*; and if digonal (sp), the process is designated *dig*. The reaction is then coded as *N*- (*exo*- or *endo*-) (*dig*- or *trig*- or *tet*-). Some examples of the use of these rules are given in Fig. 11.38. Note that although both 5-*exo-trig* and 6-*endo-trig* processes are favoured using these rules, the former proceeds at a faster rate, since it benefits from the better orbital overlap of the two.

It is important to recognise that Baldwin's rules effectively permit the prediction of the product under kinetically controlled conditions; should the reaction proceed under thermo-dynamically controlled conditions, the outcome will of course lead to the formation of the most stable product. This is illustrated in Fig. 11.39, in which the expected 5-*exo* cyclisation leads to a primary radical, but because the initial radical is (electronically) stabilised, this cyclisation is freely reversible. Under these circumstances, reversion to the starting radical then permits the alternative 6-*endo* closure which generates a more stable secondary radical, which is then quenched by hydrogen abstraction to give the cyclohexane as the major product.

More complex processes are also possible; in Fig. 11.40, a 5-*exo* cyclisation leads to a homoallyl radical capable of further rearrangement to give the cyclopropylmethyl radical. However, since this is a primary radical, further fast fragmentation generates a more stable secondary radical. In this case, however, because of the complexity of the reaction cascade, the concentration of the radical terminating tributyltin hydride determines whether product A, B or C is more likely to be obtained.

11.3 Synthetic utility

Radicals have found wide utility in modern organic synthesis, because they are highly reactive and can be reliably generated under mild thermolytic or photolytic conditions. They also

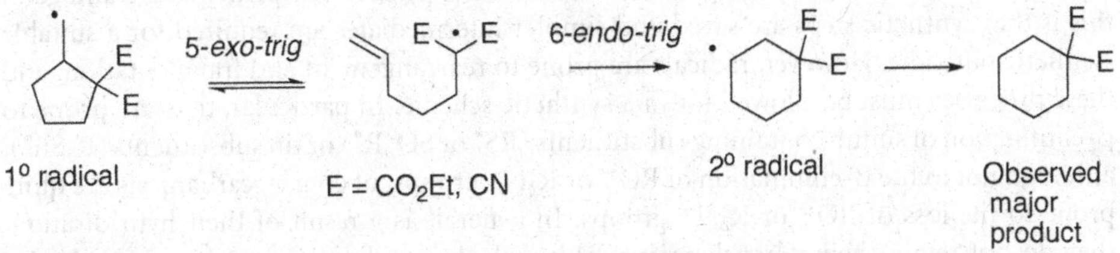

Figure 11.39 Radical cyclisation under equilibrating conditions gives thermodynamically more stable product.

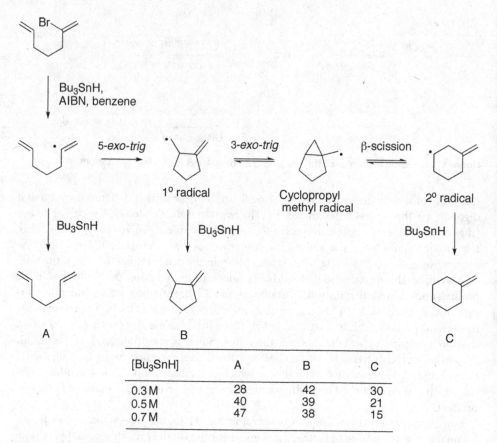

Figure 11.40 Radical cyclisation with rearrangement.

participate in well-defined reaction pathways which are different from and complementary to those of ionic species. Unlike many ionic processes, radical reactions tend to be kinetically controlled, and their substituents lead to changes in their philicity which makes for highly chemoselective and regioselective processes, as we have seen. Furthermore, they do not have large solvation spheres stabilising intermediate charge. They are therefore ideally suited to the formation of sterically hindered centres. As noted earlier, the high homolytic bond energy of oxygen–hydrogen and nitrogen–hydrogen bonds means that they tend to be inert under radical conditions (but of course not under basic conditions) and therefore do not require protection during a radical reaction sequence. The principal advantage of this is that synthetic steps are saved and simpler intermediates are required for a suitable synthetic outcome. However, radicals are prone to rearrangement and fragmentation, and these processes must be allowed for in a synthetic scheme. In particular, they are prone to β-elimination of sulfur-containing substituents ($RS^\bullet$ or $SO_nR^\bullet$) or tin substituents ($R_3Sn^\bullet$), although not to the β-elimination of RO^- or R_2N^- groups; of course, carbanions are quite prone to the loss of RO^- or R_2N^- groups. In general, as a result of their hybridisation, they do not retain their stereochemistry, although they will participate in stereoselective processes if there are other stereodirecting centres in the substrate.

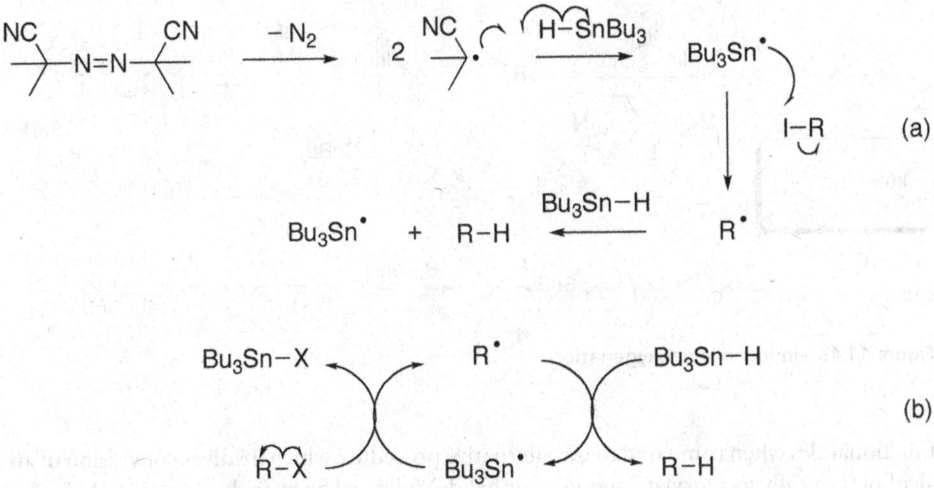

Figure 11.41 Radical reduction of an alkyl halide.

The reduction of suitable substrates, including alkyl halides, under radical conditions, offers a very useful alternative to conventional (metal hydride or hydrogenation) strategies. This reaction is conducted using the usual radical initiator, but in the presence of a suitable reducing agent, for example, Bu_3SnH or the more reactive and more environmentally acceptable $(Me_3Si)_3SiH$, reduction of the radical, which is generated rapidly, follows (Fig. 11.41a). This reaction can be represented in a cyclic form to emphasise its chain nature (Fig. 11.41b).

Although this reaction is very effective for the reduction of alkyl halides, it has been extended to a more generalised defunctionalisation process. Thus, Barton–McCombie xanthates find application for decarboxylations and dehydroxylations. In this reaction, an alcohol function is removed, firstly by conversion to the xanthate and then by treatment under radical conditions with tributyltin hydride (Fig. 11.42). Overall reduction then results; the radical deoxygenation of xanthate derivatives of alcohols offers the advantage of neutral conditions, less dependence on steric factors and compatibility with a wide range of

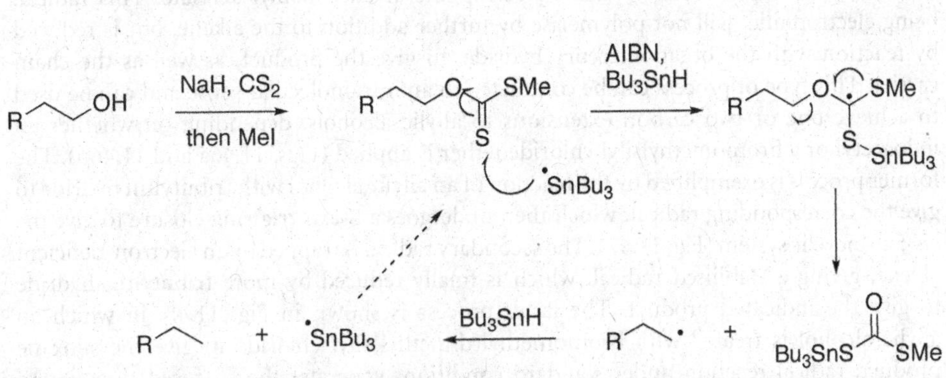

Figure 11.42 Barton–McCombie deoxygenation.

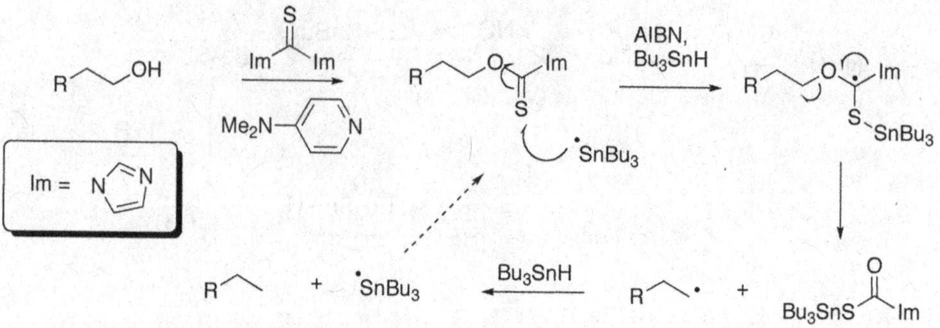

Figure 11.43 Imidazolyl deoxygenation.

functionalities when compared to the alternative procedure which involves conversion of an alcohol (typically to a tosylate, mesylate or halide) followed by hydride reduction. A similar outcome can be achieved with imidazolylthiocarbonic esters, which is especially valuable for base-sensitive alcohols. In this case, the thio derivative is prepared under especially mild conditions as shown in Fig. 11.43, and fragmentation followed by reduction then generates the alkane product.

An alternative uses the Barton pyridinethione(oxycarbonyl) (PTOC) ester system (Fig. 11.44). The PTOC system, readily generated as shown by reacting an acid chloride with the sodium salt of 2-mercaptopyridine N-oxide, is characterised by the presence of a weak N—O bond, a weak C=S bond and the potential to readily lose CO_2. Homolytic collapse, initiated by a tributylstannyl radical generated in the usual way, generates a carboxy radical which then fragments to generate an alkyl radical, which is in turn reduced by the tin hydride reagent. The aromatisation of the pyridinethione system provides a potent thermodynamic driving force for the reaction. The synthetic value of this reaction comes from the fact that the tributyltin hydride may be replaced by a number of trapping agents, and this allows a diversity of outcomes from the reaction; products include those resulting from halogenation, addition and oxygenation.

An illustration of the synthetic utility of addition reactions is shown in Fig. 11.45; in this case, generation of the nucleophilic radical using homolytic cleavage of an organomercurial is followed by rapid addition to the electrophilic alkene, methyl acrylate. This radical, being electrophilic, will not polymerise by further addition to the alkene, but is reduced by reaction with the organomercury hydride, to give the product, as well as the chain carrier. This type of process can be conducted in an intramolecular sense and can be used to achieve one or two carbon extensions to allylic alcohols, depending on whether an iodoacetyl or a bromomethylsilyl chloride tether is applied (Figs. 11.46a and 11.46b). The former process is exemplified by the reaction of an allylic alcohol with tributyltin hydride to give the corresponding radical, which then undergoes a 5-*exo-trig* ring closure to give the *cis*-5,5-bicyclic system (Fig. 11.47). The secondary radical is trapped by an electron-deficient alkene, giving a stabilised radical, which is finally reduced by more tributyltin hydride to give the indicated product. The latter process is shown in Fig. 11.48, in which an allylic alcohol is treated with bromomethyl(dimethyl)silyl chloride to give the siloxane product; radical reaction under standard conditions generates the *cis*-fused-ring system

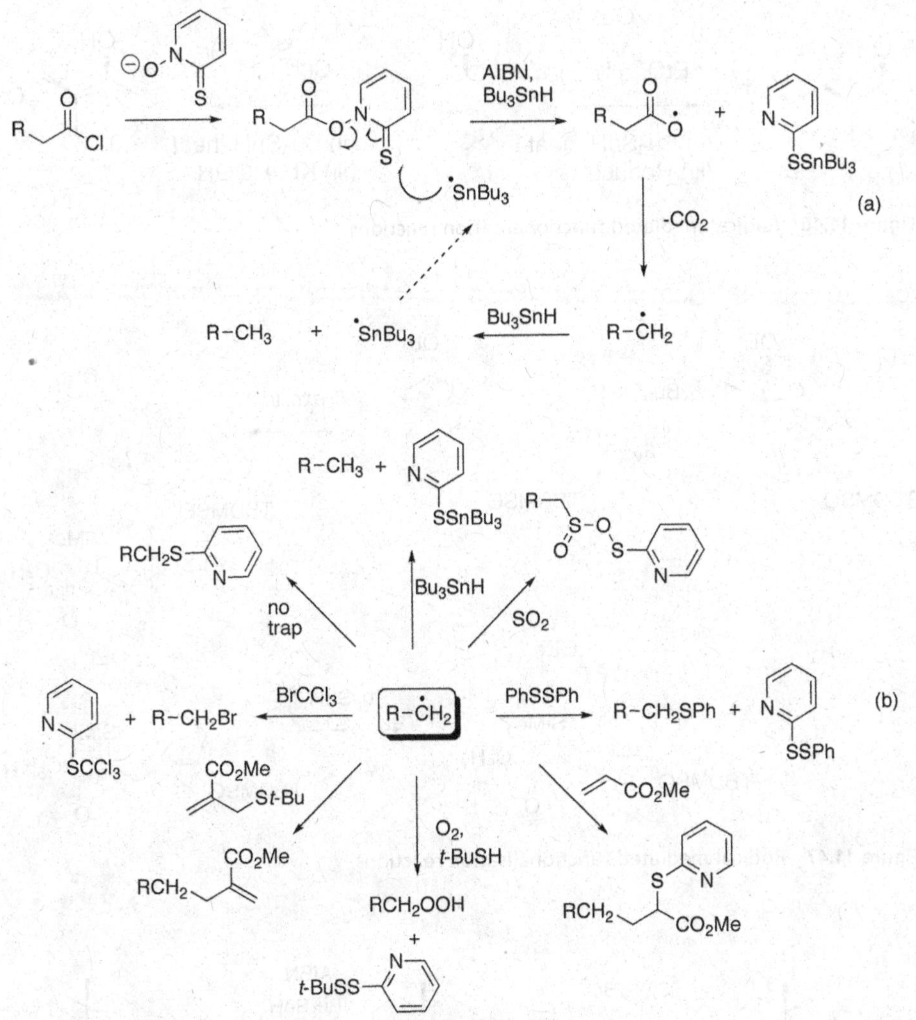

Figure 11.44 Barton PTOC ester decarboxylation: (a) general reaction; (b) some of the possible products which can be formed.

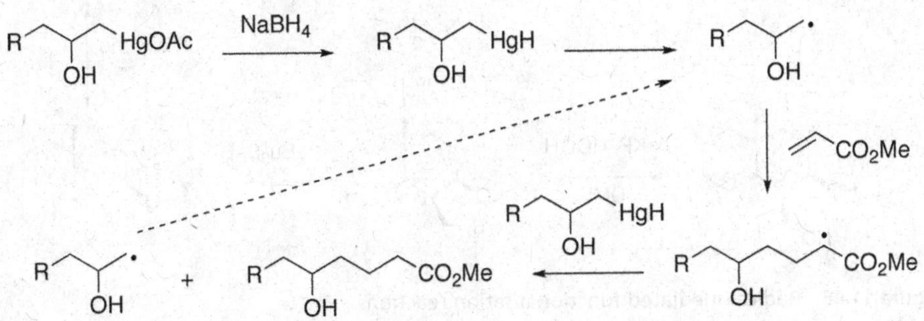

Figure 11.45 Radical conjugate addition to alkenes.

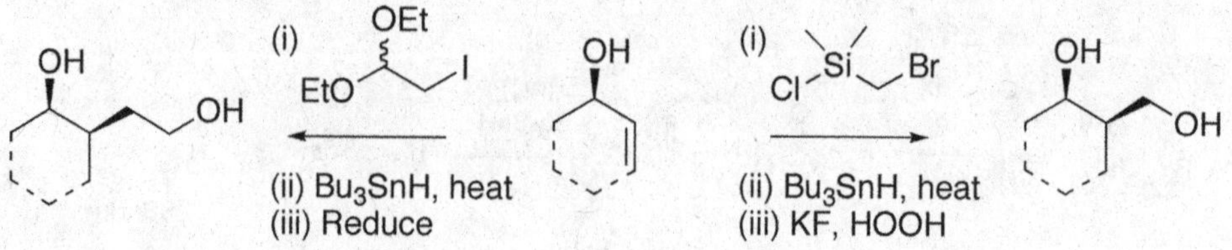

Figure 11.46 Radical-mediated functionalisation reactions.

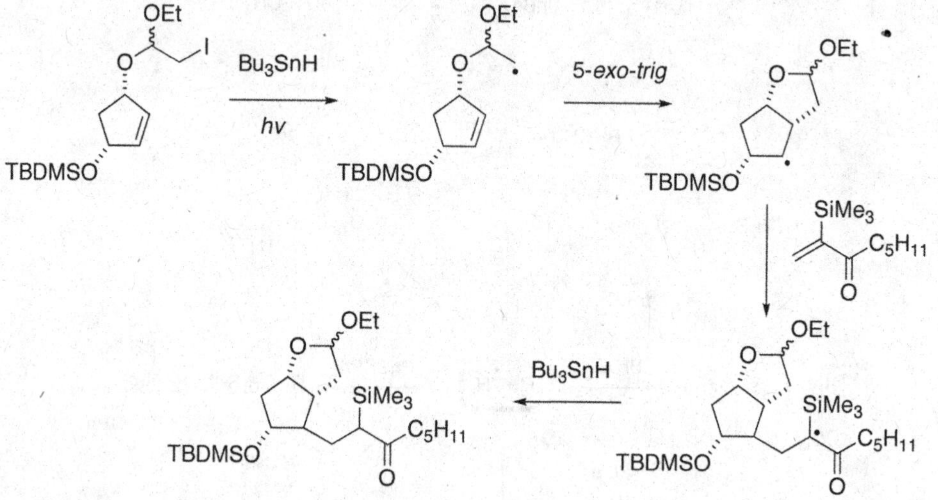

Figure 11.47 Radical-mediated functionalisation reactions.

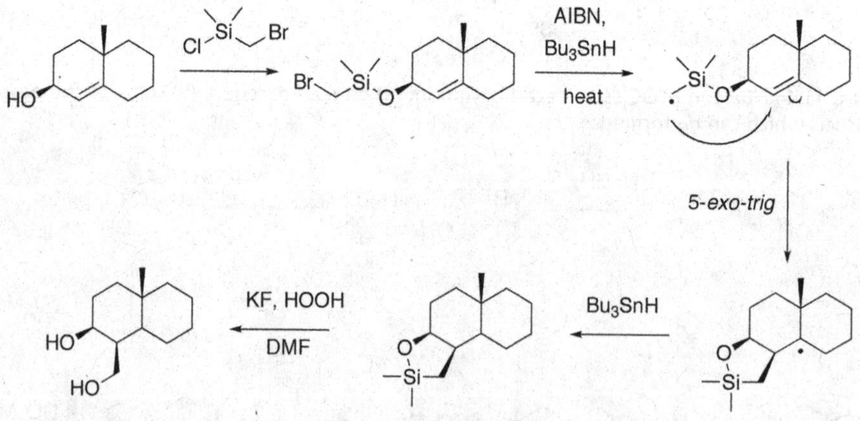

Figure 11.48 Radical-mediated functionalisation reactions.

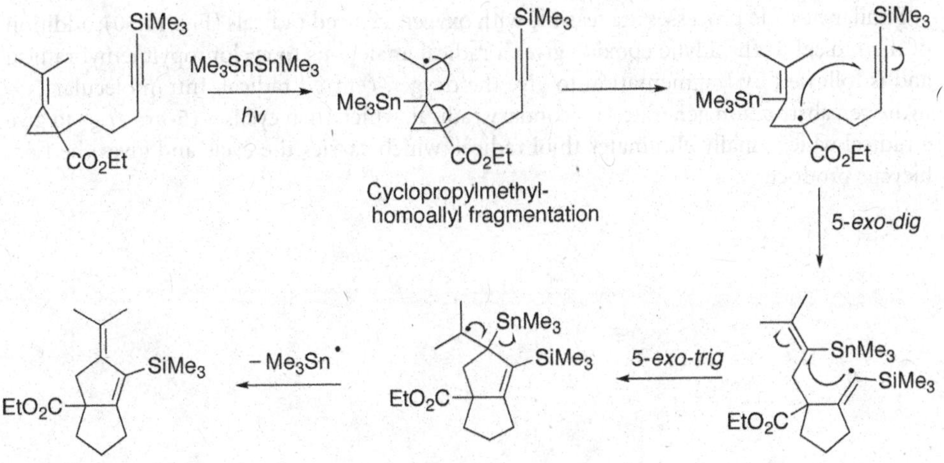

Figure 11.49 Radical-mediated functionalisation reactions.

and a secondary radical, which is reduced by tributyltin hydride to give the decalin product. Decomposition of the siloxane using Fleming–Tamao oxidation leads to the 1,3-diol product.

Even more complex ring-closing processes based on radical additions are possible (Fig. 11.49); thus, the reaction of an alkene under radical conditions (this time with the initiating radical generated by photolysis of hexamethyldistannane) leads to initial addition so as to give a tertiary radical. This radical, however, is also capable of undergoing a cyclopropylmethyl fragmentation and does so to give the homoallylic radical. This radical, in turn, is in turn capable of further 5-*exo-dig* cyclisation onto a suitably disposed alkene, generating a vinylic radical, which can cyclise further (5-*exo-trig*) to give a radical with a β-stannyl residue. This will very effectively eliminate to give the bicyclic product indicated along with trimethyltin radical, which carries the chain.

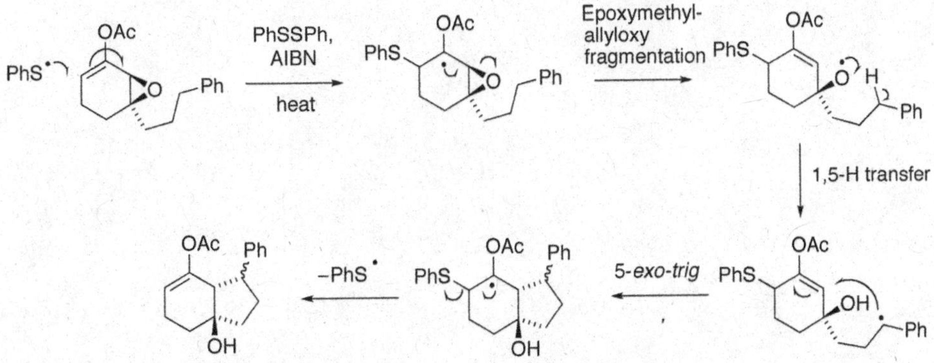

Figure 11.50 Radical-mediated functionalisation reactions.

Similar cascade processes are feasible with oxygen-centred radicals (Fig. 11.50); addition of the radical to the allylic epoxide gives a radical analogous to cyclopropylmethyl radical and is followed by fragmentation to give the oxygen-centred radical. Intramolecular 1,5-hydrogen abstraction generates a secondary radical, which then cyclises (5-*exo-trig*) to give a radical which finally eliminates thiol radical, which carries the cycle and gives the final bicyclic product.

Chapter 12
Ligand Coupling Reactions

In addition to the ionic and radical reaction mechanisms described in earlier chapters, another important class of mechanistic reaction process is that in which new bond formation occurs between ligands on a metal or metalloid, with simultaneous reduction of the metal or metalloid (Fig. 12.1). These processes have assumed enormous importance over the last three decades, because they are simple to execute, are very general, use readily available starting materials, are tolerant of a wide variety of functionality and lead to rapid construction of molecular complexity. Key to the process is the capacity for a wide variety of metals and metalloids to promote such ligand coupling; this can arise because many organic compounds (Fig. 12.1; X–Y) containing a suitable leaving group are able to add to metals in a process called oxidative addition and give rise to a product in which the X and Y units are *cis* related on the metal, and in which oxidation of the metal by two has occurred. Crucial to the success of an oxidative addition is the capacity of the metal to be oxidised by the organic substrate and that it is coordinatively unsaturated and therefore able to accept the additional ligands. The reverse process, reductive elimination, also occurs (Fig. 12.1), generating a new organic product, along with the metal in the original oxidation state, that is, reduced by two; this process, to be successful, requires that X and Y are also *cis*-related on the metal. The power of this reaction sequence comes from two important factors. Firstly, it is possible to exchange one ligand (X or Y) for another (say Z) in processes called *ligand exchange* (also called *metathesis*) and *transmetallation*, and so the product which results from ligand coupling (X–Z) is a new entity; thus, coupling of two units derived from separate starting materials is achieved. Secondly, the oxidation–reduction process of the metal is frequently reversible, and the metal is capable of cycling between low and high oxidation states during the course of the reaction, provided that a suitable oxidant is present in the reaction. Such ligand coupling processes can therefore be run in a mode which is catalytic in the metal, and this has important economic and environmental advantages. The successful development of this chemistry in recent years has in part required the development of suitable ligand systems which are capable of stabilising the reacting metal centre in both low and high oxidation states, but without removing their reactivity completely. Much is now known about such ligands, and ligand coupling has been demonstrated to be a widespread mechanism amongst transition and main group metals (Pd, Pt, Rh, Ru, Cu, Ni, Pb and Bi) and metalloids (I), and as a result, cross coupling provides a powerful synthetic strategy of substantial scope.

12.1 Palladium-mediated couplings

Ligand coupling chemistry has been most extensively developed for palladium, particularly since it can form both 16- and 18-electron ligand complexes, with a wide variety of ligands, and is commercially available in a variety of forms (including $Pd(PPh_3)_4$, $PdCl_2$, $Pd(OAc)_2$ and $PdCl_2(CH_3CN)_2$). The 16-electron configuration has a high-energy vacant orbital,

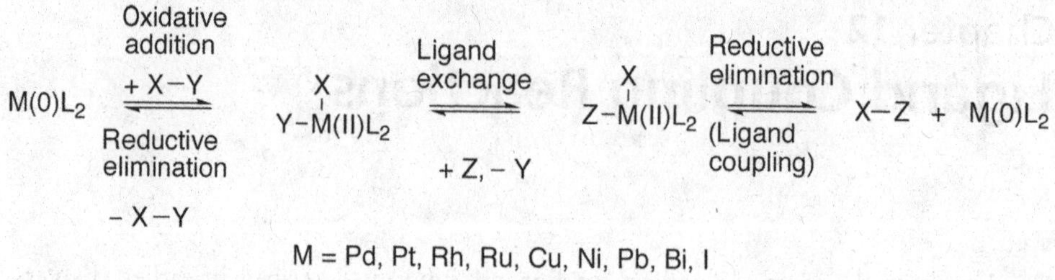

Oxidative
addition
                                        Ligand   Reductive
                                        exchange   elimination
$+ X-Y$ X   X
$M(0)L_2 \rightleftharpoons$ $\underset{|}{Y}-M(II)L_2$ $\rightleftharpoons$ $Z-\underset{|}{M}(II)L_2$ $\rightleftharpoons$ $X-Z + M(0)L_2$
Reductive   (Ligand
elimination $+ Z, - Y$ coupling)

$- X - Y$

M = Pd, Pt, Rh, Ru, Cu, Ni, Pb, Bi, I

Figure 12.1 Ligand coupling processes in organic chemistry.

capable of accepting electron density from a potential ligand, and loss of two ligands from $Pd(PPh_3)_4$ readily generates the coordinatively unsaturated $Pd(PPh_3)_2$ required for reaction. Also of significance is that the oxidation potential of $Pd(0)$ to $Pd(II)$ and of $Pd(II)$ to $Pd(IV)$ is accessible to organic oxidants and that the (reverse) reduction potential is correct for carbon–carbon bond forming processes. The reduction potentials for some common metals are shown in Table 12.1, and it can be seen that the value for palladium lies in the middle of the observed range; this corresponds with its facile oxidation and reduction chemistry. The immediate implication for this relates to the easy reduction of $Pd(II)$ to $Pd(0)$ by organic molecules, and this provides a first step in many reactions; likely mechanisms for this reduction by commonly used additives are given in Fig. 12.2.

Increasing numbers of novel ligands and catalysts are becoming commercially available, which have been optimised for wider substrate scope, ease of handling and product yields; some examples taken from a myriad of possibilities are given in Fig. 12.3. Crucial in this development has been the identification of suitable ligands which stabilise the metal, but without compromising reactivity; although this has been very successful, understanding of the correlation of ligand structure with selectivity and reactivity is far from complete. Notable has been the emergence of substituted phosphine and N-heterocyclic carbene ligands, which have proven to be excellent σ-donors, but because of the steric hindrance afforded by the adjacent substituents, these provide air- and water-stable complexes that can be easily handled in the laboratory.

12.1.1   Palladium-mediated coupling processes

Palladium has been shown to effectively oxidatively insert into a wide variety of bonds that are composed of electronegative–electropositive systems (R–X), as well as homonuclear systems

Table 12.1 Reduction potentials for common metals

Reaction	E° (V)
$Cu^{2+} + e^- \rightarrow Cu^+$	0.17
$Cu^+ + e^- \rightarrow Cu$	0.52
$Pb^{4+} + 2\,e^- \rightarrow Pb^{2+}$	1.66
$Pd^{2+} + 2\,e^- \rightarrow Pd$	0.92
$Pt^{2+} + 2\,e^- \rightarrow Pt$	1.20
$Tl^{3+} + 2\,e^- \rightarrow Tl^+$	1.26

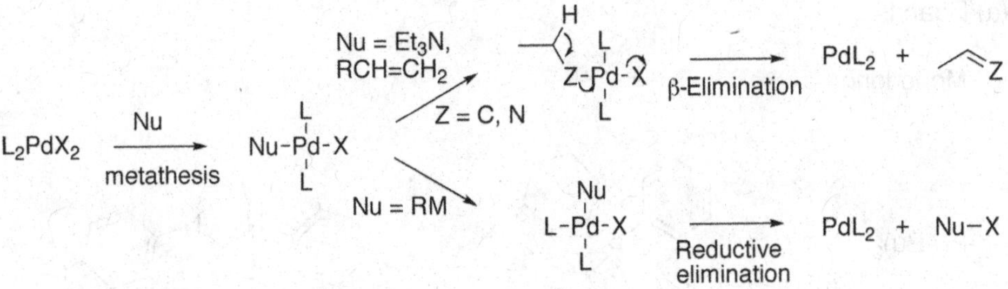

$$
\begin{array}{c}
\text{Nu = Et}_3\text{N,} \\
\text{RCH=CH}_2
\end{array}
\qquad
\begin{array}{c}
\text{H} \\
\text{—} \langle \rangle \quad \text{L} \\
\text{Z–Pd–X} \\
\text{L}
\end{array}
\xrightarrow{\beta\text{-Elimination}} \quad \text{PdL}_2 \; + \; \text{Z}
$$

$$
\text{L}_2\text{PdX}_2 \xrightarrow[\text{metathesis}]{\text{Nu}}
\begin{array}{c}
\text{L} \\
\text{Nu–Pd–X} \\
\text{L}
\end{array}
\begin{array}{c}
\text{Z = C, N}
\end{array}
$$

$$
\text{Nu = RM}
\begin{array}{c}
\text{Nu} \\
\text{L–Pd–X} \\
\text{L}
\end{array}
\xrightarrow[\substack{\text{Reductive}\\\text{elimination}}]{} \quad \text{PdL}_2 \; + \; \text{Nu–X}
$$

Figure 12.2 Mechanisms by which Pd(0) can be generated in situ from Pd(II).

containing weak X–X bond; common examples of systems susceptible to this chemistry are shown in Fig. 12.4. In the case of aryl halides, the relative reactivity for the halogens is in the order Cl < Br < I, with chloride generally too inert to be synthetically useful, although more recently developed catalyst systems are beginning to allow such substrates to be applicable. This is important because organic chlorides are generally the most readily available and therefore the most inexpensive. Some particularly effective catalysts are indicated in Fig. 12.3, and although the palladium catalyst is typically used at the level of 5–10 mol %, it has been found that significantly lower levels (less than 0.1%, and even down to the parts per million level) can give effective reaction, although the reactions can be slow under these conditions. The resulting adducts are capable of reacting with organometallic nucleophiles (Nu–M) in a cycle indicated in Fig. 12.5; initial oxidative addition of Pd(0) to the aryl or vinyl halide/triflate substrate leads to an organopalladium(II) adduct, and subsequent transmetallation with the organometallic component creates an intermediate from which reductive elimination from the metal and ligand coupling on the organic partners gives the final product. Various substrates are capable of undergoing rapid transmetallations (Fig. 12.6), and a possible mechanism for the transmetallation step, resulting in the exchange of ligands between the palladium and organometallic components, is shown in Fig. 12.7. In the course of the ligand coupling process, reduction of the metal to Pd(0) regenerates the active catalyst and allows for the operation of a catalytic cycle, beginning again by oxidative addition. This coupling process frequently operates efficiently under mild conditions for a diverse range of substrates, and for this reason as well as the fact that the reaction is

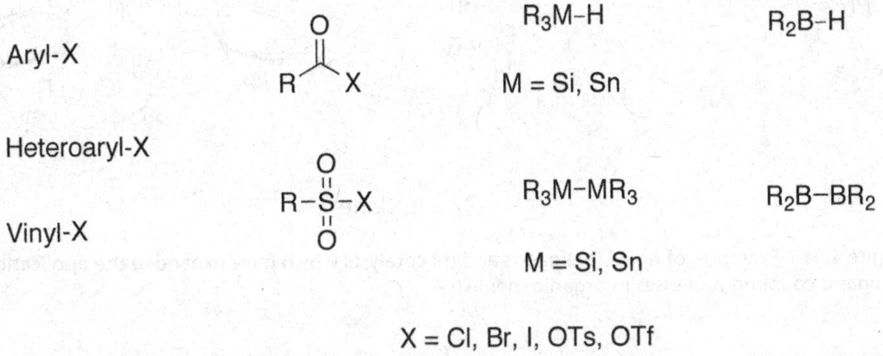

Aryl-X

$$
\begin{array}{c}
\text{O} \\
\parallel \\
\text{R} \quad \text{X}
\end{array}
$$

$$
\text{R}_3\text{M–H} \qquad\qquad \text{R}_2\text{B–H}
$$

$$
\text{M = Si, Sn}
$$

Heteroaryl-X

$$
\begin{array}{c}
\text{O} \\
\parallel \\
\text{R–S–X} \\
\parallel \\
\text{O}
\end{array}
$$

Vinyl-X

$$
\text{R}_3\text{M–MR}_3 \qquad\qquad \text{R}_2\text{B–BR}_2
$$

$$
\text{M = Si, Sn}
$$

$$
\text{X = Cl, Br, I, OTs, OTf}
$$

Figure 12.3 Substrate classes capable of undergoing oxidative addition with palladium catalysts.

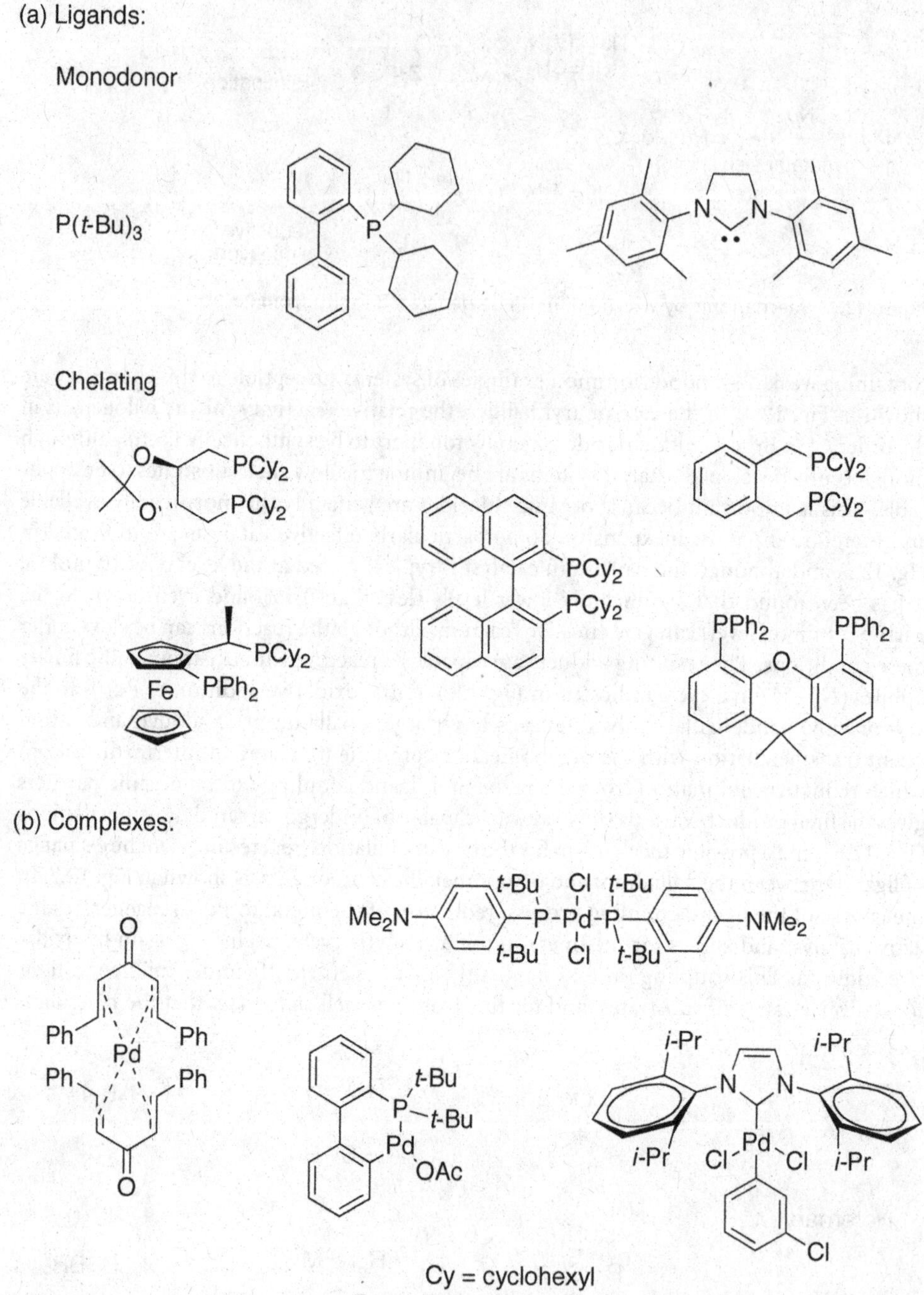

(a) Ligands:

Monodonor

P(*t*-Bu)₃

Chelating

(b) Complexes:

Cy = cyclohexyl

Figure 12.4 Examples of novel (a) ligands and (b) catalysts which have expanded the application of ligand coupling processes in organic chemistry.

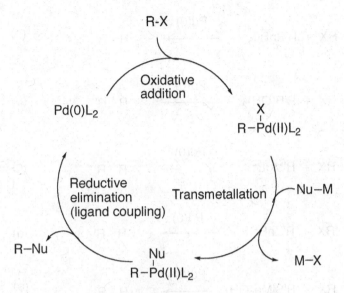

Figure 12.5 Catalytic cycle mediated by palladium.

$$R\text{-}M \qquad M = B(OH)_2, \; SiR_2OH, \; SnR_3, \; HgR, \; PbR_3, \; N_2^+$$

Figure 12.6 Substrates capable of undergoing transmetallations.

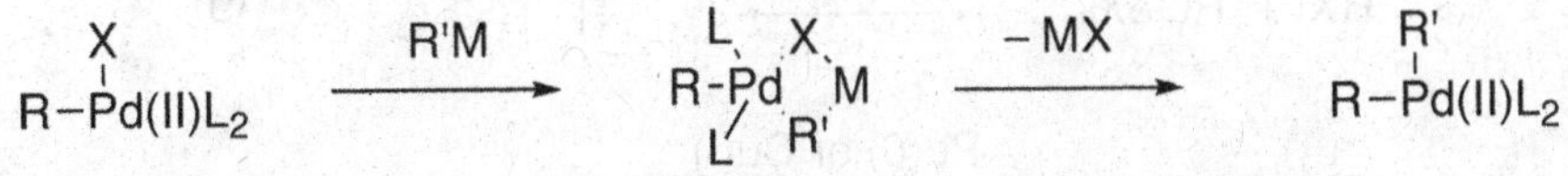

$$\overset{X}{\underset{R-Pd(II)L_2}{|}} \xrightarrow{R'M} \underset{L \diagdown R'}{\overset{L \diagup X}{R-Pd \diagup M}} \xrightarrow{-MX} \overset{R'}{\underset{R-Pd(II)L_2}{|}}$$

Figure 12.7 A mechanism for transmetallation leading to exchange of ligands.

catalytic in palladium, it has been widely embraced by the synthetic community and is used extensively in industry.

Depending on the nature of the organometallic component, the reaction is called the Stille (NuM = organotin derivative), Suzuki–Miyaura (NuM = organoboron), Corriu–Kumada–Tamao (NuM = organomagnesium), Negishi (NuM = organozinc), Sonagashira (NuM = organocopper) or Denmark coupling reaction (Fig. 12.8). These organometallic reagents have been chosen largely because of their ready availability and functional group tolerance. There are disadvantages with the use of certain organometallic derivatives in these reactions; organostannanes are not ideal because of their toxicity and the difficulty of removal of tin by-product residues, and organomagnesiums and organozincs because of their instability and (sometimes) difficulty in formation. Organoborons have therefore generally been much preferred, especially since they are readily prepared but also boron waste products are easily removed by washing with water. However, the reaction may use other organometals as sources of nucleophiles; recently, the use of organosilanols has been reported by Denmark. Other more esoteric nucleophiles can be used, and these include organolead and organogermanium compounds, as well as aryldiazonium salts.

$$RX + R'SnBu_3 \xrightarrow{Pd(0)} R - R' \qquad (a)$$

$$RX + R'B(OH)_2 \xrightarrow{Pd(0)} R - R' \qquad (b)$$

$$RX + R'MgBr \xrightarrow{Pd(0)} R - R' \qquad (c)$$

$$RX + R'ZnBr \xrightarrow{Pd(0)} R - R' \qquad (d)$$

$$RX + R'SiMe_2OH \xrightarrow{Pd(0)} R - R' \qquad (e)$$

$$RX + RPb(OAc)_3 \xrightarrow{Pd(0)} R - R' \qquad (f)$$

$$RX + RGeX_3 \xrightarrow{Pd(0)} R - R' \qquad (g)$$

$$RX + \;\equiv\!\!-R' \xrightarrow{Pd(0)\ or\ Cu(I)} R\!-\!\!\equiv\!\!-R' \qquad (h)$$

$$X = Cl,\ Br,\ I$$
$$R = aryl,\ vinyl$$
$$R' = aryl,\ vinyl$$

Figure 12.8 (a) Stille, (b) Suzuki–Miyaura, (c) Corriu–Kumada–Tamao, (d) Negishi, (e) Denmark, (f) organolead, (g) organogermanium and (h) Sonagashira ligand coupling processes in organic chemistry.

The Suzuki–Miyaura process is the palladium-catalysed cross coupling of an aryl halide (chloride, bromide or iodide) or triflate with an organic-derived boronic acid or boronate for the transmetallation. This reaction is of great utility, since organoborons are available by several methods, as shown in Fig. 12.9a–12.9e, and this provides wide substrate scope. The boronic acid preferentially contains a C–sp^2–B bond (aryl or alkenyl boronate) but is also applicable to certain types of alkylboronic acids; thus, 9-borabicyclononane (9-BBN) derivatives will effectively couple with alkenyl and aryl halides and tosylates to give the corresponding alkyl coupled products (Fig. 12.10). Recently, the scope of this reaction has been

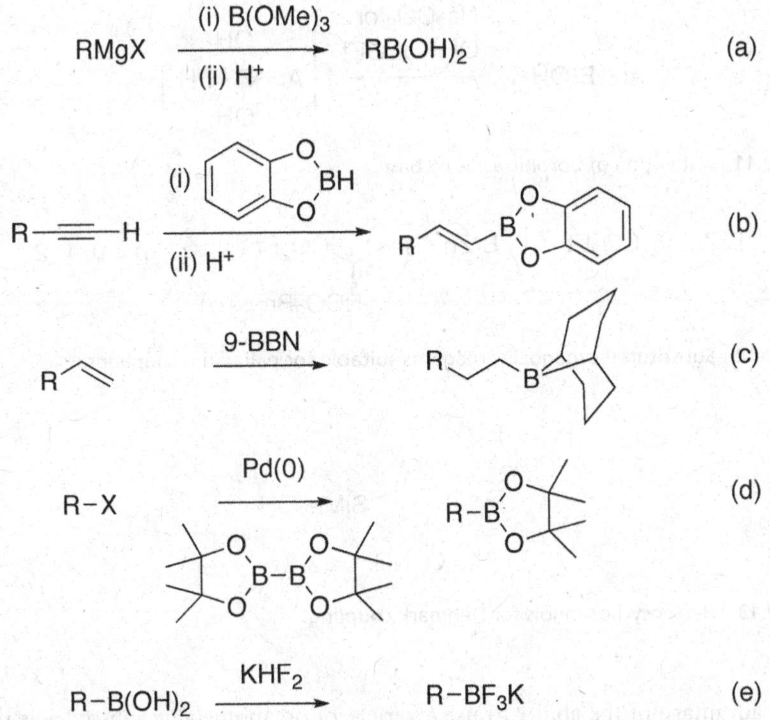

$$RMgX \xrightarrow[\text{(ii) } H^+]{\text{(i) } B(OMe)_3} RB(OH)_2 \qquad \text{(a)}$$

$$R-\!\!\equiv\!\!-H \xrightarrow[\text{(ii) } H^+]{\text{(i)}} \qquad \text{(b)}$$

$$R\diagup\!\!\diagdown \xrightarrow{\text{9-BBN}} \qquad \text{(c)}$$

$$R-X \xrightarrow{Pd(0)} \qquad \text{(d)}$$

$$R-B(OH)_2 \xrightarrow{KHF_2} R-BF_3K \qquad \text{(e)}$$

Figure 12.9 Common methods for the preparation of organoboron derivatives suitable for palladium-mediated couplings.

extended to include potassium organotrifluoroborates, which are generally more readily available, indefinitely stable in the air and therefore easy to handle as compared to the corresponding boronates; they are prepared by treating the required organoboronic acid or ester with potassium hydrogen fluoride (Fig. 12.9e). Furthermore, organoborates can conveniently be prepared in a very efficient manner by the palladium-mediated coupling of organohalides with bis(pinacolato)boron (Fig. 12.9d). The Suzuki–Miyaura coupling reaction is catalysed by base, typically an alkali metal carbonate or potassium phosphate, which generates the ate complex, as shown in Fig. 12.11, and which is more reactive to exchange with the palladium complex because of its negative charge than the starting boronic acid. This activation is further assisted if water is used as the solvent, by solvating the negatively charged species, although phase-transfer catalysts are also often needed under these conditions to solvate the reacting species.

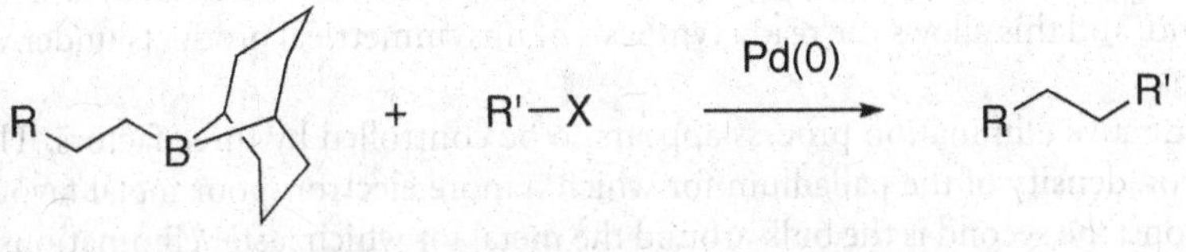

$$R\diagup\!\!\diagdown\!\!B\text{<} + R'-X \xrightarrow{Pd(0)} R\diagup\!\!\diagdown R'$$

Figure 12.10 Suzuki couplings of alkylboronic acid derivatives.

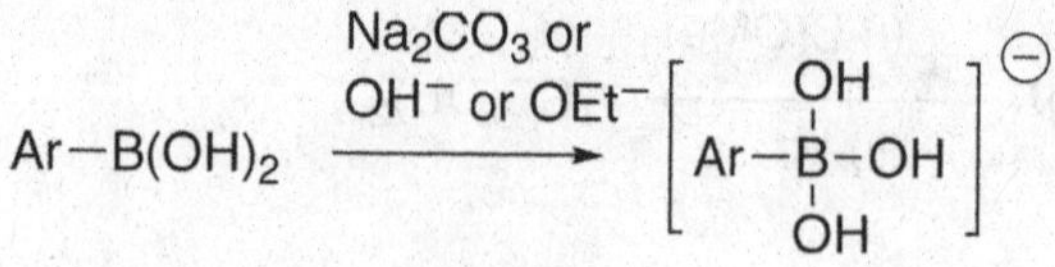

Figure 12.11 Activation of boronic acids by base.

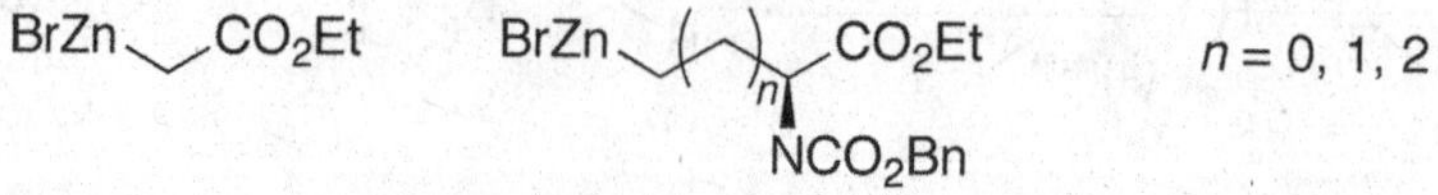

$n = 0, 1, 2$

Figure 12.12 Substituted organozinc reagents suitable for palladium couplings.

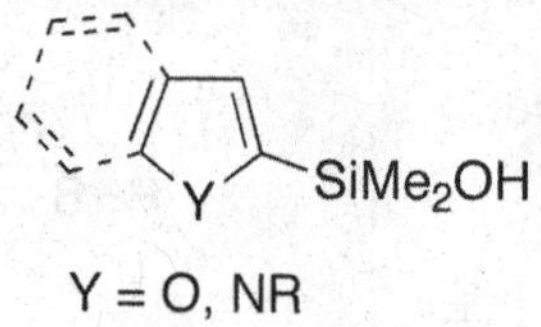

$Y = O, NR$

Figure 12.13 Heterocyclic silanols for Denmark coupling.

A key advantage of the ability to use a variety of organometallic substrates is that the scope of functionality which can be tolerated in the reaction is expanded; the applicability of Reformatsky and Jackson organozinc reagents is notable, since such reagents tolerate a range of electrophilic functions not acceptable in, for example, a Grignard reagent (Fig. 12.12).

In the case of Denmark coupling (Fig. 12.8e), alkenyl, aryl and heteroarylsilanols react with aryl halides in the presence of a base and a palladium catalyst (Fig. 12.13). The resulting polydimethylsiloxane by-products are readily removed from the reaction mixture by aqueous washing. This approach has the significant advantage that 2-substituted heteroaryl substrates react efficiently under coupling conditions.

These processes are being continually expanded in scope by careful development of reaction conditions. For example, the formation of sp^3 coupled species is possible; thus, methylation of aryl halides under palladium-mediated coupling conditions is now possible, as shown in Fig. 12.14. Furthermore, since the order of leaving group ability has been found to be I > Br > OTf > Cl in these couplings, it is possible to conduct sequential processes, in which the more reactive leaving group is substituted first by using a limiting amount of one boronic acid component and the second group is substituted only upon addition of an excess of a second component. Thus, as shown in Fig. 12.15, sequential coupling processes can be achieved, firstly, by reacting a more reactive C–I bond followed by a less reactive C–Br bond, and this allows the ready synthesis of unsymmetrical products under very mild conditions.

The reductive elimination process appears to be controlled by three factors: The first is the electron density of the palladium for which a more electron-poor metal favours faster eliminations; the second is the bulk around the metal for which faster eliminations are promoted by a bulky ligand; and the third is the nature of the remaining ligands on the metal, and

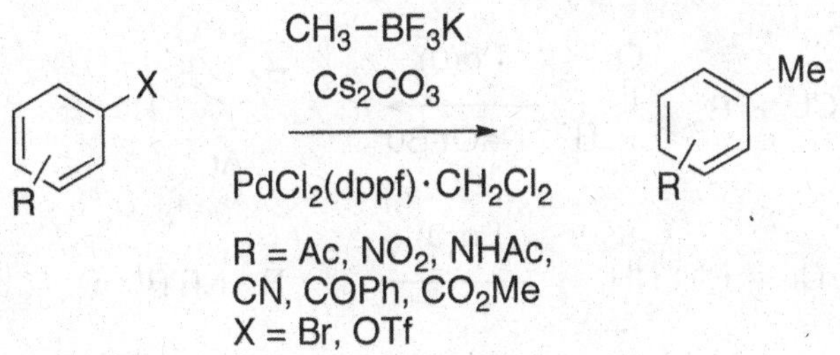

$$CH_3-BF_3K$$
$$Cs_2CO_3$$

$$PdCl_2(dppf) \cdot CH_2Cl_2$$

R = Ac, NO$_2$, NHAc,
CN, COPh, CO$_2$Me
X = Br, OTf

Figure 12.14 Methylation of aryl halides using organoborons by palladium coupling.

π-acceptor ligands favour reductive elimination. Overall, the outcome of all of these processes is similar, and palladium catalysts generally permit the coupling of aryl and vinyl halides and related substrates with organometallic nucleophiles. The synthetic value in these processes is that a leaving group (such as halogen, triflate, tosylate or mesylate) directly attached to an sp^2 centre is substituted; such a process is generally difficult to achieve using more conventional reactions, although examples include addition–elimination at a carbon–carbon double bond or aromatic ring substituted with strongly electron-withdrawing groups (see Chapter 9, S$_N$Ar and the benzyne reaction).

The aryl halide has traditionally been needed to be a bromide or iodide (although iodides are more reactive, they are less commonly available than bromides) since chlorides are relatively unreactive, but the development of new catalysts has permitted the widespread application of aryl chlorides (Fig. 12.16). Furthermore, the coupling process has been applied to soft nucleophiles in place of the harder organometallic systems, and the work of Hartwig and Buchwald has been significant in the establishment of conditions which allow coupling with oxygen, sulfur, phosphorus, boron and silicon nucleophiles, as well as soft carbon nucleophiles such as enolates (Figs. 12.16a–12.16c). Crucial in the successful outcome of this work has been the development of effective ligands for palladium which allow reversible cycling between the Pd(0) and Pd(II) oxidation states.

It is noteworthy that the Suzuki and other similar reactions can be achieved using other transition metal catalysts, including nickel, platinum, copper and ruthenium, but these approaches have not been so widely developed as the palladium coupling process.

12.1.2 Heck coupling

The Heck reaction involves coupling of an alkenyl halide or triflate with an alkene in a reaction catalysed by palladium(II) acetate and triphenylphosphine (Figs. 12.17a–12.17d);

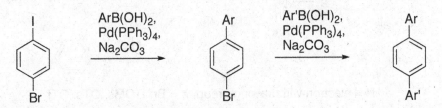

ArB(OH)$_2$,
Pd(PPh$_3$)$_4$,
Na$_2$CO$_3$

Ar'B(OH)$_2$,
Pd(PPh$_3$)$_4$,
Na$_2$CO$_3$

Figure 12.15 Sequential coupling in Suzuki couplings.

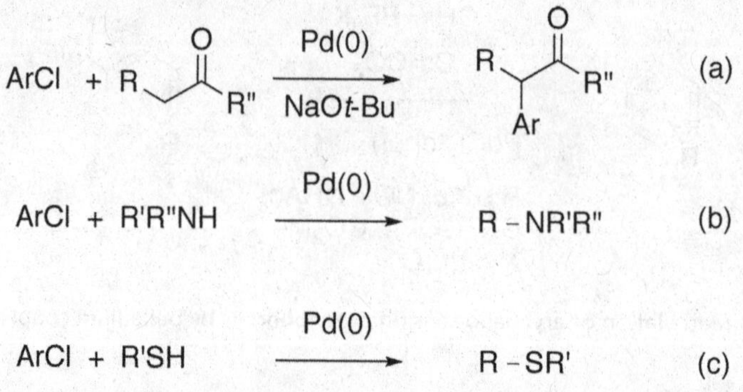

Figure 12.16 (a) Hartwig arylation, (b) Buchwald–Hartwig amination and (c) Hartwig thiolation ligand coupling processes in organic chemistry.

in the case of geometrically defined vinylic substrates, the reaction proceeds with retention of configuration. One equivalent of a base, normally triethylamine, is also required, since it reduces the Pd(II) to the catalytically active Pd(0) species and also neutralises strong acid (HX) produced in the course of the reaction. Initial π-complexation of the palladium to the organic halide is followed by oxidative addition. The σ-palladium complex then

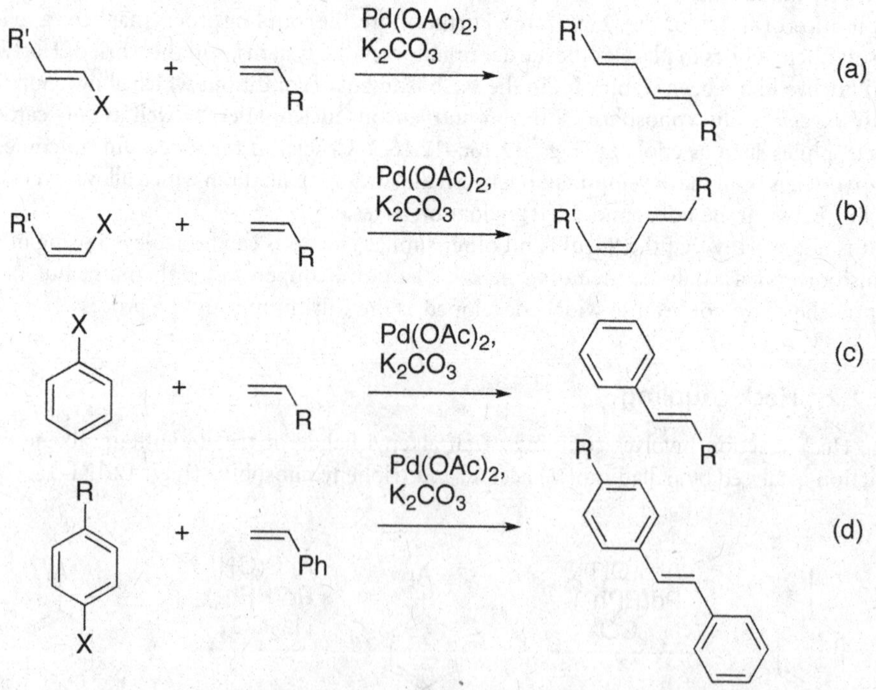

R = electron-withdrawing group, X = Br, I, OMs, OTs, OTf

Figure 12.17 Heck-type coupling processes.

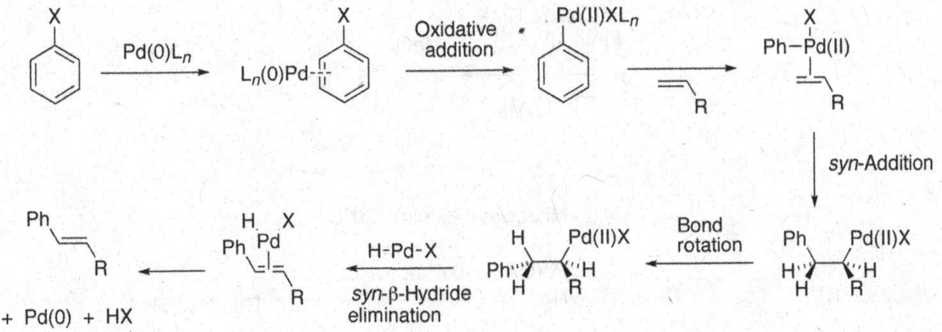

Figure 12.18 Mechanism for the Heck coupling process.

coordinates to and adds across the carbon–carbon double bond containing component to form a σ-bonded complex, in a *syn*-selective and regioselective carbopalladation process, as shown in Fig. 12.18. In order for this intermediate to eliminate palladium hydride, it requires a bond rotation to occur so that the palladium and the hydrogen are *syn*-related; this then enforces an arrangement leading to the *trans*-product and determines the stereochemical outcome. Significantly, the reaction will also proceed with retention of configuration at the vinyl halide component; the reaction is therefore stereospecific. These β-hydride eliminations from orgnopalladium species are very rapid and will occur without difficulty in an appropriate substrate. The reaction is regioselective, and double bonds substituted with electron-releasing groups tend to couple adjacent to the releasing group, while electron-withdrawing groups substitute at the terminal position of the alkene (Fig. 12.19).

These organometallic processes have in general been found to be very sensitive to microwave acceleration, principally as a result of the high receptivity of the metal catalysts to microwave energy; for example, the Heck, Suzuki, Sonagashira and Buchwald–Hartwig amination reactions have all been found to be substantially accelerated in this way (Figs. 12.20a–12.20e).

12.1.3 Allylic coupling processes

Allyl acetates and carbonates will readily react with Pd(0) and a nucleophile, typically a β-dicarbonyl derivative (Fig. 12.21), to give the product of allylic substitution; the reaction is called the Tsuji–Trost allylation. Mechanistically, the reaction is thought to proceed via initial π-allyl palladium cation formation (an η³ complex), which is then intercepted by the nucleophile to give the allyl product; unlike the ligand coupling processes described above, which require a *syn*-reductive elimination, in this case the reaction can proceed with overall inversion or retention, that is, by nucleophilic attack either *cis*- or *trans*- to the palladium, depending on whether a soft (e.g. β-dicarbonyl) or a hard nucleophile (organozinc) is used respectively, giving the products as shown in Figs. 12.22a and 12.22b.

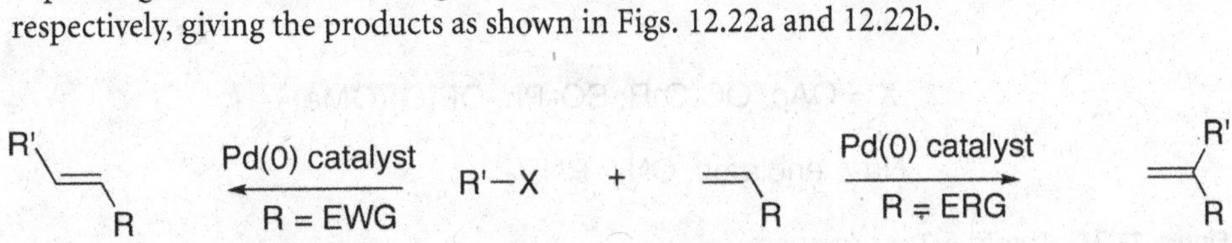

Figure 12.19 Regioselectivity of the Heck coupling process.

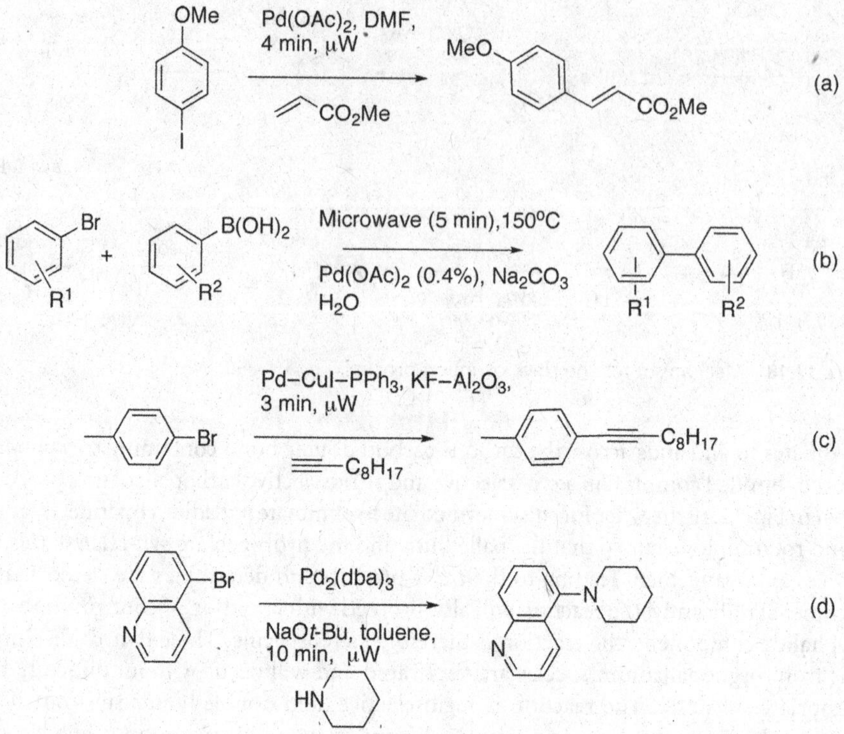

Figure 12.20 Examples of microwave-accelerated palladium-catalysed couplings.

The reaction can be conducted in such a way as to generate cyclic products (Fig. 12.23). If this system is modified as a 2(trimethylsilylmethyl)allyl carbonate, it is possible to generate a 1,3-dipole equivalent which will add to alkenes (Fig. 12.23a). An unusual but valuable variant of this process is the reaction of methylene cyclopropanes with unsaturated esters, leading to cyclopentane derivatives (Fig. 12.23b) via π-allyl intermediate systems; interaction of Pd(0) with the methylene cyclopropane leads to the π-allyl intermediate which is intercepted by the other alkene to give a cyclopentane product. This reaction can be conducted in an intramolecular sense too and be used to access highly functionalised bicyclic ring systems (Fig. 12.23c) as well as heterocyclic systems (Fig. 12.23d).

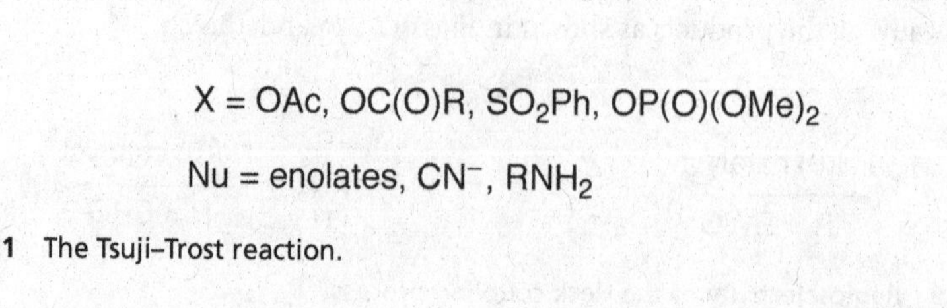

$$X = OAc, OC(O)R, SO_2Ph, OP(O)(OMe)_2$$

$$Nu = \text{enolates, } CN^-, RNH_2$$

Figure 12.21 The Tsuji–Trost reaction.

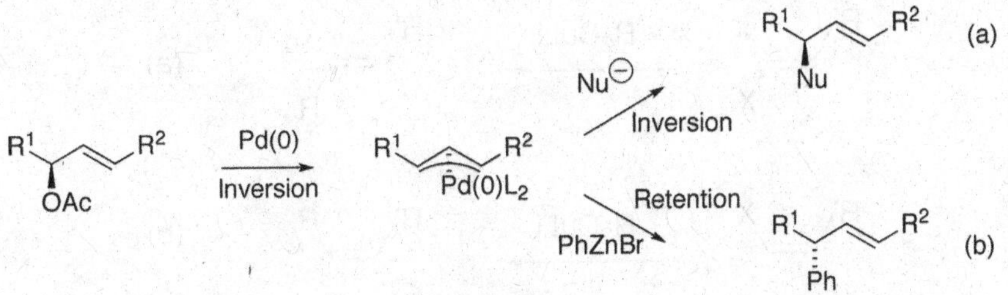

Figure 12.22 Reactions of π-allyl palladium species proceed with (a) overall retention or (b) inversion depending on the reacting nucleophile.

12.2 Ligand coupling processes mediated by other elements

The myriad of coupling processes which are mediated by palladium makes it an extraordinarily valuable metal. However, it is neither cheap nor non-toxic, although the ease with

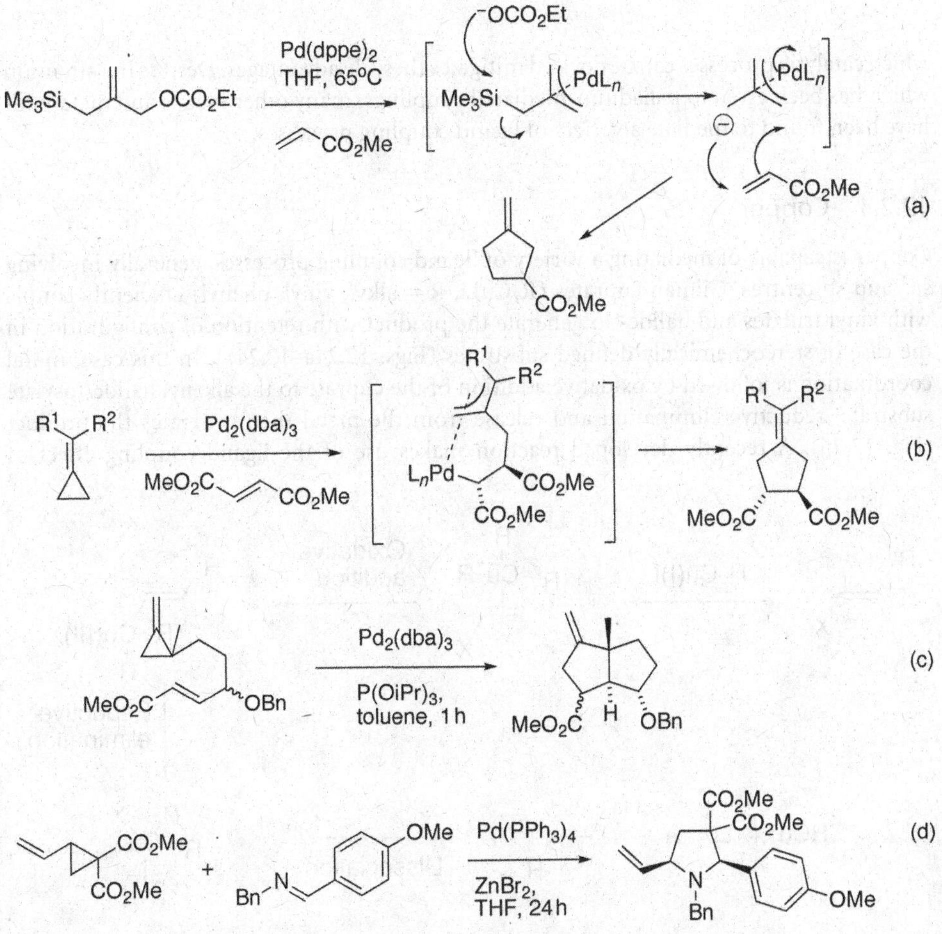

Figure 12.23 Insertion reactions of π-allyl palladium species.

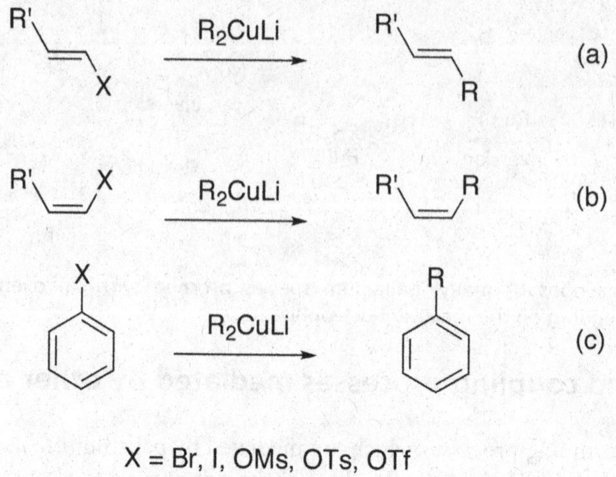

X = Br, I, OMs, OTs, OTf

Figure 12.24 Coupling reactions of Gilman cuprates.

which catalytic processes can be devised mitigates these disadvantages. Despite the attention which has been given to palladium-mediated couplings, many other metals and metalloids have been found to mediate a variety of ligand coupling processes.

12.2.1 Copper

Copper is capable of mediating a variety of ligand coupling processes, generally involving sp^2 and sp centres. Gilman cuprates (R_2CuLi, R = alkyl, vinyl, phenyl) efficiently couple with vinyl triflates and halides to generate the product with retention of configuration in the case of stereochemically defined substrates (Figs. 12.24a–12.24c). In this case, initial coordination is followed by oxidative addition of the cuprate to the alkenyl halide/tosylate substrate; reductive elimination and release from the metal then generates the product (Fig. 12.25). A recently developed reaction makes use of the ligand coupling effect of

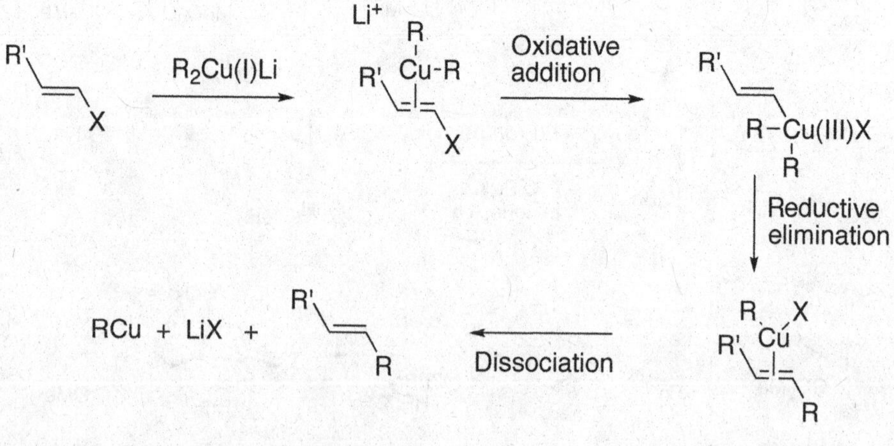

Figure 12.25 Mechanism for the Gilman coupling process.

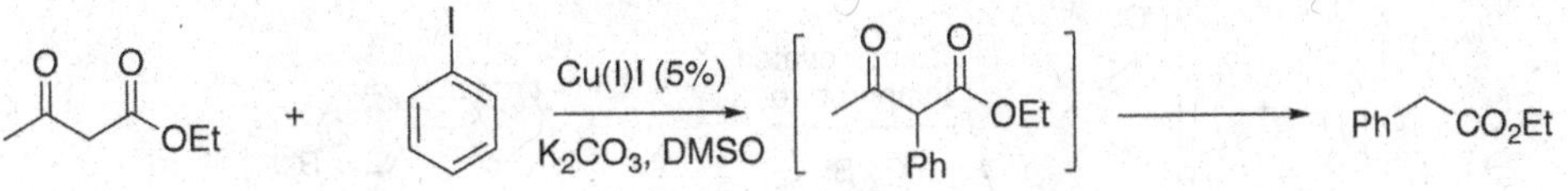

Figure 12.26 Copper(I)-mediated coupling process of β-dicarbonyl compounds.

copper(I) to arylate dicarbonyl compounds; the arylated acetoacetate ester is deacylated under the reaction conditions resulting in the generation of 2-arylacetic acid esters, and this process constitutes a mild method for the direct arylation of carboxylate esters. This reaction is interesting because copper replaces the more commonly used metal – palladium – for such ligand coupling processes (Fig. 12.26).

Copper also mediates two specific types of coupling, called the Ullman coupling reaction and the Cadiot–Chodkiewicz coupling reaction. These lead to the formation of biaryls and bis(acetylenic) derivatives respectively (Figs. 12.27a and 12.27b). In the former process, oxidative addition with the more reactive aryl halide leads to an intermediate arylcopper intermediate, which then ligand exchanges with the less reactive halide and then reductively eliminates to generate the unsymmetrical biaryl species. Homocoupling is also possible using this process, leading to a symmetrical biaryl product. In the latter reaction, coupling of an acetylene and an acetylenic halide leads to the formation of an unsymmetrical bisalkyne. Mechanistically, these processes are very similar, and the key bond-forming processes are driven by a ligand coupling with concomitant reduction of the metal. It is also possible to couple aryl or alkenyl halides and triflates with acetylenes under copper(I) and palladium(0) conditions, in reactions called Cacchi or Sonogashira–Hagihara couplings (Figs. 12.27c and 12.27d). The reactions typically require an excess of amine base, whose function is to generate ammonium acetylide, which in turn is converted to a copper acetylide (Fig. 12.28). This nucleophilic entity is capable of ligand exchange at palladium, thereby generating a species set for the required ligand coupling and ultimately product formation. The ligand exchange also releases copper(I), which is capable of re-entering the catalytic cycle, allowing the formation of more copper acetylide. This reaction scheme may also be applied to the synthesis of 1,3-eneynes (Figs. 12.27d and 12.27e); in this case, the reaction is stereospecific, since the configuration of the starting alkenyl substrate is preserved in the product. Stereospecific couplings of organostannanes with organoiodides using a copper(I) thiophenecarboxylate catalyst have also been reported (Fig. 12.27f)..

12.2.2 Magnesium

Alkyl, vinyl and aryl Grignard reagents, in the presence of a nickel(0) catalyst, will couple with alkenyl halides or aryl halides or triflates to give the corresponding product (Fig. 12.29), in which retention of configuration is obtained. The nickel catalyst is most conveniently added as a nickel(II) salt, which is efficiently reduced to the required nickel(0) species by excess organometallic reagent. A likely mechanism follows a familiar pattern (Fig. 12.30): Initial coordination of the catalyst to the substrate generates a π-complex, which then undergoes oxidative addition. Metathesis with the Grignard reagent then gives an intermediate set-up to reductively eliminate, thereby giving the product. A dissociation step then releases the catalyst, ready for re-entering the cycle.

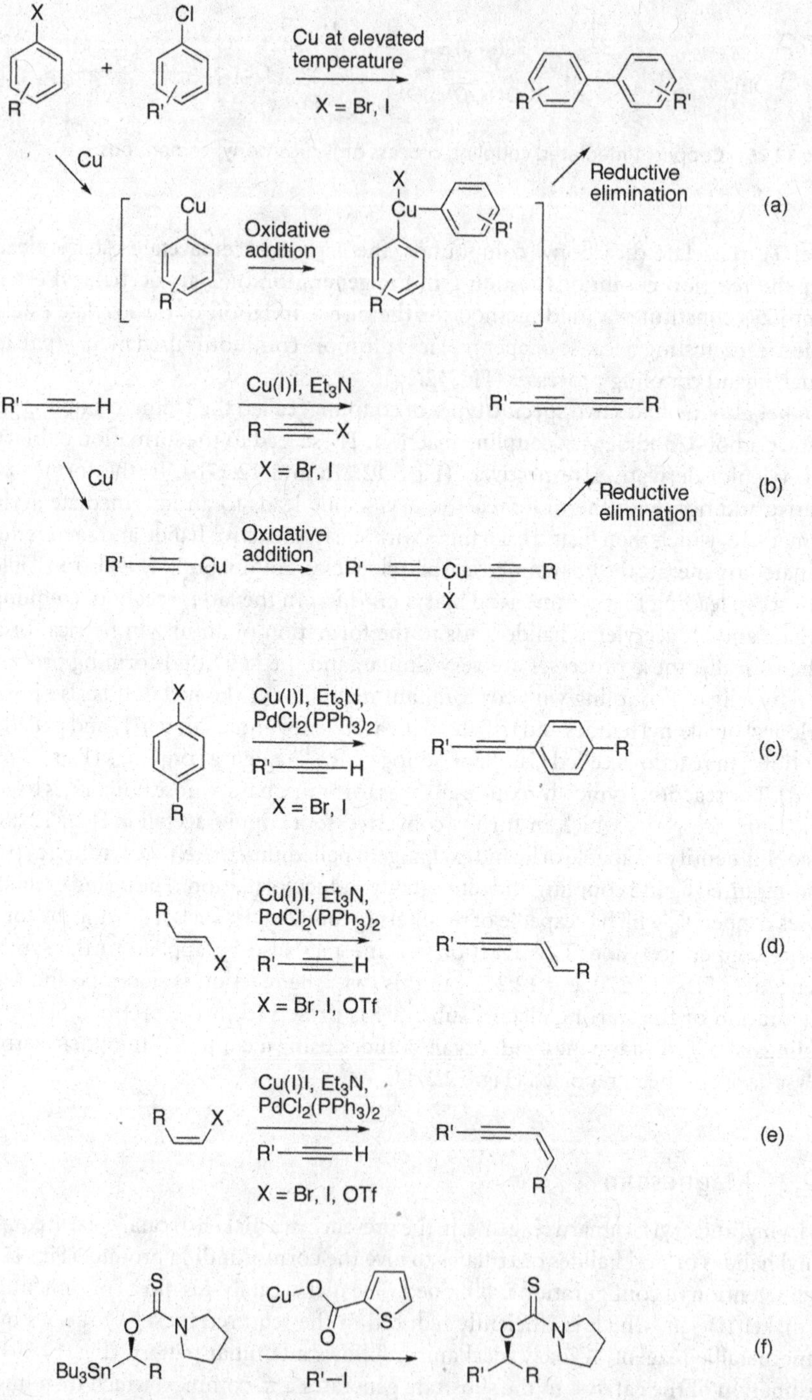

Figure 12.27 Coupling processes mediated by copper: (a) Ullman coupling, (b) Cadiot–Chodkiewicz coupling, (c) Sonogashira-type coupling and (d–f) stereospecific enyne and alcohol synthesis.

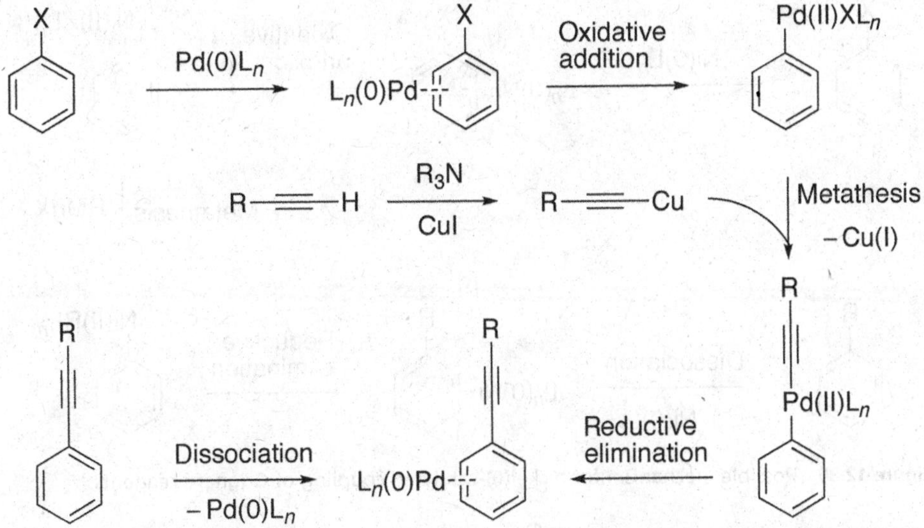

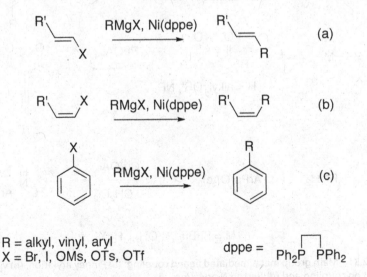

Figure 12.28 Possible mechanism for palladium(0)-catalysed coupling of copper acetylides.

12.2.3 Lead

Lead(IV) is known to mediate a range of ligand coupling reactions, which are noteworthy since they facilitate reactions which have only recently become accessible to palladium-mediated processes (Fig. 12.31). The first process to be discovered was Pinhey arylation (Fig. 12.31a), and this involves the treatment of an aryllead(IV) triacetate with a β-dicarbonyl compound in the presence of pyridine, which has been found to be an important σ-donor ligand for the facilitation of the reaction. A likely mechanism for the process is shown in

R = alkyl, vinyl, aryl
X = Br, I, OMs, OTs, OTf

dppe = Ph$_2$P PPh$_2$

Figure 12.29 Nickel(0)-catalysed coupling reactions of Grignard reagents.

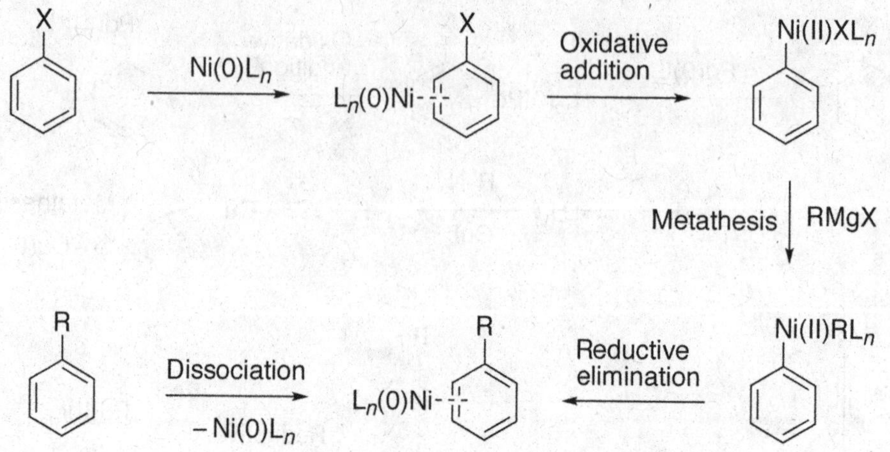

Figure 12.30 Possible mechanism for nickel(0)-catalysed coupling of Grignard reagents.

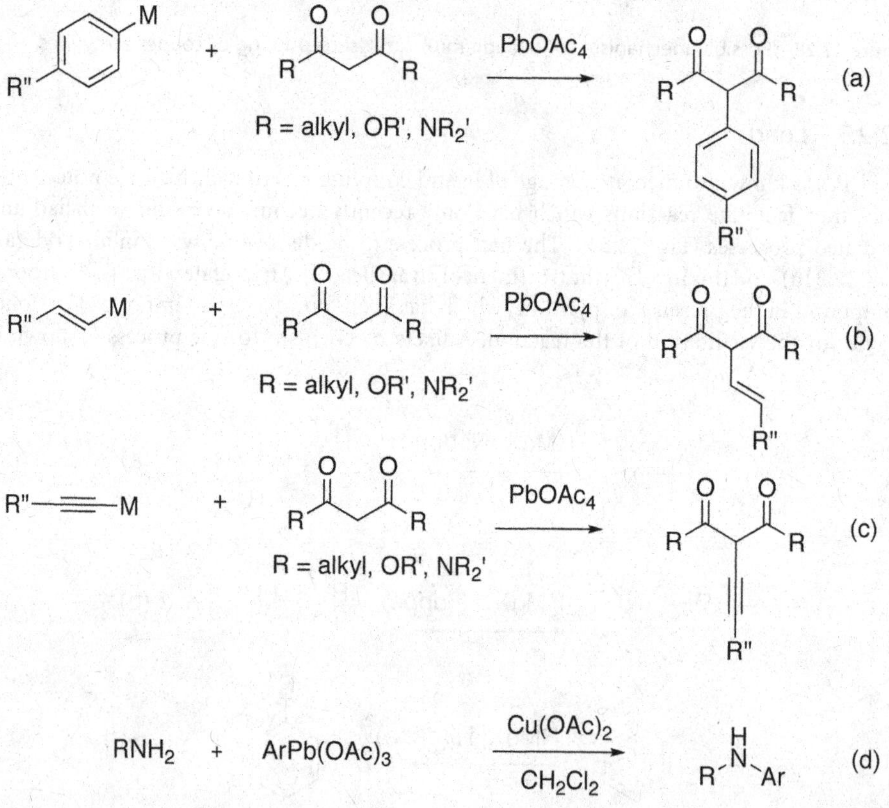

M = SnBu$_3$, B(OH)$_2$, HgX

Figure 12.31 Main group metal mediated ligand coupling: (a) Pinhey arylation, (b) vinylation, (c) alkynylation coupling and (d) Barton N-arylation.

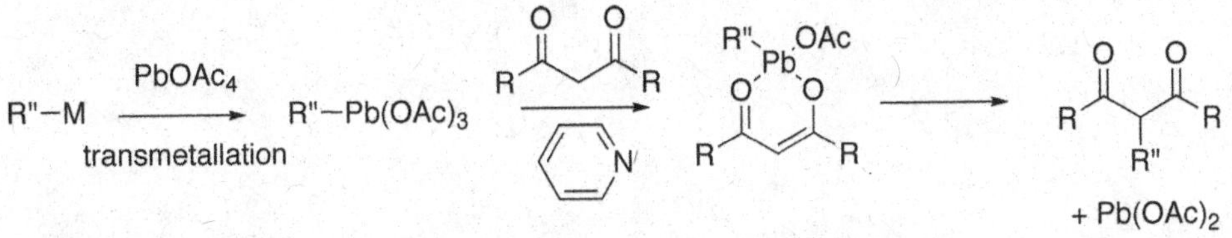

Figure 12.32 Possible mechanism for lead(IV)-mediated ligand coupling.

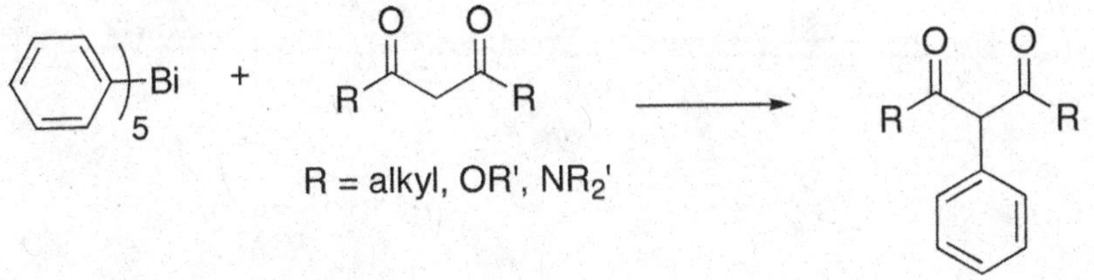

R = alkyl, OR', NR$_2$'

Figure 12.33 Bismuth-mediated arylations of β-dicarbonyl substrates.

Fig. 12.32; although a very effective reaction, giving high yields of product arising from an otherwise inaccessible arylcation intermediate under mild conditions, the reaction is nonetheless stoichiometric in lead. The required aryllead(IV) species, which are isolable and storable, are most readily available by transmetallation of the corresponding arylmercury, stannane or boronic acids. Later developments of this process gave rise to vinylation (Fig. 12.31b) and alkynylation (Fig. 12.31c), which proceed in a mechanistically analogous manner, although in this case the intermediate vinyllead and alkynyllead species are too unstable to isolate, but nonetheless may be generated and reacted successfully in situ. These latter reactions are of particular interest, since the intermediate lead species behave like vinyl and alkynyl cations, species which are otherwise virtually inaccessible by more conventional routes. Of interest is that palladium equivalents of this process have only very recently become viable, largely as a result of ligand design which has expanded the scope of possible substrates (see Section 12.1.1). The scope of this reaction has been further expanded away from soft dicarbonyl nucleophiles by the demonstration that amine nucleophiles were also applicable; thus, it is possible to arylate amines directly using aryllead triacetates in the presence of copper(II) (Fig. 12.31d).

Noteworthy is that other main group elements mediate similar processes; pentaphenyl bismuth will efficiently phenylate β-dicarbonyl systems, but the synthetic scope of this process is restricted by the fact that there are only a limited number of routes available for the synthesis of penta-substituted bismuth compounds (Fig. 12.33).

Chapter 13
Pericyclic Reactions

Thus far, the existence of charged (carbocations, carbanions) and uncharged (carbenes, radicals) intermediates provides an excellent explanation of a wide variety of chemical processes. However, there are some reactions which do not proceed via the formation of observable intermediates and for which a separate mechanistic rationale must be made. A reaction mechanism in which electrons are redistributed by moving in a closed ring of atoms, but without the formation of intermediates (i.e. the reaction proceeds through a transition state), is called *pericyclic reaction*, which may be formulated in a general equation as shown in Fig. 13.1. Because these reactions proceed without the formation of intermediates, they are sometimes erroneously referred to as 'no-mechanism' reactions. However, they are readily understood by a consideration of the molecular orbitals (MOs) which are involved in the reaction (called the frontier molecular orbitals (FMOs)), and this is described in more detail in the following sections. It is worth noting that other means for rationalising these reactions have been developed, including aromatic transition state theory and orbital symmetry correlation, but these will not be considered here.

13.1   Molecular orbitals and the FMO approach

The MOs for extended π-systems can be derived by a consideration of the atomic orbitals (AOs) of which they are comprised; the simplest analysis uses a linear combination of atomic orbitals (the so-called LCAO method). Thus, we consider that an MO is derived by a simple algebraic sum of its constituent AOs; each AO in an extended π-system is added or subtracted to generate a series of MOs in which successive orbitals are either aligned or non-aligned in phase. For an MO comprising n AOs, there will be n MOs. This is illustrated in Fig. 13.2 for ethene, for which two MOs (π and π^*) are generated from two p orbitals; butadiene, for which four MOs (ψ_1, ψ_2, ψ_3, ψ_4) are generated from four p orbitals; and hexatriene, for which six MOs (ψ_1, ψ_2, ψ_3, ψ_4, ψ_5, ψ_6) are generated from six p orbitals. With the MOs drawn, it is possible to locate the electrons of the compound by filling orbitals according to the Hund rule and the Aufbau principle. The orbital of highest energy containing electrons is called the highest occupied molecular orbital (HOMO) and the orbital of lowest energy not containing electrons is called the lowest unoccupied molecular orbital (LUMO); in any reaction, these are the most likely to interact, with the former providing the nucleophilic component and the latter the electrophilic component.

Inspection of the MOs for these molecules indicates that (i) the nth orbital has $n-1$ nodes symmetrically distributed around the midpoint; (ii) the orbitals are alternately symmetric and antisymmetric with regard to a mirror plane at the midpoint; (iii) the orbitals are alternately antisymmetric and symmetric with regard to a twofold rotational axis of symmetry at the midpoint.

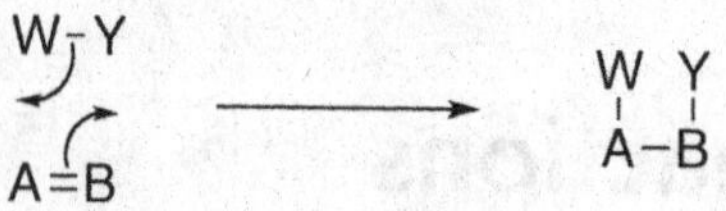

$$W\overset{\frown}{-}Y \qquad\qquad \longrightarrow \qquad\qquad \begin{matrix} W & Y \\ | & | \\ A & -B \end{matrix}$$
$$A\overset{\frown}{=}B$$

Figure 13.1 Pericyclic reactions.

This analysis is also applicable to systems with odd numbers of atoms: Here one MO is non-bonding (i.e. energetically neutral) and has nodes at the nuclei (this is why electron deficiency or electron density lies at the end of the molecule). Using this analysis, the results for the cation, radical and anions of allyl and pentadienyl systems are shown in Fig. 13.3.

With this information in hand, it is possible to develop an FMO approach for understanding reactions: This considers the interactions of HOMO (electron donor) and LUMO (electron acceptor) to determine the feasibility of any given reaction; remember that bonding arises only when orbitals of the same phase overlap. The easiest path (i.e. lowest E_{act}) for a reaction is one which maximises the bonding during the reaction process; since bonding results from orbital overlap, we require maximum overlap which can only be achieved when orbitals are of the same phase. Thus, concerted reactions can be either *symmetry allowed* or *symmetry forbidden*. This can easily be decided by a consideration of the FMOs, which contain the 'valence electrons' of the molecules. This approach is particularly valuable when it comes to understanding different types of pericyclic reactions.

13.2 Pericyclic reactions

Pericyclic reactions are concerted (concerted = simultaneous bond making/breaking) reactions that proceed via cyclic transition states. They are characterised by having a very small

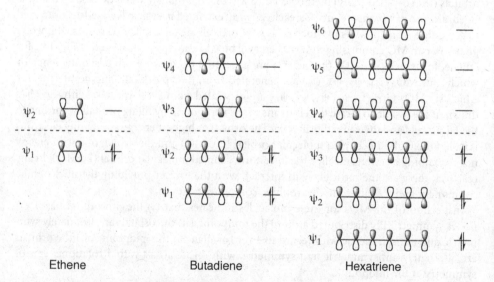

Figure 13.2 Molecular orbitals for ethene, butadiene and hexatriene.

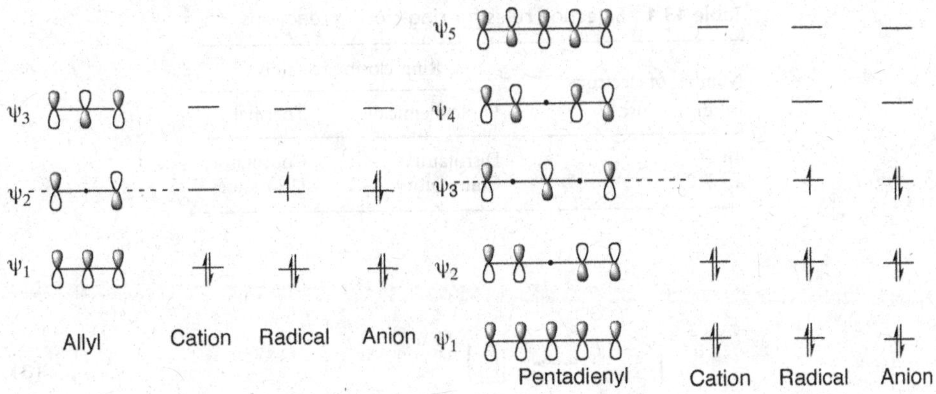

Figure 13.3 Molecular orbitals for allyl and pentadienyl cation, radical and anion systems.

change in entropy ($\Delta S \sim 0$) in the course of the reaction, and they are often highly stereo-specific and reversible. There are several key types, and these are described below.

13.2.1 Electrocyclic reactions

These are pericyclic reactions that involve ring opening or ring closure, by bond forma-tion/cleavage between the ends of a linear system of π-electrons; important examples are the ring closure of butadiene and hexatriene (Figs. 13.4a and 13.4b respectively). The reac-tion is of greatest interest when there are substituents on the polyene termini, since in this case the substituents can end up either *cis*- or *trans*-related, and this is determined by the number of π-electrons involved in the ring closure (Figs. 13.4c and 13.4d). The mode of ring closure (conrotatory, in which the substituents rotate in the same direction, or disrotatory, in which they rotate in opposite directions) can be understood by a consideration of the interaction of the FMOs, in particular the HOMO of the polyene; ring closure is *symmetry allowed* only when orbital overlap of the same phase can occur.

1. Butadiene $\rightarrow$ Cyclobutene
    In this case, the ring closure involves four π-electrons in the diene and under ther-mal conditions, conrotatory orbital movement of the HOMO ψ_2 leads to a net bond-ing arrangement and the configuration of the substituents reflects this movement (Fig. 13.4c). Under photochemical conditions, electron excitation makes ψ_3 the HOMO, and it is disrotatory orbital movement that leads to a net bonding arrangement.
2. Hexatriene $\rightarrow$ Cyclohexadiene
    Simple extension of the butadiene case indicates that with six π-electrons in the triene, the HOMO is now ψ_3 and so thermal ring closure requires a disrotatory movement to lead to appropriate orbital overlap and a bonding outcome. On the other hand, electron excitation makes ψ_4 the HOMO, and it is conrotatory orbital movement that now leads to a net bonding arrangement.

These results can be generalised for any system containing n π-electrons in the ring-closing step to give the selection rules shown in Table 13.1.

Table 13.1 Selection rules for ring-closing reactions

Number of electrons in ring closure	Ring-closing reaction	
	Photochemical	Thermal
$4n$	Disrotatory	Conrotatory
$4n+2$	Conrotatory	Disrotatory

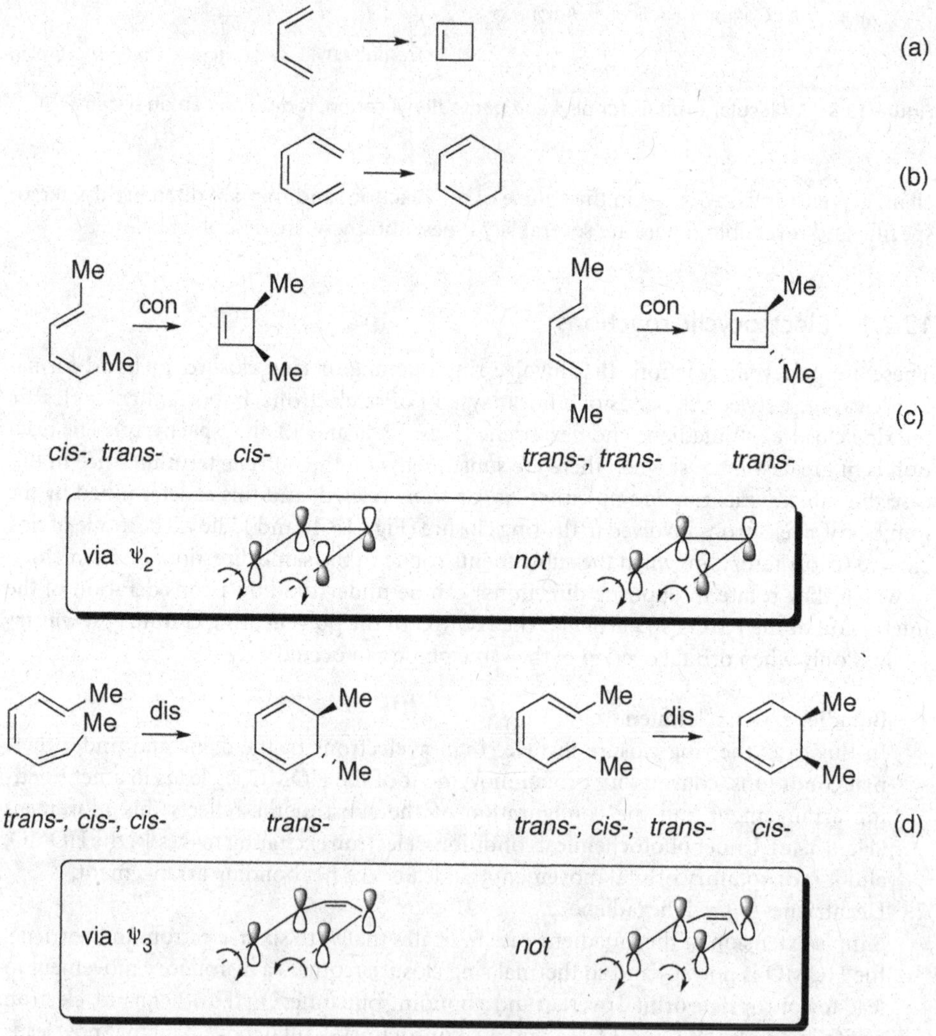

(a)
(b)
Me
con
Me
Me
Me
Me
con
Me
Me
Me
(c)
cis-, trans-
cis-
trans-, trans-
trans-
via Ψ2
not
Me
Me
dis
Me
Me
Me
dis
Me
Me
Me
trans-, cis-, cis-
trans-
trans-, cis-, trans-
cis-
(d)
via Ψ3
not

Figure 13.4 Electrocyclic reactions.

13.2.2 Cycloaddition reactions

These are pericyclic reactions in which two unsaturated molecules react to form one cyclic molecule; the π-electrons are used to form two new σ-bonds. In order to assign an appropriate descriptor to a cycloaddition process, we need to assign (i) the type of orbitals directly participating in the bond-breaking/forming process, which may be σ π, or even AOs (designated ω); (ii) the number of electrons (n) in each of these participating orbitals; (iii) whether these orbitals are reacting in a suprafacial (s, on the same face) or antarafacial (a, on opposite faces) manner. This allows the assignment of a descriptor; thus, [orbital-n-facial] such as [π4s + π2s] or [π2s + ω2a].

Rules devised by Woodward and Hofmann can be conveniently used to determine when cycloaddition is allowed. In order to establish whether a reaction is allowed, it is necessary firstly to examine the reactive components and identify the $(4q + 2)$ and $(4r)$ systems, where q and r are integers; for example, an alkene is a two π-electron system and is therefore a $(4q + 2)$ system with $q = 0$. A diene, being a four π-electron system, is therefore a $(4r)$ system with $r = 1$. With this information in hand, it is necessary to draw out the reacting components and the reaction mechanism, identify those π-bonds (and only those π-bonds) that are involved in the reaction, to label the reactive components using their $(4q + 2)$ and $(4r)$ descriptors and to establish whether the process is suprafacial (s) or antarafacial (a); in the example shown, this allows identification of the π2s and π4s components (Fig. 13.5). Then, all that is necessary is to count the number of $(4q + 2)$s and $(4r)$a components (or $(4q + 2)$a and $(4r)$s ones). The Woodward–Hofmann rules state that a thermal pericyclic reaction is symmetry allowed when the total number of $(4q + 2)$s and $(4r)$a components is odd, and a photochemical pericyclic reaction is symmetry allowed when the total number of $(4q + 2)$s and $(4r)$a components is even (note that q and r are integers). An example is shown in Fig. 13.5; here, the number of $(4q + 2)$s components is one and $(4r)$a is none, giving a total of one, meaning that the Diels–Alder reaction is an allowed process. It is possible to generalise the stereochemical outcome of these reactions using the Woodward–Hofmann rules, as shown in Table 13.2.

13.2.2.1 *Diels–Alder reaction*

This reaction involves the addition of a diene with a dienophile to give a cyclohexene derivative (Fig. 13.6a) by a [4 + 2]-cycloaddition. The Diels–Alder reaction is a [π4s + π2s], and since both components react in a suprafacial manner, and there is only one $(4q + 2)$ suprafacial component, the reaction is thermally allowed by the

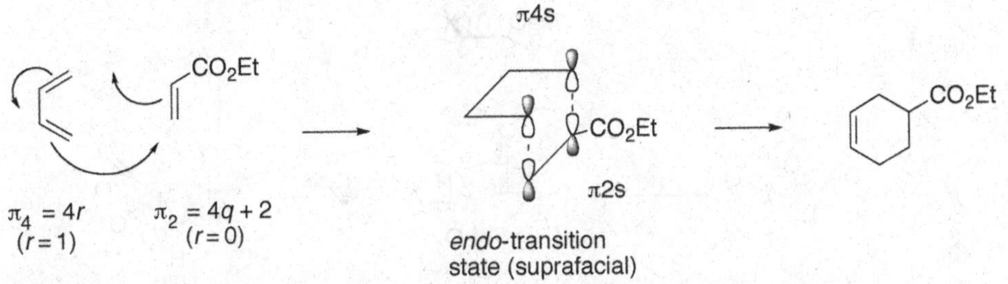

$\pi_4 = 4r$ $\pi_2 = 4q + 2$
$(r = 1)$ $(r = 0)$

endo-transition state (suprafacial)

Figure 13.5 Woodward–Hofmann rules for cycloaddition reactions.

Table 13.2 Woodward–Hofmann rules for
the facial selectivity of $[i + j]$-cycloadditions

$i + j$	Thermal	Photochemical
$4n$	s, a (or a, s)	s, s (or a, a)
$4n + 2$	s, s (or a, a)	s, a (or a, s)

Woodward–Hofmann rules. FMO analysis indicates that the dominant interaction is HOMO
of the diene (ψ_2) and LUMO of the dienophile (π_2) (Fig. 13.6b), and this interaction is promoted by electron-releasing substituents (including alkyl, oxygen and nitrogen) on the
diene and electron-withdrawing substituents (especially resonance-withdrawing groups,
such as ester, amide, nitrile and nitro groups) on the dienophile. Such electron-releasing or
-withdrawing groups may have very significant effects on the rate of reaction, as shown in
Figs. 13.7a and 13.7b respectively; electron-withdrawing groups on the dienophile

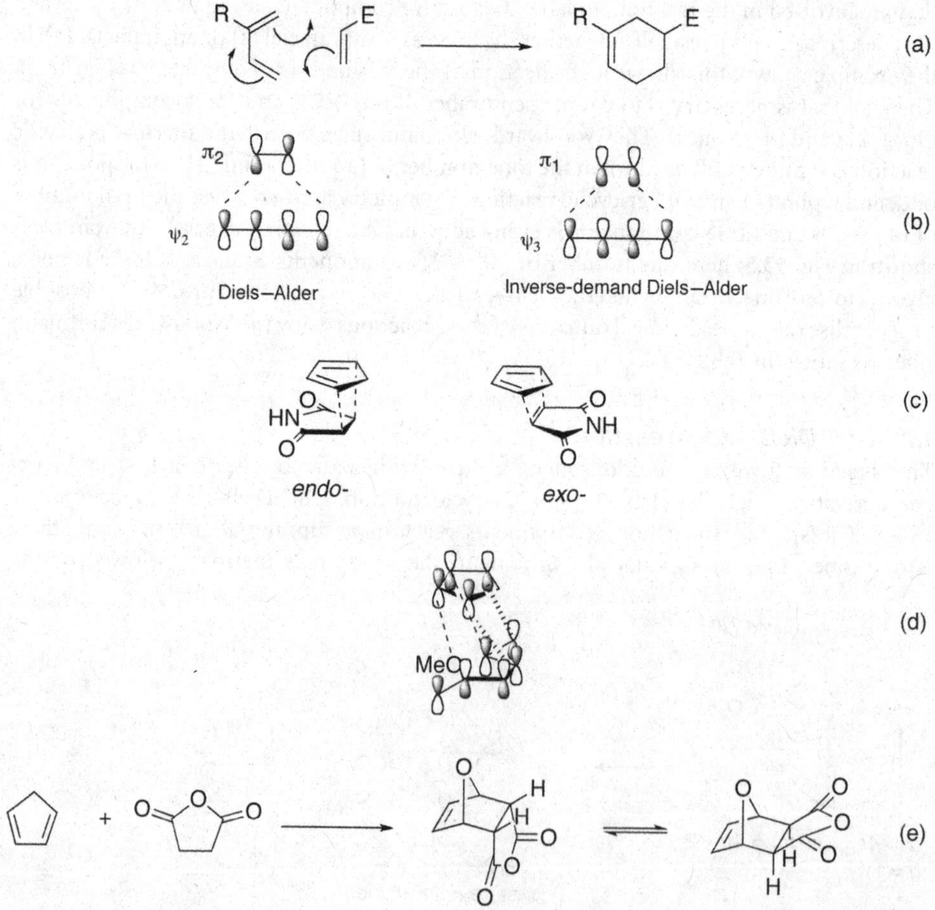

Figure 13.6 [4 + 2]-Cycloaddition reactions.

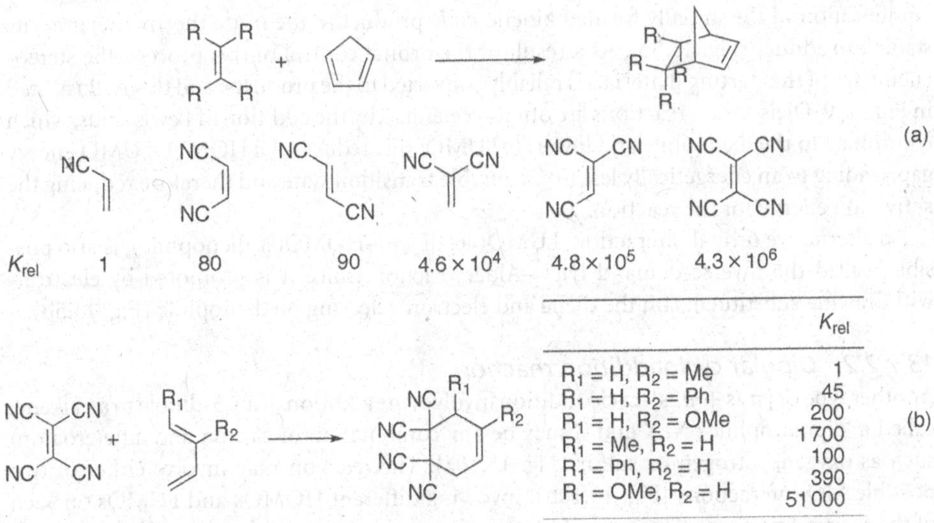

	K_{rel}
R_1 = H, R_2 = Me	1
R_1 = H, R_2 = Ph	45
R_1 = H, R_2 = OMe	200
R_1 = Me, R_2 = H	1700
R_1 = Ph, R_2 = H	100
	390
R_1 = OMe, R_2 = H	51000

Figure 13.7 Electronic effects on the Diels–Alder reaction.

activate the reaction relative to an unsubstituted alkene, and electron-releasing substituents on the diene also activate, with multiple substituents leading to increasing activation effects. Resonance-donating or -releasing groups have a much stronger activating effect than inductively modifying groups. Substituents may also exert significant effects on the regioselectivity of the process; the regioselectivity is most simply understood by considering the effects of substituents on the favoured canonical forms of the reacting partners. Thus, electron-donating groups at the 1- or 2-positions of the diene function promote the regiochemistry shown in Figs. 13.8a and 13.8b respectively, giving the ortho or para products; the Diels–Alder reaction has in this sense been referred to as ortho and para directing. There are several important characteristics: The diene must exist in an *s-cis* conformation; the addition is stereospecifically *syn* with respect to the alkene dienophile (suprafacial); *endo* additions are kinetically preferred (Fig. 13.6c) as in the required orbital interaction; favourable secondary orbital interactions between the HOMO and the LUMO of adjacent functionality in the transition state are possible (Fig. 13.6d). In some cases, the reaction can be reversible, and this can lead to

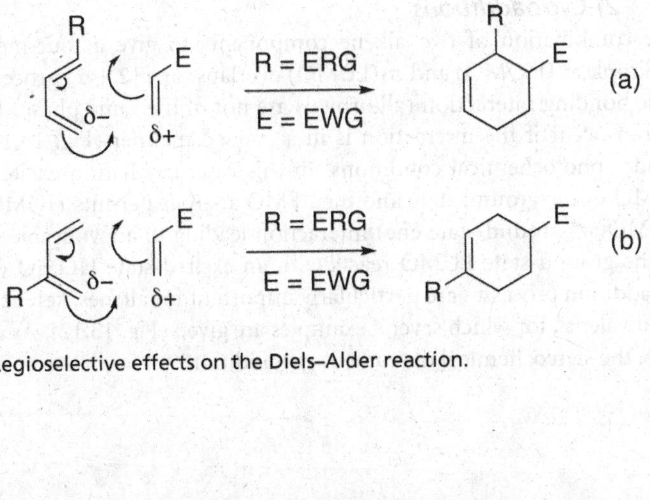

Figure 13.8 Regioselective effects on the Diels–Alder reaction.

equilibration of the initially formed kinetic *endo* product to the more thermodynamically stable *exo* adduct (Fig. 13.5e). As a result of the orbital control of this process, the stereochemistry of the starting materials is reliably converted to the products, and this is illustrated in Fig. 13.9. Diels–Alder reactions are often accelerated by the addition of Lewis acids, which coordinate to the dienophile and lower its LUMO; this reduces the HOMO–LUMO energy gap, leading to an energetically less unfavourable transition state and therefore reducing the activation energy for the reaction.

An alternative orbital interaction, LUMO(ψ_3 diene)–HOMO(π dienophile), is also possible, called the inverse-demand Diels–Alder reaction, since it is promoted by electron-withdrawing substituents on the diene and electron-releasing on dienophile (Fig. 13.5b).

13.2.2.2   Dipolar cycloaddition reaction

Another type of $[\pi 4s + \pi 2s]$-cycloaddition involves the addition of a 1,3-dipole to an alkene, called a 'dipolarophile'; X, Y and Z may be any combination of carbon and a heteroatom such as oxygen, nitrogen or sulfur (Fig. 13.10a). The reaction may involve either of two possible MO interactions (Fig. 13.10b), involving different HOMOs and LUMOs on each of the reacting partners. This type of reaction is possible for a wide variety of 1,3-dipoles.

There are a number of important examples of these processes, including the cycloaddition of diazomethane to give pyrazolines, azides to give triazoles, nitrile oxides to give isoxazolines and ozone to give molozonides (Figs. 13.10c–13.10f). This last reaction, the cleavage of a carbon–carbon double bond to give two carbonyl compounds using ozone (O_3), was used historically for the elucidation of the structure of organic compounds, but its main use today is in synthesis (Figs. 13.10g–13.10i). The reaction proceeds by spontaneous rearrangement of the intermediate molozonide to the corresponding ozonide, which is not isolated for reasons of both convenience and safety (since these compounds are often explosive). Depending on whether this intermediate is treated under reducing, neutral or oxidising conditions, it is in fact possible to access two carbonyl compounds (Fig. 13.10g), two carboxylic acid products (Fig. 13.10h) or two alcohol products (Fig. 13.10i).

The reaction of alkenes with osmium tetroxide, and of alkaline permanganate, gives *syn*-1,2-diols. Both of these reactions proceed by initial cycloaddition of the metal oxide to the π-bond of the alkene; for this reason, oxygen adds to the carbon–carbon double bond from the same side, and this determines the stereochemical outcome as *syn*; this reaction is therefore complementary to the opening of epoxides. The diol is formed by subsequent hydrolysis by reaction with water (Fig. 13.10j).

13.2.2.3   [2 + 2]-Cycloadditions

These are the combination of two alkene components to give a four-membered ring (Fig 13.11a). Simple π_1(HOMO) and π_2(LUMO) overlaps for a [2 + 2] process do not lead to a favourable bonding interaction (all orbitals are not of the same phase) (Fig. 13.11b), but bonding can occur if the interaction is in a crossed manner (Fig. 13.11c) or if it is conducted under photochemical conditions. In this case, irradiation excites an electron from the HOMO of the ground state and then FMO analysis permits HOMO(π^*excited-state ene)–LUMO(π^* ground-state ene) interaction leading to a favourable reaction outcome. Thus, the ground-state LUMO reacts with an excited-state HOMO (Fig. 13.11d). [2 + 2]-Cycloaddition processes are particularly important for allenes, ketenes and hetero-substituted equivalents, for which several examples are given (Fig. 13.12). As a result of the orbital overlap, the stereochemical outcome is *syn*-addition.

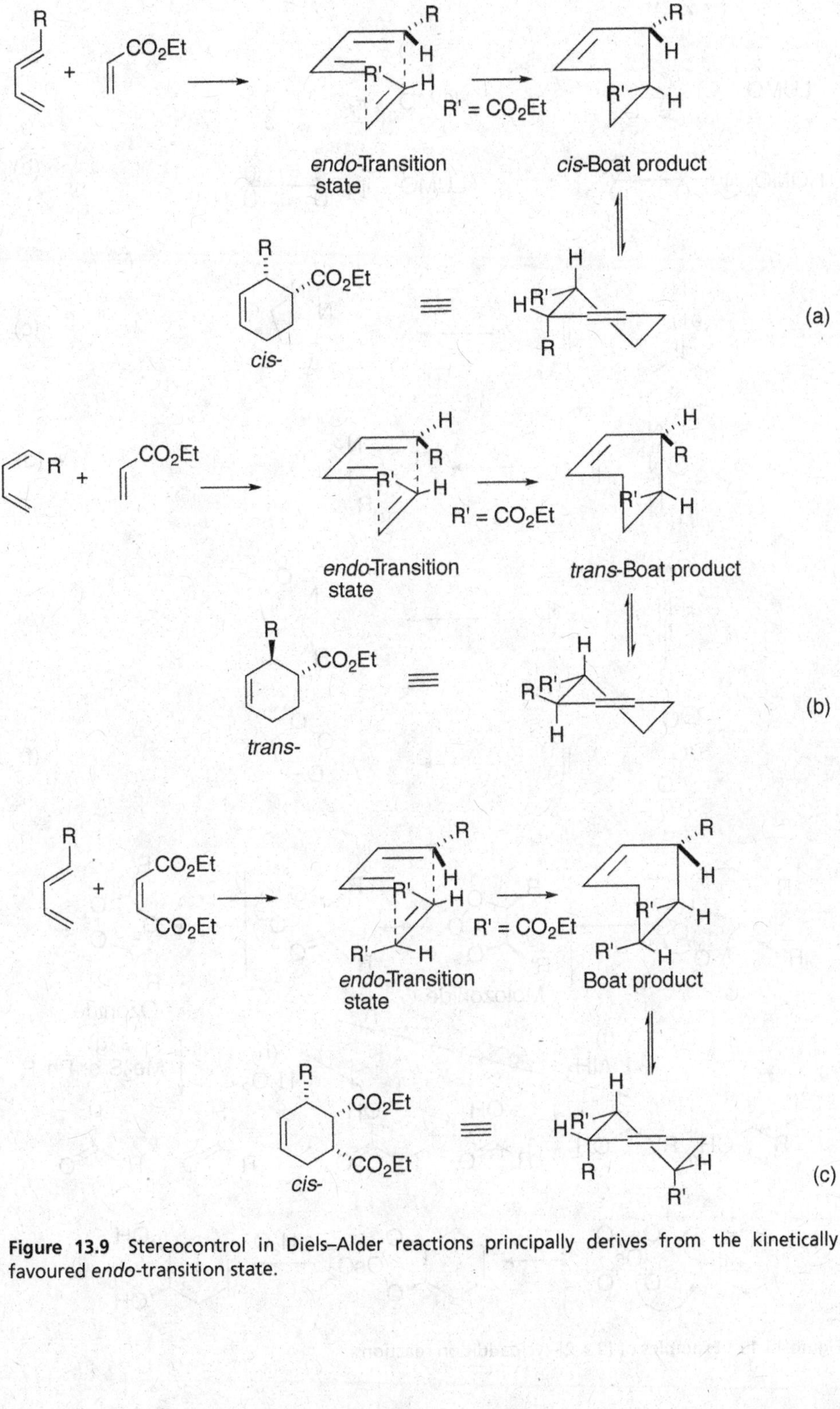

Figure 13.9 Stereocontrol in Diels–Alder reactions principally derives from the kinetically favoured *endo*-transition state.

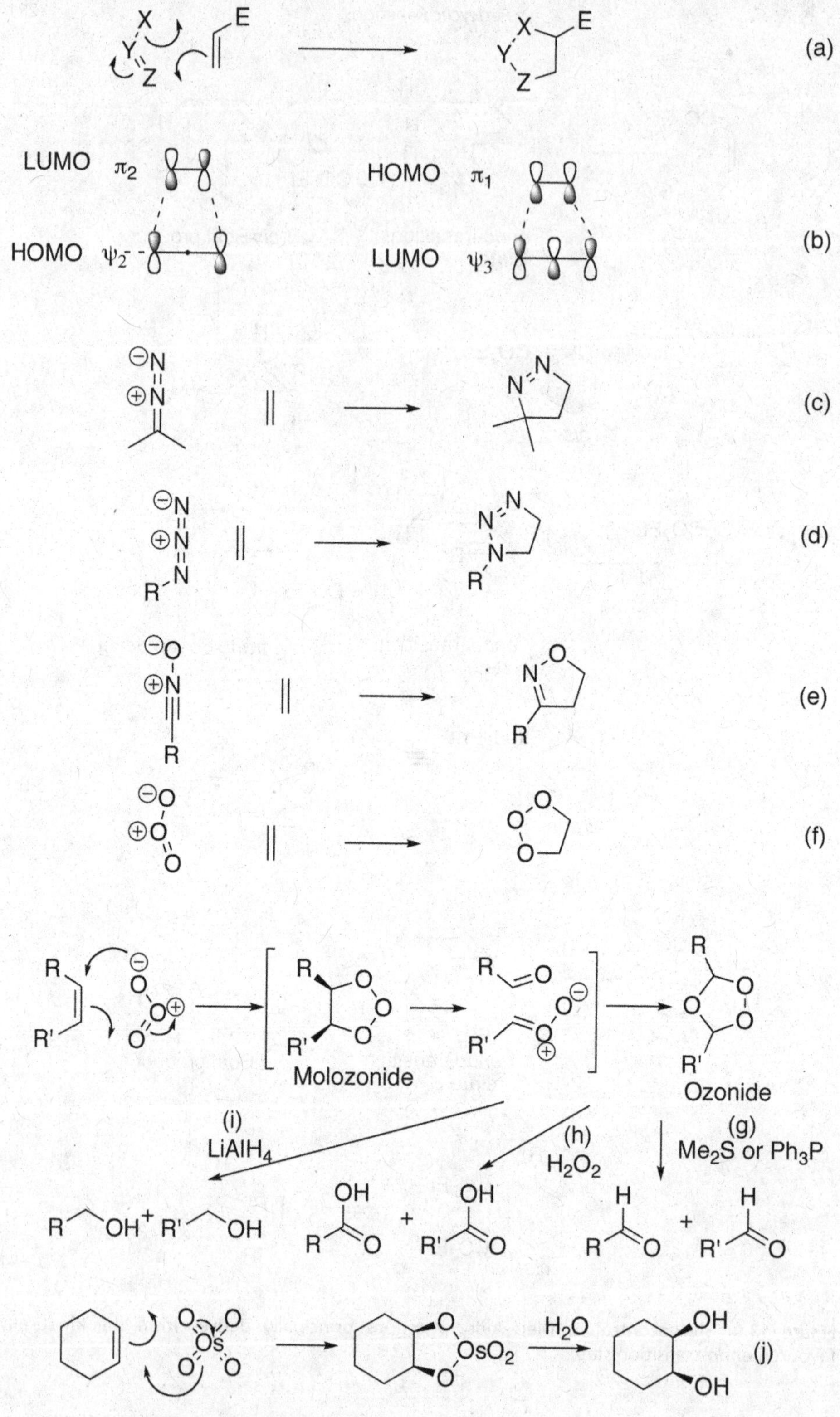

Figure 13.10 Examples of [3 + 2]-cycloaddition reactions.

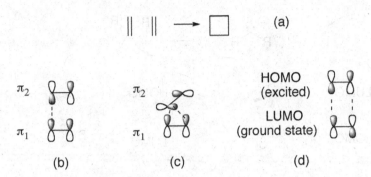

Figure 13.11 [2 + 2]-cycloaddition reactions.

13.2.2.4 Cheletropic reactions

A final type of cycloaddition (or its reverse) processes are called cheletropic reactions. These are reactions in which two σ-bonds are made to a single atom and are typically processes such as carbene addition (a [π2s + ω2a] reaction (Fig. 13.13a)), arising by one of two possible MO interactions shown involving side-on attack of the carbene (Fig. 13.13b), or sulfur extrusion (a [π4s + ω2s] reaction), involving head-on attack of the carbene (Figs. 13.13c and 13.13d).

13.2.3 Sigmatropic reactions

Sigmatropic rearrangements are pericyclic reactions which proceed by a migration, in an uncatalysed intramolecular process, of a σ-bond, adjacent to one or more π-systems, to a new position in the molecule, and in which the π-system becomes reorganised. The process is

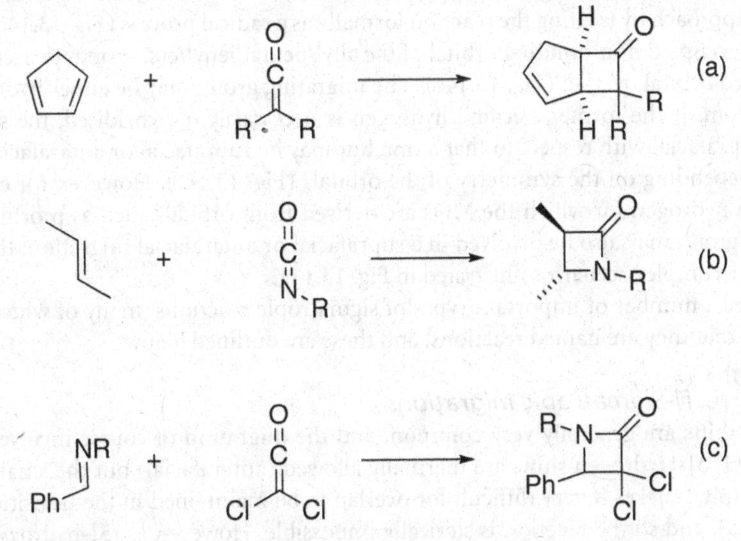

Figure 13.12 Synthetic applications of [2 + 2]-cycloaddition reactions.

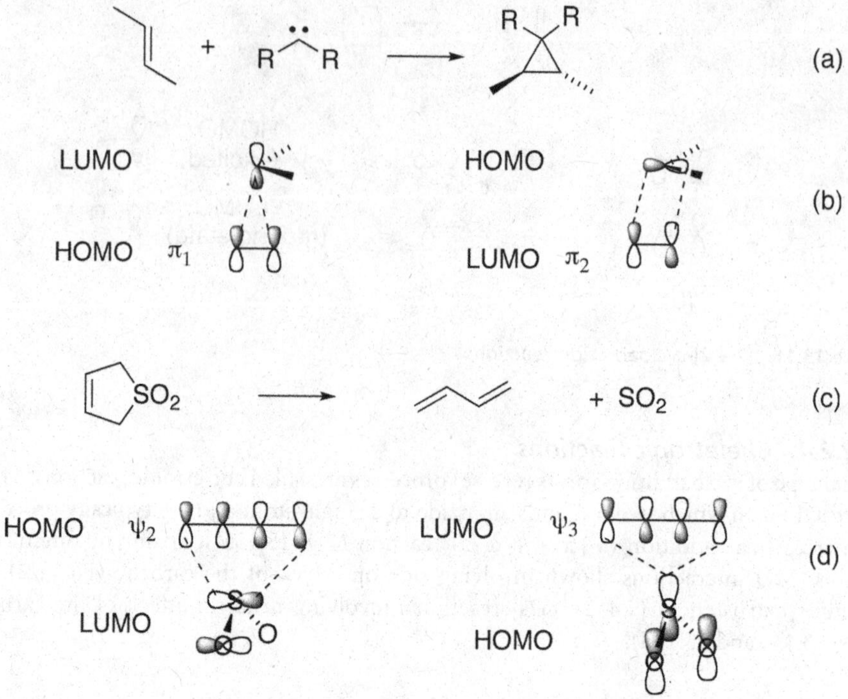

Figure 13.13 Cheletropic cycloaddition reactions.

illustrated in general in Fig. 13.14a and described by an $[i, j]$ descriptor, with each terminus of the bond being broken given the number 1 and the position of the new bond being formed identified by counting around the chain from the broken bond. Examples of [1, 3] and [3, 3] processes are shown in Figs. 13.14b and 13.14c. In the former case, it is possible to use the FMO approach by treating the reaction formally as a radical process (Fig. 13.14d), using the singly occupied non-bonding orbital of the allyl/pentadienyl/etc. system to account for the required orbital overlap (Fig. 13.14e). The migrating group may be either hydrogen or another atom. If the former, because hydrogen is necessarily σ-hybridised, the shift can only be suprafacial with respect to that atom, but may be suprafacial or antarafacial to the polyene, depending on the symmetry of the orbitals (Fig. 13.14f). However, for elements other than hydrogen for which the MOs are derived from orbitals such as p orbitals, the migrating group may also be involved in a suprafacial or antarafacial migration; this leads to the more complex scenarios illustrated in Fig. 13.14g.

There are a number of important types of sigmatropic reactions, many of which are so distinctive that they are named reactions, and these are outlined below.

13.2.3.1 [i, j]-Sigmatropic migrations

Hydrogen shifts are generally very common, and the migration of course involves the H 1s orbital. [1, 3]-Hydrogen shifts are thermally allowed (antarafacial) but the small size of the H 1s orbital makes it very difficult for overlap to be maintained in the transition state (Fig. 13.15a), and so the reaction is sterically impossible. However, [1, 5]-hydrogen shifts are thermally allowed (suprafacial) since the H 1s can readily overlap simultaneously with

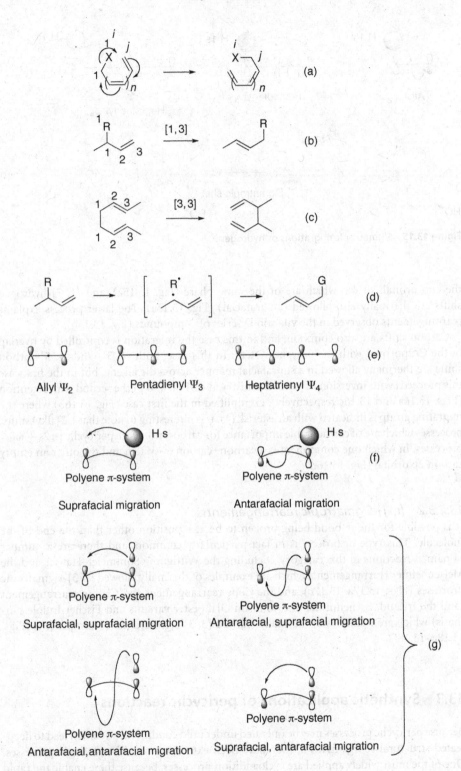

Figure 13.14 Sigmatropic reactions.

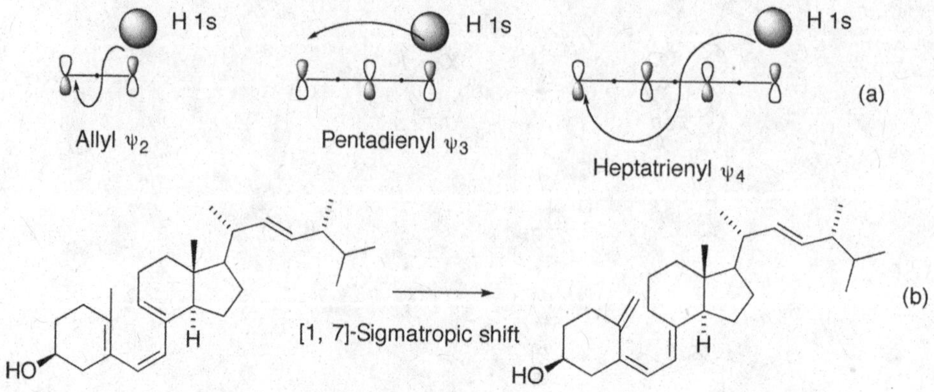

Allyl ψ_2 Pentadienyl ψ_3

Heptatrienyl ψ_4 (a)

[1, 7]-Sigmatropic shift   (b)

Figure 13.15 Sigmatropic migrations of hydrogen.

the end orbitals of ψ_3, which are of the same phase (Fig. 13.15a), and [1, 7]-hydrogen shifts are thermally also allowed (antarafacial) (Fig. 13.15a). The latter process explains rearrangements observed in the vitamin D series of compounds (Fig. 13.15b).

Carbon shifts are also common, and in this case the migration is controlled by overlap of the C 2p orbital with the polyene system. In this case, both [1, 3]- and [1, 5]-carbon shifts are thermally allowed in a suprafacial manner across the alkene, but in the first case this proceeds with inversion of stereochemistry at carbon and in the second with retention (Figs. 13.16a and 13.16c respectively), exemplified in the first case (Fig. 13.16b) where the migrating group is indicated with an asterisk (*). It is interesting to note that 1,2-alkyl shifts, processes which are of considerable importance for carbocations, are pericyclic [$\sigma2s + \omega0s$] processes, in which one component is a carbon–carbon σ-system and the other an empty carbon 2p orbital (Fig. 13.16d).

13.2.3.2 [i, j]-Sigmatropic rearrangements
It is possible for the σ-bond being broken to be at a position other than the end of the molecule. This type of process is in fact particularly common, and there are a number of named reactions in this category, including the Wittig, the Sommelet–Hauser and the Meisenheimer rearrangements which are examples of thermally allowed [2, 3]-sigmatropic processes (Figs. 13.17a–13.17c), and the Cope rearrangement and Claisen rearrangement (and the Ireland, Eschenmoser and Johnson Orthoester variants and Fischer Indole synthesis) which are examples of thermally allowed [3, 3]-sigmatropic rearrangements (Figs. 13.18a–13.18f).

13.3 Synthetic applications of pericyclic reactions

Because pericyclic processes may be operated under mild conditions, and often lead to deep-seated structural reorganisation, they have been extensively used in synthetic processes. One of the most widely applied are cycloaddition processes, because these enable the rapid

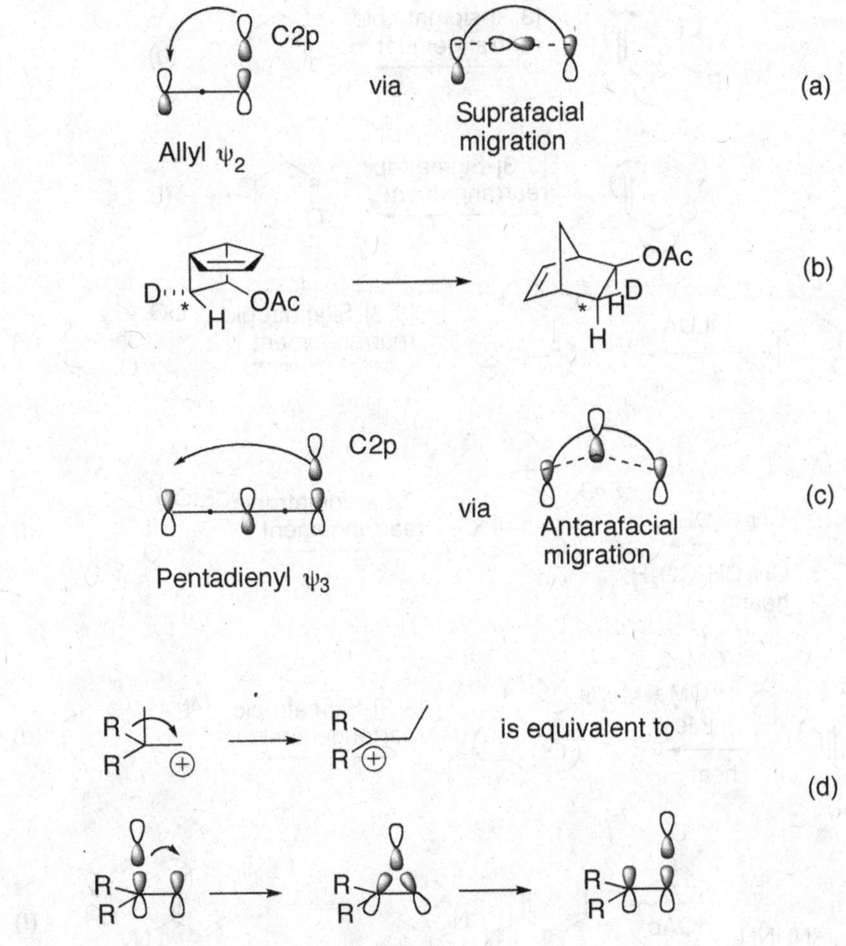

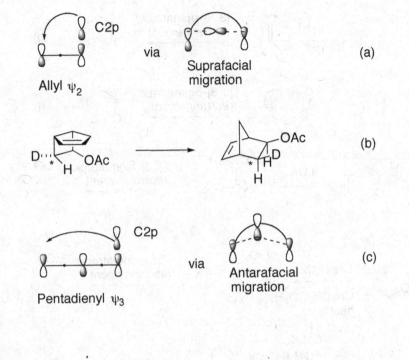

Figure 13.16 Sigmatropic migrations of carbon.

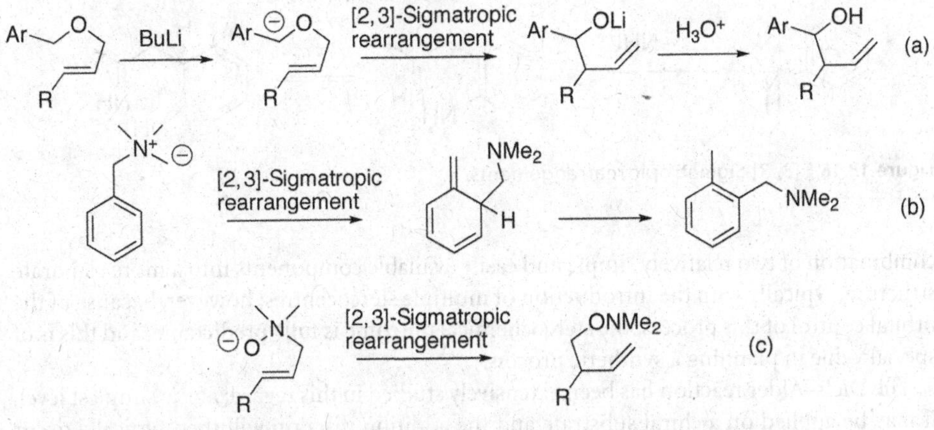

Figure 13.17 [2, 3]-Sigmatropic rearrangements.

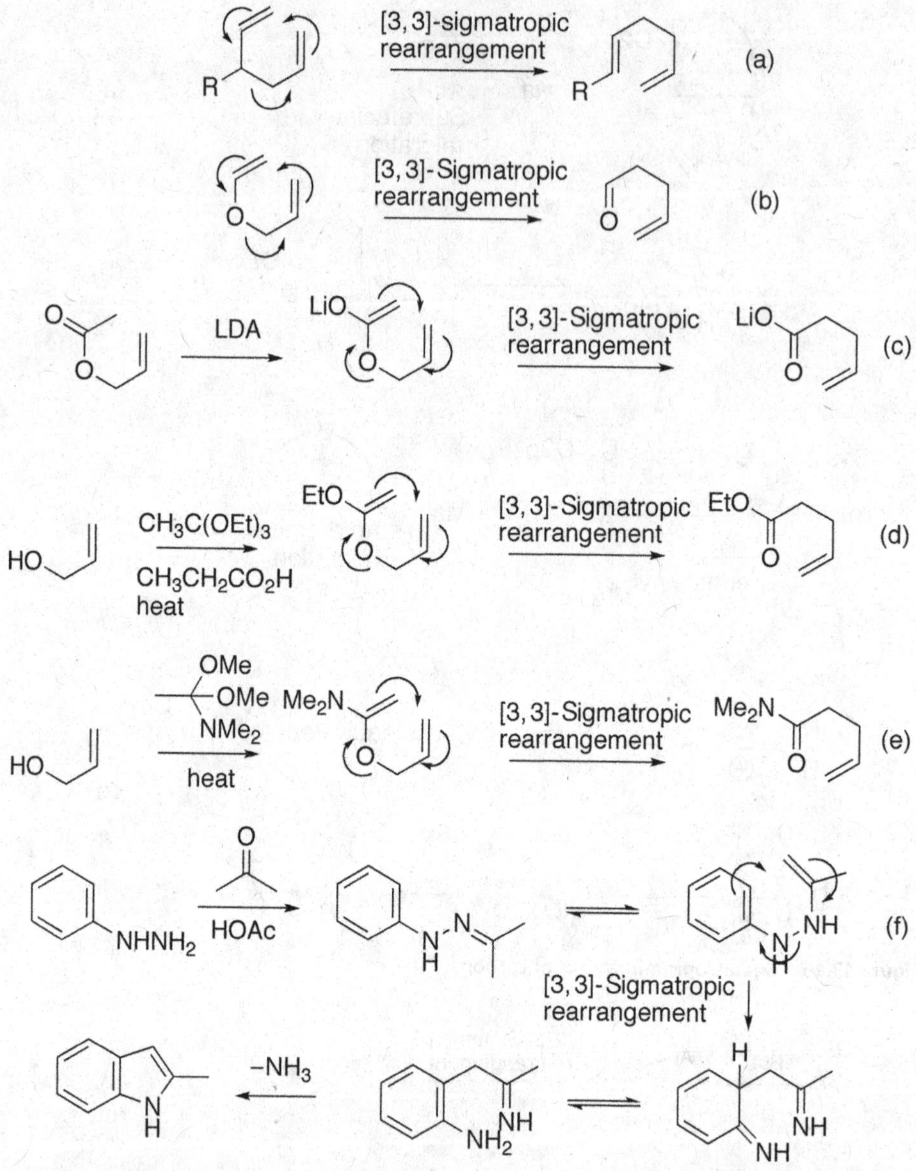

Figure 13.18   [3, 3]-Sigmatropic rearrangements.

combination of two relatively simple and easily available components into a more elaborate structure, typically with the introduction of multiple stereocentres; however, because of the orbital control of this process, the stereochemical outcome is fully predictable, and this is of special value in planning a synthetic process.

The Diels–Alder reaction has been extensively studied in this regard; at the simplest level, it may be applied on a chiral substrate and the addition reaction will then typically occur to the least hindered face of the molecule (Fig. 13.19a); in the case shown, this attack is

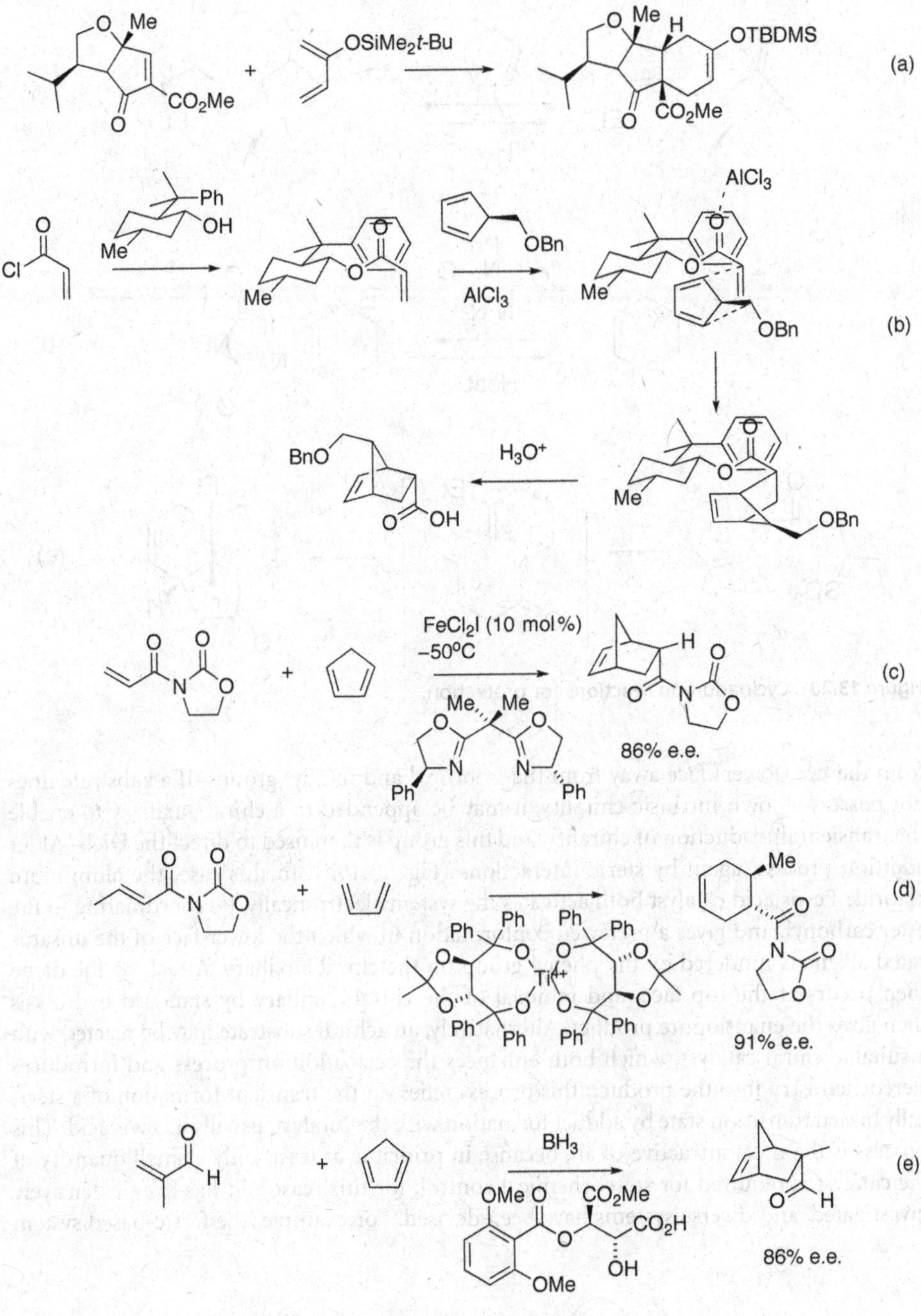

Figure 13.19 Stereocontrolled cycloaddition reactions.

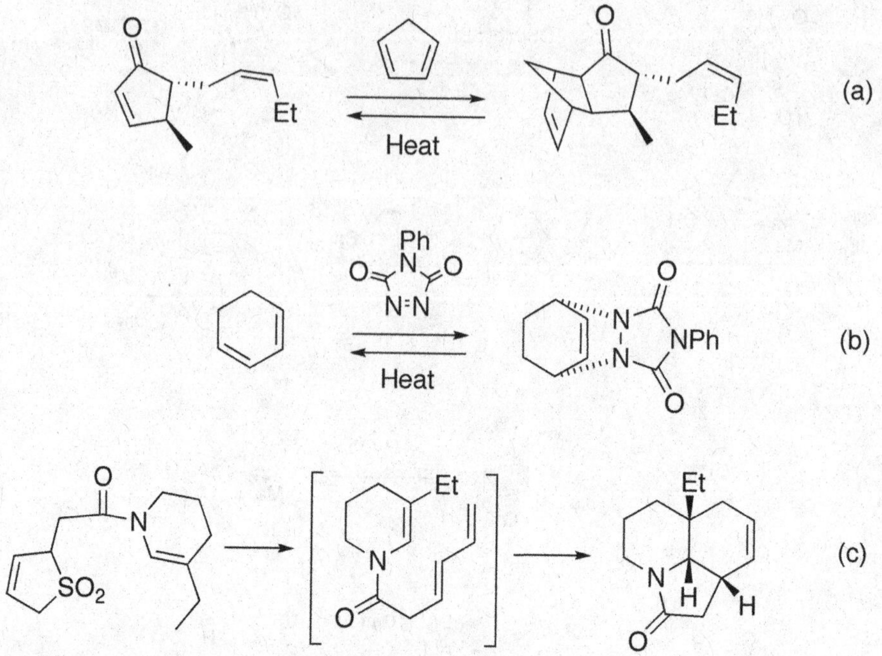

Figure 13.20 Cycloaddition reactions for protection.

from the *exo* (lower) face away from the isopropyl and methyl groups. If a substrate does not possess its own intrinsic chirality, it may be appended to a chiral auxiliary to enable the transient introduction of chirality, and this group is then used to direct the Diels–Alder addition process, again by steric interactions (Fig. 13.19b). In this case, the aluminium chloride Lewis acid catalyst both activates the system electronically, by coordinating to the ester carbonyl, and gives a preferred conformation in which the lower face of the unsaturated alkene is hindered by the phenyl group on the chiral auxiliary. Attack by the diene then occurs at the top face, and removal of the chiral auxiliary by standard hydrolysis then gives the enantiopure product. Alternatively, an achiral substrate may be reacted with a suitable chiral catalyst, which both enhances the cycloaddition process and introduces stereochemistry into the product; this process relies on the transient formation of a sterically biased transition state by adduct formation with the catalyst, usually a Lewis acid. This process is the most attractive of all, because in principle at least, only a small quantity of the catalyst is required for stereochemical control; for this reason, it has been extensively investigated, and diverse systems have been devised. For example, the ferric-based system

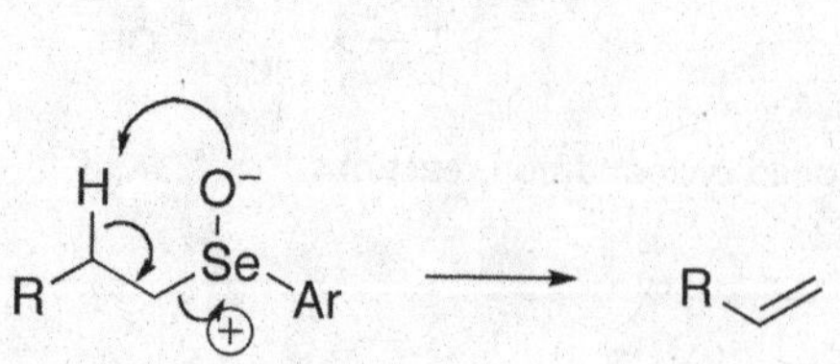

Figure 13.21 [2, 3]-Sigmatropic elimination reactions.

is particularly effective (Fig. 13.19c), not least because this Lewis acid markedly acceler-
ates the addition reaction, enabling it to be run at very low temperature, where the energy
difference between the diastereomeric transition states is maximised. Titanium-derived systems can also be effective (Fig. 13.19d) and so can chiral Lewis acid oxaborolidine systems
(Fig. 13.19e).

Cycloaddition processes, and their reverse, have been very effectively used for the protection of certain functional groups; this makes use of the fact that pericyclic processes are
readily reversible by appropriate modification of reaction conditions. Using anthracene, cyclopentadiene, dimethylfulvene or 1-methoxy-1,3-cyclohexadiene to react with alkenes by a
cycloaddition process, it is possible to protect that alkene and release it at a later stage by thermolysis. Note that this approach allows selective reaction at the more electron-poor alkene,
as a result of the intrinsic electronics of the Diels–Alder process (Fig. 13.20a). Electron-rich
alkenes may be protected using triazoline derivatives, which are readily formed by cycloaddition, but may be reversed by thermolysis in a high-boiling solvent, such as 1,3,5-collidine
(Fig. 13.20b). Similarly, cyclic sulfones may be used to generate diene systems by thermolysis;
this process is often used to generate reactive diene systems in situ for immediate trapping,
often in intramolecular processes (Fig. 13.20c).

The formation of $\alpha\beta$-unsaturated carbonyl compounds from carbonyls via sulfoxides or
selenoxides (Fig. 13.21) by [2, 3]-sigmatropic elimination is a very important process for
the introduction of unsaturation; the required intermediates are readily prepared as shown
in Chapter 8 (Fig. 8.31).

Index